For

Steven Michael

Carolyn Kearns

Laura Ann

Ann Odell

Kate Arminè

Cormac Michéal

Armen H. Zemanian

Graphs and Networks

Transfinite and Nonstandard

Birkhäuser
Boston • Basel • Berlin

Armen H. Zemanian
Department of Electrical Engineering
State University of New York at Stony Brook
Stony Brook, NY 11794-2350
U.S.A.

Library of Congress Cataloging-in-Publication Data
Zemanian, A. H. (Armen H.)
Graphs and networks : transfinite and nonstandard / Armen H. Zemanian.
p. cm.
Includes bibliographical references and indexes.
ISBN 0-8176-4292-7 (acid-free paper)
1. Graphy theory. 2. Transfinite numbers. I. Title.

QA166.Z447 2004
511'.5-dc22 2004043745
CIP

AMS Subject Classifications: 94C05, 03H05, 05C40, 05C12, 03H10, 04A10

ISBN 0-8176-4292-7 Printed on acid-free paper.

Printed in the United States of America. (TXQ/SB)

9 8 7 6 5 4 3 2 1 SPIN 10980231

Birkhäuser is a part of *Springer Science+Business Media*

www.birkhauser.com

Contents

Preface

Scientia Gratiā Scientiae

It is now thirteen years since the first book that discusses transfinite graphs and electrical networks appeared [50]. This was followed by two more books [51] and [54] which compiled results from an ongoing research effort on that subject. Why then is a fourth book, this one, being offered? Simply because still more has been achieved beyond that appearing in those prior books. An exposition of these more recent results is the purpose of this book.

The idea of transfiniteness for graphs and networks appeared as virgin research territory about seventeen years ago. Notwithstanding the progress that has since been achieved, much more remains to be done—or so it appears. Many conclusions concerning conventionally infinite graphs and networks can be reformulated as open problems for transfinite graphs and networks. Furthermore, questions peculiar to transfinite concepts for graphs and networks can be suggested. Indeed, these two considerations have inspired the new results displayed herein.

There are presently four nonequivalent ways of introducing transfiniteness for graphs and networks. The original method is discussed in [50], with [48] being the earliest reconnaissance into that approach. It views the extremities of conventionally infinite graphs as equivalence classes, called "tips," of one-way infinite paths that are pairwise viewed as "equivalent" if they are pairwise eventually identical; connections between those tips are implemented by "1-nodes," these being transfinite generalizations of conventional nodes. Such ideas can be extended to higher ranks of transfiniteness and are examined at length in [51]. A concise exposition of this approach is presented in Chapter 2.

It turns out that transfinite connectedness is not in general a transitive binary relation between nodes. Specifically, one node may be connected to another node through a transfinite path, and the second node may be connected to a third node through another transfinite path, but there may be no transfinite path connecting the first and third nodes. This pathology can be avoided by imposing some additional restrictions on the transfinite graph. On the other hand, this difficulty disappears if the extremities (now called "walk-tips" or simply "wtips") of a graph are represented by equivalence classes of one-way infinite walks, rather than paths, with the walk-based tips being used to construct walk-based transfinite nodes. This radically expands the realm of transfinite graphs but also introduces more complicated—indeed, strange—structures as walk-based transfinite graphs. The restrictions necessitated by the non-transitivity of path-connectedness are no longer needed for walk-connectedness. Chapter 5 herein presents this walk-based approach to transfinite graphs.

The path-based and walk-based transfinite graphs are strictly graph-theoretic constructs. Nonetheless, by assigning electrical parameters to the branches, transfinite networks and their electrical behavior can be explored.

For instance, a different approach, due to B.D. Calvert, becomes feasible when the branches are assigned positive resistances and yields another theory for transfinite electrical networks. The sums of branch resistances in paths provide path lengths, which lead to a metric for the set of nodes of a conventionally infinite, locally finite graph. That metric space is discrete. However, its completion may have limit points, and these can be interpreted as the infinite extremities of the network. Connections between those limit points then yield a transfinite network. This leads to a substantial theory for transfinite electrical networks. Moreover, it can be extended to higher ranks of transfiniteness wherein each rank has its own metric, lower-ranked metrics being stronger than higher-ranked metrics. This version of transfinite networks is examined at some length in the book [54], but it is not discussed herein. We mention it now since it is one of the four extant methods for constructing transfinite networks.

The fourth and radically different approach to transfinite graphs is provided by nonstandard analysis. The nodes of a conventionally infinite graph along with the real numbers are taken as the individuals for a superstructure of sets, from which ultrapower constructions yield "nonstandard nodes," "nonstandard branches," and thereby "nonstandard graphs." This is explored in Chapter 8, wherein "nonstandard transfinite graphs and networks" are also proposed. Standard theorems concerning the existence and uniqueness of current-voltage regimes in transfinite electrical networks can be lifted into a nonstandard setting with now hyperreal currents and voltages.

Actually, hyperreal currents and voltages can be introduced into transfinite networks without enlarging those networks in a nonstandard way; this is the subject of Chapter 6. The idea is to view the transfinite network as the end result of an expanding sequence of finite networks that fill out the transfinite network in such a way that the connections provided by the transfinite nodes in the original network are maintained throughout the expansion procedure. This yields a sequence of currents in each branch, which can be identified as a hyperreal current for that branch, and similarly for the branch and node voltages. Thus, the transfinite network obtains a hyperreal current-voltage regime, which satisfies Kirchhoff's laws and Ohm's law, again in a particular nonstandard way. Here, however, the transfinite network will have many different hyperreal current-voltage regimes due to the many different ways of filling out the network with sequences of finite networks. Another advantage is that transfinite networks having inductors and capacitors in addition to resistors can now be admitted. (All the prior approaches to transfinite networks are restricted to purely resistive networks.) Thus, we can now examine hyperreal-valued transients in hyperreal time, as is done in Chapter 7. As a particular result of this sort, hyperreal transients on artificial and distributed, transfinite, transmission lines and cables are established.

Let us also mention that a fifth method of defining transfinite graphs can be devised by starting with the "ends" of conventionally infinite graphs, as defined by Halin [25] and others. However, the restriction of the infinite extremities of conventional graphs to their ends turns out to be too crude for many purposes. For example, a one-way infinite ladder has only one end, but connections to different extremities represented by the tips within that single end are needed for electrical analysis. This issue is discussed in Sec. 2.9.

As for the other chapters of this book, some symbols and notations are specified and transfinite ranks are explicated in Chapter 1. Chapter 2 is a concise presentation of the earliest construction of transfinite graphs, a topic that is explored at some length in [51]. Chapter 2 provides enough information to make this book understandable independently of that prior book. In Chapter 3, some issues regarding transfinite connectedness are examined, and transfinite graphs are related to hypergraphs in a certain way.

Furthermore, distances between nodes have played an important role in graph theory. Those distances are usually natural numbers. Chapter 4 extends that subject to transfinite graphs by allowing those distances to be transfinite ordinals as well. A variety of standard results are extended transfinitely in this way.

This then is the scope of our book. But perhaps a few words about some general properties and peculiarities of transfinite graphs and networks might be worthwhile. Their definitions can be quite complicated. For example, a path in a conventional graph can be defined in a sentence or two, whereas the definition of a transfinite path of natural-number rank requires half a page along with another page or so of some preliminary specifications of terms used in that definition. Moreover, since paths are defined recursively along increasing transfinite ranks, with the arrow-rank paths having still another distinctive definition, the whole structure is rather cumbersome. To be sure, a concise characterization of transfinite paths is given in Sec. 2.7, but it requires that the lengthy definitions of transfinite paths be established first. This complexity is typical in all of the four different kinds of transfinite graphs and networks. Would that there were a more concise definition of transfinite paths. Perhaps there is, but it is yet to be devised.

One last comment: Transfinite numbers were introduced by Cantor into mathematics over 100 years ago, and this led to a major development in the set-theoretic foundations of mathematics. This has expanded the scope of a variety of subjects in modern mathematics. In contrast, graph theory has remained on "this side of infinity," notwithstanding the substantial body of results concerning conventionally infinite graphs. Undoubtedly, the advent of transfinite graphs will not by any means have the same impact as have transfinite numbers. The subject is for the most part isolated from other areas of mathematics—in contrast to conventional graph theory, but perhaps this may change. It would change if practical applications to engineering and science were to be found, but nothing of this nature exists presently. All that can be claimed is that transfinite graphs and networks comprise a new area of pure mathematics, hence the motto at the top of this Preface. Let us simply note that advancements in theory do at times reappear, perhaps unexpectedly, in real-world situations, and let us hope that such will happen for transfinite graphs.

Stony Brook, New York *Armen H. Zemanian*

Graphs and Networks

1

Some Preliminaries

The mathematical notation and terminology used in this book are conventional. However some concepts are symbolized in varying ways throughout the literature. To avoid ambiguity, we specify some of our usage in the next section. In Sec. 1.2, we discuss the ranks of transfiniteness; these play an important role in our theory of transfinite graphs and networks.

1.1 Concerning Symbols and Terminology

Almost all of our symbols are standard. In this initial section, we list a few items of information that may be helpful later on. Included are some comments on those few mathematical symbols that differ from those used in our prior books [50], [51], [54] or need to be specified for some other reason. One might skip this section and refer to it if and when the need arises.

- Although we concisely present in Chapter 2 all that is essential for a comprehension of this book, one may wish to refer to the prior books [50], [51], [54] ([51] is the most pertinent of these), which present more extensive discussions by means of examples and explanations. In this regard, one should take note of the Errata for those books available at the URL:

  http://www.ee.sunysb.edu/~zeman

  under "Books."
- A set, that foundational concept for mathematics [17], has the property that all its elements are distinct; moreover, in this book a set, such as a set of branches, can have any cardinality. A sequence, on the other hand, is a mapping from a set of integers into a set and has thereby an order inherited from those integers; thus, its range elements can repeat in the sequence, but that range is countable.

In our prior books, we used the same notation for sets and sequences. It often did not matter which entity was intended, but when it did matter, the context clarified the intent. In this book, we will use different symbols for these two concepts.

A set is denoted by braces $\{\cdots\}$, and a sequence by angle brackets $\langle\cdots\rangle$. Thus, $\{a, b, \ldots, y, z\}$ denotes a set of distinct elements $a, b, \ldots, y, z$, where the ellipses indicate possibly infinitely many elements and with no order necessarily imposed. Sometimes, we write $\{a_i\}_{i\in I}$ or $\{a_i : i \in I\}$ for a set whose elements a_i are indexed by $i \in I$, with I being a set of any cardinality. Still more generally, we may write $\{a \in A : P\}$ for a subset of elements in given set A determined by some property P. We also distinguish between the adjectives "countable" and "denumerable" for a set. A set is *finite* if it has finitely many elements, *denumerable* if it has cardinality $\aleph_0$ (i.e., its elements can be placed in a one-to-one correspondence with the natural numbers), and *countable* if it is either finite or denumerable.

On the other hand, we use $\langle a_i : i \in I\rangle$ to indicate a sequence with (not necessarily distinct) elements a_i and with a countable, totally ordered, index set I, or we might simply list the elements, as in $\langle a, b, c, \ldots\rangle$ for example.

Here too, it will not matter whether we use the set symbol or the sequence symbol whenever the elements at hand are all distinct, are countably many, and have an (explicit or implicit) order.

- $\mathbb{N}$ denotes the set of natural numbers $\{0, 1, 2, \ldots\}$, $\mathbb{R}$ the set of real numbers, and $\mathbb{R}_+$ the set of nonnegative real numbers. For other symbols, see the Index of Symbols for the pages where they are defined.
- Most of the facts concerning ordinal and cardinal numbers that we shall need are listed in [51, Appendix A]. Occasionally, we refer to [1] for other results concerning such numbers. The cardinality of a set A is denoted either by $\overline{\overline{A}}$ or by card A.
- A *partition* of a set A is a set $\{A_i : i \in I\}$ of subsets A_i of A such that $A = \bigcup_{i\in I} A_i$, $A_i \neq \emptyset$, and $A_i \bigcap A_j = \emptyset$ if $i \neq j$. This is a standard concept, but we state it here just for specificity.
- In this book, we use the notation $A \subset B$ to denote that A is a strictly proper subset of B, and we use $A \subseteq B$ when $A = B$ is a possibility. This differs from the notation used in our prior books wherein $A \subset B$ allowed $A = B$.
- Instead of $A - B$, we use the set-difference notation $A \setminus B$ to denote the set of elements in A that are not in B.

1.2 Ranks of Transfiniteness

Throughout this book we will be discussing a variety of objects such as "tips," "nodes," "paths," "sections," "graphs," and "networks." They will appear in a hierarchical structure of transfiniteness. The position of any such object in that structure will be designated by an assigned *rank* ρ, which will be displayed by referring to the object as a *ρ-object.* ρ will be either a countable ordinal or an entity that precedes a countable limit ordinal or simply -1 for the initial rank. In fact, ranks form the totally ordered set

$$\mathcal{R} = \{-1, 0, 1, 2, \ldots, \vec{\omega}, \omega, \omega+1, \ldots, \vec{\omega\cdot 2}, \omega\cdot 2, \omega\cdot 2+1, \ldots\}$$

where ω denotes the first infinite ordinal, as usual, and is the first limit ordinal. $\mathcal{R}$ consists of all the countable ordinals along with -1 initially and also with a rank $\vec{\lambda}$ inserted before each countable limit ordinal λ. We call the initial rank -1 the *elementary rank* or (-1)*-rank* and call the ranks $\vec{\lambda}$ *arrow ranks* or simply *arrows*; these are not ordinals.[1] All the other ranks are countable ordinals. $\vec{\omega}$ is the first arrow; it is larger than every natural-number rank and less than ω. $\mathcal{R}$ is a *well-ordered* set; that is, it is totally ordered, and every subset has a least member. Ranks will be symbolized with lower-case Greek letters with possibly arrows attached. Also, there will be occasions where we use $\rho - 1$ to symbolize the predecessor of some arbitrary ordinal rank ρ; in the event that ρ is a limit ordinal λ, $\lambda - 1$ will be taken to designate $\vec{\lambda}$. Thus, $\vec{\lambda} + 1$ will mean λ. In particular, we have $\omega - 1 = \vec{\omega}$ and $\vec{\omega} + 1 = \omega$. The same convention works for the elementary rank -1; namely, $0 - 1 = -1$ and $-1 + 1 = 0$.

We can view all this as a notational convenience, and this is how it was used in the prior books [50], [51], [54]. However, for the purposes of this book, it is expedient to define the arrows mathematically. We do so by taking arrows to be certain equivalence classes of nondecreasing sequences of countable ordinals.

With λ being a countable limit ordinal, we define the *arrow* $\vec{\lambda}$ as an equivalence class of all nondecreasing sequences of ordinals, where two such sequences $\langle \alpha_k : k \in \mathbb{N} \rangle$ and $\langle \beta_k : k \in \mathbb{N} \rangle$ are taken to be equivalent if, for each γ less than λ, there exists a natural number k_0 such that $\gamma < \alpha_k, \beta_k < \lambda$ for all $k > k_0$. The axioms of an equivalence relationship are clearly satisfied. Each such sequence $\langle \alpha_k : k \in \mathbb{N} \rangle$ is a representative of $\vec{\lambda}$, and we say that $\langle \alpha_k : k \in \mathbb{N} \rangle$ *reaches* $\vec{\lambda}$. According to this definition, each representative sequence of $\vec{\lambda}$ is *persistently increasing* in the sense that every member of that sequence is less than some subsequent member of the sequence.

Note that this equivalence class of nondecreasing sequences is different from the set of ordinals less than λ. The latter is λ itself by the definition of ordinals. Also, we are distinguishing this equivalence class from the limit λ of any such sequence in the equivalence class [1, pp. 165–166].

[1] In our prior works [51] and [54] we also called the elementary rank the "$\vec{0}$-*rank*." In this work we shall not do so; the first arrow rank will be $\vec{\omega}$.

2

Transfinite Graphs

In this chapter we define transfinite graphs and list a number of their properties. We shall be fairly concise, for most of what is presented herein has been discussed at some length in [51] along with a variety of illustrative examples.[1] Our purpose here is to provide enough information to relieve the reader from any need to consult that prior work.

2.1 Branches or Synonymously (-1)-Graphs

Our graphs are arranged in a hierarchy of transfiniteness indexed by the ranks specified in Sec. 1.2. They are defined recursively in accordance with those ranks. At the most elementary level, we have the (-1)-*graph*, defined as a two-element set, which we always refer to as a *branch* and whose two elements are called *elementary tips*; at times, these are also called (-1)-*tips* when they occur as special cases of tips in general. Elementary tips are typically denoted as t^{-1}.

Graphs of higher ranks will consist of sets of branches along with tips and nodes of various ranks defined recursively, but it will always be understood that the branches of any such graph are pairwise disjoint. Tips and nodes of ranks 0 and higher will be sets too, whereas elementary tips are the *individuals* for our transfinite graphs, that is, they are not sets because they have no elements.[2]

To conform with subsequent terminology, we say that a branch b *traverses* (but does not "embrace") each of its two elementary tips and that b is a *representative* of each of its tips. We may also view b as an *endless* (-1)-*path*, again to conform with endless paths of higher ranks. For a similar reason, we also view a branch as a (-1)-*section*.

[1] An earlier version was given in [50], and a simplified exposition can be found in [54] under the restriction that all nodes are "pristine."

[2] In Chapter 8, we use 0-nodes instead of elementary tips as the individuals for the nonstandard 0-graphs discussed therein, but a virtually equivalent theory for nonstandard 0-graphs based upon elementary tips as the individuals is also available [58].

2.2 0-Graphs

A 0-*graph* is a graph in the customary sense, but we shall define it in an unusual way, again to conform with the definitions of transfinite graphs. $\mathcal{B}$ will always denote a (finite or infinite) set of pairwise disjoint branches. Let $\mathcal{T}^{-1}$ be the union of all the branches in $\mathcal{B}$; thus, $\mathcal{T}^{-1}$ is a set of elementary tips. Partition $\mathcal{T}^{-1}$ in any arbitrary fashion. Each set of the partition is called a 0-*node*, and the set of 0-nodes is denoted by $\mathcal{X}^0$. With $x^0 \in \mathcal{X}^0$, we say that the elementary tips in x^0 are *shorted together by* x^0. If x^0 is a singleton, its sole elementary tip is said to be *open*. We also say that x^0 *embraces* itself and each of its elementary tips. The *degree* of a 0-node is its cardinality. If a 0-node x^0 contains an elementary tip t^{-1} of a branch b, we say that x^0 and b are *incident* to each other and that b *reaches* x^0 *through* t^{-1}. We also say that x^0 is a 0-node *of* b. (According to this construction, there are no "isolated" 0-nodes because every 0-node has at least one incident branch.) A *self-loop* is a branch whose two elementary tips are shorted together; thus, a self-loop is incident to just one 0-node. If two branches are incident to the same 0-node, we say that they are *adjacent*. Also, two branches that are not self-loops are said to be *in parallel* and are called *parallel branches* if they are incident to the same two 0-nodes. Also, two branches are said to be *in series* if they are incident to the same 0-node of cardinality 2.

Given a set $\mathcal{B}$ of branches and a set $\mathcal{X}^0$ of 0-nodes constructed as above, we define a 0-*graph* $\mathcal{G}^0$ as the doublet

$$\mathcal{G}^0 = \{\mathcal{B}, \mathcal{X}^0\}. \tag{2.1}$$

$\mathcal{G}^0$ is called either *finite, infinite, countable, uncountable,* or *of a* (particular) *cardinality* if its branch set $\mathcal{B}$ has one of these properties respectively. Also, if the degree of every one of the 0-nodes is finite, $\mathcal{G}^0$ is called *locally finite*.

We now define a "branch-induced subgraph" of $\mathcal{G}^0$; we shall refer to it simply as a "0-subgraph." Let $\mathcal{B}_s$ be a subset of $\mathcal{B}$. $\mathcal{X}_s^0$ will be the subset of $\mathcal{X}^0$ consisting of those 0-nodes, each having at least one elementary tip belonging to a branch of $\mathcal{B}_s$. Then, the *0-subgraph* $\mathcal{G}_s^0$ *of* $\mathcal{G}^0$ *induced by* $\mathcal{B}_s$ (or *induced by the branches of* $\mathcal{B}_s$) is the doublet

$$\mathcal{G}_s^0 = \{\mathcal{B}_s, \mathcal{X}_s^0\}. \tag{2.2}$$

We can also define a 0-subgraph $\mathcal{G}_{ss} = \{\mathcal{B}_{ss}, \mathcal{X}_{ss}^0\}$ of $\mathcal{G}_s^0$ by choosing a subset $\mathcal{B}_{ss}$ of $\mathcal{B}_s$ and defining $\mathcal{X}_{ss}^0$ from $\mathcal{B}_{ss}$ as $\mathcal{X}_s^0$ was defined from $\mathcal{B}_s$.

Note that $\mathcal{G}_s^0$ need not be a 0-graph because $\mathcal{X}_s^0$ may contain a 0-node having an elementary tip belonging to a branch not in $\mathcal{B}_s$. To overcome this anomaly, we *reduce* such 0-nodes as follows. We remove from each $x^0 \in \mathcal{X}_s^0$ every elementary tip belonging to a branch not in $\mathcal{B}_s$; this yields a *reduced* 0-node x_r^0 all of whose elementary tips belong to branches in $\mathcal{B}_s$. Let $\mathcal{X}_r^0$ be the set of such reduced 0-nodes. Then, the *reduced 0-graph* $\mathcal{G}_r^0$ *of* $\mathcal{G}^0$ *induced by* $\mathcal{B}_s$ is the doublet

$$\mathcal{G}_r^0 = \{\mathcal{B}_s, \mathcal{X}_r^0\}.$$

This conforms with our prior definition of a 0-graph; that is, $\mathcal{G}_r^0$ is a 0-graph.

We say that a 0-subgraph $\mathcal{G}_s^0$ *embraces* itself, its 0-subgraphs, its 0-nodes, its branches, and the elementary tips of its branches. Each of those entities is also said to be *in* or *of* $\mathcal{G}_s^0$. A similar terminology is used for a reduced 0-graph.

The *union* (resp. *intersection*) of two 0-subgraphs is the 0-subgraph induced by the union (resp. intersection) of their branch sets. Such an intersection does not exist if those two branch sets are disjoint. Two 0-subgraphs are said to *meet* or to be *incident* if they embrace a common 0-node. Otherwise, they are called *disjoint*, but in this case we also say that they are *totally disjoint* in order to again conform with some subsequent terminology. A similar terminology is used for reduced 0-graphs.

Next, let $\mathcal{B} = \bigcup_{k \in K} \mathcal{B}_k$ be a partitioning of the branch set $\mathcal{B}$ of $\mathcal{G}^0$ into the subsets $\mathcal{B}_k$. Let $\mathcal{G}_k^0$ be the subgraph of $\mathcal{G}^0$ induced by $\mathcal{B}_k$. Then, the set $\{\mathcal{G}_k^0\}_{k \in K}$ is called a *partitioning* of $\mathcal{G}^0$.

A *trivial 0-path* is a singleton set $\{x^0\}$ consisting of one 0-node x^0. A *nontrivial 0-path* P^0 in $\mathcal{G}^0$ is an alternating sequence of 0-nodes and branches with at least one branch and two 0-nodes:

$$P^0 = \langle \ldots, x_m^0, b_m, x_{m+1}^0, b_{m+1}, \ldots \rangle \tag{2.3}$$

in which no 0-node and no branch appear more than once and moreover every branch and 0-node that are adjacent in the sequence (2.3) are incident in the 0-graph $\mathcal{G}^0$. The indices $\ldots, m, m+1, \ldots$ traverse a consecutive sequence of integers (i.e., they do not extend to the transfinite ordinals). If (2.3) terminates on either side, it terminates at a 0-node. A nontrivial 0-path is a special case of a 0-subgraph; it is induced by its branches.

An *orientation* can be assigned to P^0 in accordance with increasing or decreasing indices. We will also speak of a *tracing* of P^0 as a progression through its elements in accordance with a chosen orientation. P^0 is called *finite* or *one-ended* or *endless* if (2.3) is respectively a finite or one-way infinite or two-way infinite sequence. A finite 0-path is also called *two-ended.* A trivial 0-path is considered to be a special case of a finite 0-path. A 0-*loop* is a nontrivial finite 0-path except that its two terminal elements are required to be the same 0-node.[3] A self-loop is thus a special kind of 0-loop. We can assign an *orientation* to a 0-loop in accordance with increasing or decreasing indices as before, and similarly for a *tracing* of it.

A 0-path or 0-loop is said to *embrace* itself, all its elements, and all the elements embraced by its elements.

Our next objective is to define the "infinite extremities" of a 0-graph $\mathcal{G}^0$. Such an extremity occurs when $\mathcal{G}^0$ contains at least one one-ended 0-path. Two one-ended 0-paths in $\mathcal{G}^0$ will be called *equivalent* if they are identical except for at most finitely many branches and 0-nodes. This is truly an equivalence relationship, and it partitions the set of one-ended 0-paths in $\mathcal{G}^0$ into equivalence classes. Each such equivalence class is called a 0-*tip* and is taken to be an *infinite extremity* of $\mathcal{G}^0$. Any one-ended path in the equivalence class is a *representative* of the 0-tip.

[3]More precisely, it is a circulant sequence satisfying the adjacency-incidence requirement of a 0-path. Which 0-node is chosen as the terminal node when writing (2.3) is immaterial to its definition.

We can now define the *infinite extremities* of a 0-subgraph $\mathcal{G}_s^0$ in the same way, but now the one-ended paths comprising any 0-tip of $\mathcal{G}_s^0$ must be in $\mathcal{G}_s^0$. Thus, a one-ended (resp. endless) 0-path in $\mathcal{G}^0$ has exactly one (resp. two) 0-tips. $\mathcal{G}_s^0$ is said to *traverse* each of its 0-tips and also each of the elementary tips of its branches.

2.3 1-Graphs

Given a 0-graph $\mathcal{G}^0 = \{\mathcal{B}, \mathcal{X}^0\}$ having at least one 0-tip, let $\mathcal{T}^0$ be its set of 0-tips. Partition $\mathcal{T}^0$ in any arbitrary fashion into subsets $\mathcal{T}_\tau^0$; then, $\mathcal{T}^0 = \bigcup_\tau \mathcal{T}_\tau^0$, where τ is the index for the partition, each $\mathcal{T}_\tau^0$ is nonempty, and $\mathcal{T}_{\tau_1}^0 \cap \mathcal{T}_{\tau_2}^0 = \emptyset$ if $\tau_1 \neq \tau_2$. Also, for each τ, let $\mathcal{X}_\tau^0$ denote either the empty set or a singleton $\{x^0\}$ consisting of exactly one 0-node x^0 of $\mathcal{G}^0$; we also require that $\mathcal{X}_{\tau_1} \cap \mathcal{X}_{\tau_2}^0 = \emptyset$ if $\tau_1 \neq \tau_2$. Then, for each τ, we define a 1-*node* as the set $x_\tau^1 = \mathcal{T}_\tau^0 \cup \mathcal{X}_\tau^0$. Thus, a 1-node x^1 contains at least one 0-tip and at most one 0-node, and the members of x^1 will not appear in any other 1-node.

A 1-node is said to *embrace* itself, all its members, and all the elementary tips in its 0-node if it contains a 0-node. If a 1-node x^1 does contain a 0-node x^0, we call x^0 the *exceptional element* of x^1. If x^1 is a singleton, its one and only 0-tip is said to be *open*. If x^1 is a nonsingleton, its embraced elements are said to be *shorted together* by x^1, and its 0-tips are called *nonopen*. A 0-node that is not a member of a 1-node is called a *maximal* 0-node. Also, the *degree* of x^1 is the cardinality of its set of embraced tips (that is, all its 0-tips and all its embraced elementary tips).

A 1-*graph* $\mathcal{G}^1$ is defined as follows: With $\mathcal{B}$ being a given set of branches, with $\mathcal{X}^0$ being the set of 0-nodes, and with $\mathcal{X}^1$ being the set of 1-nodes, we define $\mathcal{G}^1$ as the triplet

$$\mathcal{G}^1 = \{\mathcal{B}, \mathcal{X}^0, \mathcal{X}^1\}. \tag{2.4}$$

The doublet $\{\mathcal{B}, \mathcal{X}^0\}$ is the 0-*graph of* $\mathcal{G}^1$.

As for a " branch-induced subgraph" of $\mathcal{G}^1$, let $\mathcal{B}_s$ be a nonempty subset of $\mathcal{B}$, let $\mathcal{X}_s^0$ be the set of 0-nodes such that each is incident to at least one branch of $\mathcal{B}_s$, and let $\mathcal{X}_s^1$ be the set of 1-nodes in $\mathcal{X}^1$ such that each has at least one 0-tip with a representative whose branches are all in $\mathcal{B}_s$. Then, the *subgraph of* $\mathcal{G}^1$ *induced by* $\mathcal{B}_s$ (or *by the branches of* $\mathcal{B}_s$) is the triplet $\mathcal{G}_s^1 = \{\mathcal{B}_s, \mathcal{X}_s^0, \mathcal{X}_s^1\}$ if $\mathcal{X}_s^1$ is nonempty, and it is the doublet $\mathcal{G}_s^0 = \{\mathcal{B}_s, \mathcal{X}_s^0\}$ if $\mathcal{X}_s^1$ is empty. In a similar way, we can define a subgraph of a subgraph.

Here too, a subgraph need not be a graph because its nodes may contain tips that are not "traversed" by it (that is, it may have a tip none of whose representative paths lie entirely in the subgraph). But, once again we can overcome this anomaly by reducing its nodes as follows. From each 0-node or 1-node of the subgraph remove every tip that is not traversed by the subgraph (i.e., that does not have a representative with all its branches in $\mathcal{B}_s$), to get thereby a *reduced* node. (Nodes that do not need reduction will still be called reduced nodes.) With $\mathcal{X}_r^0$ (resp. $\mathcal{X}_r^1$) being the set of all such reduced nodes of rank 0 (resp. rank 1), the *reduced graph induced by* $\mathcal{B}_s$ is either the 1-graph $\{\mathcal{B}_s, \mathcal{X}_r^0, \mathcal{X}_r^1\}$ if $\mathcal{X}_r^1$ is nonempty or is the 0-graph $\{\mathcal{B}_s, \mathcal{X}_r^0\}$ if $\mathcal{X}_r^1$ is empty.

A subgraph or reduced graph is said to *embrace* itself, its branches, its nodes, and all the tips embraced by its nodes. Two subgraphs, or a subgraph and a node, or two nodes are called *totally disjoint* if they do not embrace a common node.

Let us explicate some more terminology regarding branches, tips, paths, and subgraphs. (This terminology will hold for graphs of higher ranks as well.) A subgraph is said to *traverse* a tip if the subgraph contains a representative of the tip in the sense that all the branches of the representative lie in the subgraph. Two subgraphs are said to be *incident* if they traverse tips that are shorted to each other. If x is a node that contains those traversed tips, we also say that the subgraphs *meet* at x. (Note that x belongs to both subgraphs.) This also applies to two-ended paths since these are special cases of subgraphs. Similarly, a subgraph and a node are said to be *incident* if the subgraph traverses a tip embraced by the node.

However, incidence and meeting are different concepts for one-ended and endless 0-paths. Indeed, two 0-paths may be *incident* because a 0-tip of one and a 0-tip of the other are shorted together by a 1-node, but they will not meet because there is no 0-node in both paths contained in the 1-node. (Note that a one-ended 0-path is not a subgraph of $\mathcal{G}^1$ because the 1-node containing its 0-tip is not part of the 0-path, and similarly an endless 0-path is not a subgraph of $\mathcal{G}^1$ either.) Here too, we say that a 0-path is *incident* to a 1-node if the 0-path traverses a 0-tip contained in the 1-node, in which case we also say that the 0-path *reaches* the 1-node through that 0-tip. A 0-path P^0 *passes through* each of its embraced elements and also through every 1-node that embraces one of its 0-nodes—except for any terminal node of P^0 if such exists. P^0 is said to be *terminally incident* to a 1-node x^1 if it terminates at a 0-node in x^1 or if it traverses a 0-tip in x^1.

Now, consider subgraphs, paths, and nodes. Any two such entities are called *totally disjoint* if they do not embrace a common node. Otherwise, they are said to *meet*, as before.

A *trivial 1-path* is a singleton $\{x^1\}$ consisting of one 1-node x^1. A *nontrivial 1-path* P^1 is an alternating sequence of three or more elements:

$$P^1 = \langle \ldots, x_m^1, P_m^0, x_{m+1}^1, P_{m+1}^0, \ldots \rangle \tag{2.5}$$

containing 1-nodes and 0-paths, where the indices $\ldots, m, \ldots$ are restricted to a consecutive set of integers and where the conditions stated in the next three paragraphs (a), (b), and (c) are satisfied.

(a) All the paths displayed in the right-hand side of (2.5) are nontrivial 0-paths and all the nodes displayed therein are 1-nodes except possibly when (2.5) terminates on the left and/or right. In the latter case, each terminal element is either a 0-node or a 1-node.

(b) Every 0-path is terminally incident to the two nodes adjacent to it in (2.5) but is otherwise totally disjoint from those nodes in the sense that its nonterminal 0-nodes are not embraced by those adjacent nodes. (It follows that, if P^1 terminates on either side at a 0-node, then that 0-node is a terminal node of the adjacent 0-path.)

(c) Every two nonadjacent elements in the right-hand side of (2.5) are totally disjoint.

The requirement (c) implies that, for each nonterminal 1-node x^1_{m+1}, at least one (perhaps both) of P^0_m and P^0_{m+1} reaches x^1_{m+1} through a 0-tip.

A nontrivial 1-path in $\mathcal{G}^1$ is a special case of a subgraph, as is a two-ended 0-path. Now, however, a one-ended or endless 0-path P^0 in $\mathcal{G}^1$ is not, as was noted above; indeed, for each 0-tip of P^0, one must append the 1-node reached by that 0-tip in order to get a subgraph.

Here too, an *orientation* can be assigned to P^1 in accordance with either increasing or decreasing indices m, and we speak of a *tracing* of P^1 as a progression along one of those two possible orientations. P^1 is called *two-ended* or *one-ended* or *endless* if the sequence (2.5) terminates on both sides or terminates on just one side or does not terminate on either side, respectively.

A 1-*loop* is a nontrivial two-ended 1-path except that one of its terminal nodes embraces the other terminal node.

As with a 0-path, a 1-path is said to *traverse* a 0-tip (resp. elementary tip) if it embraces all the branches of a representative of the 0-tip (resp. the branch for the elementary tip).

Now, assume that $\mathcal{G}^1$ contains at least one one-ended 1-path. Two one-ended 1-paths will be called *equivalent* if they are identical except for at most finitely many 1-nodes and intervening 0-paths in their sequential representations (2.5). This defines an equivalence relationship, which partitions the set of all one-ended 1-paths in $\mathcal{G}^1$ into equivalence classes, called 1-*tips*. The 1-*tips* of a subgraph of $\mathcal{G}^1$ are defined in the same way with however the proviso that the one-ended 1-paths are required to be in the subgraph. Thus, a one-ended (resp. endless) 1-path has exactly one (resp. two) 1-tips. A subgraph is said to *traverse* the elementary tips, 0-tips, and 1-tips induced by its branches.

2.4 μ-Graphs

Throughout this book, μ will denote any natural number, but in this section, we take it to be a positive one. Having defined 0-graphs and 1-graphs, we now define recursively transfinite graphs of any natural-number rank, as follows. Assume that, for each $\rho = 0, 1, \ldots, \mu-1$, the ρ-graphs $\mathcal{G}^\rho = \{\mathcal{B}, \mathcal{X}^0, \ldots, \mathcal{X}^\rho\}$ have been defined along with their concomitant ideas such as ρ-nodes x^ρ and ρ-paths P^ρ:

$$P^\rho \;=\; \langle \ldots, x^\rho_m, P^{\alpha_m}_m, x^\rho_{m+1}, P^{\alpha_{m+1}}_{m+1}, \ldots \rangle \tag{2.6}$$

where, for each m, $0 \le \alpha_m < \rho$ and $P^{\alpha_m}_m$ is a nontrivial α_m-path. We now present a recursive argument that will establish them for all natural numbers.

Consider a $(\mu-1)$-graph $\mathcal{G}^{\mu-1} = \{\mathcal{B}, \mathcal{X}^0, \ldots, \mathcal{X}^{\mu-1}\}$, and assume it has at least one one-ended $(\mu-1)$-path. Thus, we have at least one $(\mu-1)$-tip, defined as follows (to be specific now). Partition the set of all one-ended $(\mu-1)$-paths in

$\mathcal{G}^{\mu-1}$ into equivalence classes by taking two one-ended paths to be *equivalent* if their representative sequences (such as (2.6) with $\rho = \mu - 1$) are identical except for at most finitely many $(\mu - 1)$-nodes and their intervening α_m-paths. Each equivalence class $t^{\mu-1}$ is called a *$(\mu-1)$-tip*. A *representative* of $t^{\mu-1}$ is any one of the one-ended paths in that equivalence class.

Now, partition the set $\mathcal{T}^{\mu-1}$ of all $(\mu - 1)$-tips into subsets $\mathcal{T}_\tau^{\mu-1}$, where τ is the index for the partition. Thus, $\mathcal{T}^{\mu-1} = \bigcup_\tau \mathcal{T}_\tau^{\mu-1}$, where $\mathcal{T}_\tau^{\mu-1} \neq \emptyset$ for every τ and $\mathcal{T}_{\tau_1}^{\mu-1} \bigcup \mathcal{T}_{\tau_2}^{\mu-1} = \emptyset$ if $\tau_1 \neq \tau_2$. Also, for each τ, let $\mathcal{X}_\tau^{\mu-1}$ be either the empty set or a singleton whose sole member is an α-node x_τ^α, where $0 \leq \alpha \leq \mu - 1$. Then the set,

$$x_\tau^\mu = \mathcal{T}_\tau^{\mu-1} \cup \mathcal{X}_\tau^{\mu-1} \tag{2.7}$$

is defined to be a *μ-node* so long as the following condition holds: Whenever $\mathcal{X}_\tau^{\mu-1}$ is nonempty, its sole member x_τ^α is not a member of any other β-node ($\alpha < \beta \leq \mu-1$). μ is the *rank* (of transfiniteness) of x_τ^μ.

Thus, a μ-node x^μ contains at least one $(\mu - 1)$-tip and at most one node of rank no larger than $\mu - 1$, called the *exceptional element* of x^μ when it exists. That exceptional element x^{α_1} may in turn contain an exceptional element x^{α_2}, which again may contain an exceptional element x^{α_3}, and so on finitely many times; thus, $\alpha_1 > \alpha_2 > \alpha_3 > \cdots > \alpha_p$. We say that x^μ *embraces* itself, all its elements, and all the elements embraced by its exceptional element x^{α_1} (specifically, x^μ embraces $x^\mu, x^{\alpha_1}, x^{\alpha_2}, \ldots, x^{\alpha_p}$ and their tips of ranks $\mu - 1, \alpha_1 - 1, \alpha_2 - 1, \ldots, \alpha_p - 1$). If x^μ is a nonsingleton, all its embraced elements are said to be *shorted together by x^μ*. If however x^μ is a singleton, its single $(\mu - 1)$-tip is called *open* (*nonopen* otherwise). A fact [51, Lemma 2.2-1] that is of some importance is the following.

Lemma 2.4-1. *If x_a^α and x_b^β are an α-node and a β-node with $0 \leq \alpha \leq \beta$ and if x_a^α and x_b^β embrace a common node, then x_b^β embraces x_a^α. If in addition $\alpha = \beta$, then $x_a^\alpha = x_b^\beta$.*

A *μ-graph* $\mathcal{G}^\mu$ of *rank* μ is by definition a $(\mu + 2)$-tuplet:

$$\mathcal{G}^\mu = \{\mathcal{B}, \mathcal{X}^0, \ldots, \mathcal{X}^\mu\} \tag{2.8}$$

where $\mathcal{B}$ is a set of branches and, for each $\rho = 0, \ldots, \mu$, $\mathcal{X}^\rho$ is the nonempty set of all the ρ-nodes. The subset $\mathcal{G}^\rho = \{\mathcal{B}, \mathcal{X}^0, \ldots, \mathcal{X}^\rho\}$ is called the *ρ-graph* of $\mathcal{G}^\mu$.

An α-node x^α is said to be *maximal* if x^α is not embraced by a node of higher rank; otherwise, x^α is called *nonmaximal*. The maximal nodes of $\mathcal{G}^\mu$ partition all the tips of $\mathcal{G}^\mu$; specifically, each subset of the partition consists of all the tips embraced by the maximal node.

The *degree* of a node (whether maximal or not) is the cardinality of the set of all its embraced tips.

Next, let $\mathcal{B}_s$ be a nonempty subset of $\mathcal{B}$. With $\mathcal{G}^\mu$ being defined by (2.8) and for each $\lambda = 0, \ldots, \mu$, let $\mathcal{X}_s^\lambda$ be the subset of $\mathcal{X}^\lambda$ consisting of all λ-nodes x^λ, each containing at least one $(\lambda - 1)$-tip with a representative all of whose branches are in $\mathcal{B}_s$. For $\lambda \geq 1$, $\mathcal{X}_s^\lambda$ may be empty, but there will be some maximum rank γ $(0 \leq \gamma \leq \mu)$ for which all the $\mathcal{X}_s^\lambda$ $(\lambda = 0, \ldots, \gamma)$ are nonempty; moreover, if $\gamma < \mu$, then all the $\mathcal{X}_s^\lambda$ $(\lambda = \gamma + 1, \ldots, \mu)$ will be empty. The *subgraph* $\mathcal{G}_s^\gamma$ of $\mathcal{G}^\mu$ *induced by* $\mathcal{B}_s$ (or *by the branches of* $\mathcal{B}_s$) is by definition the set

$$\mathcal{G}_s^\gamma = \{\mathcal{B}_s, \mathcal{X}_s^0, \ldots, \mathcal{X}_s^\gamma\}. \tag{2.9}$$

In the same way, we can define a subgraph $\mathcal{G}_{ss}$ of a subgraph $\mathcal{G}_s^\gamma$ induced by a subset $\mathcal{B}_{ss}$ of $\mathcal{B}_s$.

As before, a subgraph need not be a graph because some node of some $\mathcal{X}_s^\lambda$ $(0 \leq \lambda \leq \gamma)$ may contain a tip none of whose representatives have all their branches residing only in $\mathcal{B}_s$. But, we can identify a unique graph related to $\mathcal{G}_s^\gamma$ by removing every tip in every node of every $\mathcal{X}_s^\lambda$ having the property that every representative of the tip in question has some branches not in $\mathcal{B}_s$ (that is, the tip has no representative with all its branches in $\mathcal{B}_s$). In this way, we *reduce* every such node x^λ of each $\mathcal{X}_s^\lambda$ to get a *reduced node* $x_r^{\lambda_r}$; its rank λ_r is either equal to or less than λ. (Nodes not requiring reduction will also be called reduced nodes.) With $\mathcal{X}_r^0, \ldots, \mathcal{X}_r^\rho$ being the sets of reduced nodes of ranks $0, \ldots, \rho$ $(\rho \leq \gamma)$,

$$\mathcal{G}_r^\rho = \{\mathcal{B}_s, \mathcal{X}_r^0, \ldots, \mathcal{X}_r^\rho\} \tag{2.10}$$

is called the *reduced graph of* $\mathcal{G}^\mu$ *induced by* $\mathcal{B}_s$. $\mathcal{G}_r^\rho$ is a ρ-graph. The idea of a reduced graph simplifies the definition of a path, as will be seen in Sec. 2.7.

A set of subgraphs of a subgraph $\mathcal{G}_s^\gamma$ (possibly $\mathcal{G}_s^\gamma = \mathcal{G}^\mu$) is said to *partition* $\mathcal{G}_s^\gamma$ if the branch sets of the former subgraphs comprise a partition of the branch set of $\mathcal{G}_s^\gamma$.

A node or a subgraph or a reduced graph is said to *embrace* itself, its branches, its nodes (whether maximal or nonmaximal), and all the tips in all its nodes. On the other hand, a subgraph is said to *traverse* a tip if that subgraph embraces a representative of the tip, that is, if it embraces all the branches and nodes of that representative path; in this case, we also say that the tip is *in* or *belongs to* the subgraph. For a subgraph, a traversed tip will be embraced, but an embraced tip need not be traversed because none of its representatives entirely lie in the subgraph. (In a reduced graph, a tip is embraced if and only if it is traversed.) A subgraph $\mathcal{G}$ is said to be *incident to* or to *reach* a tip, or a node, or a branch, or another subgraph if $\mathcal{G}$ traverses a tip that is shorted to a tip or node of the latter entity, in which case we also say that the latter entity is *incident to* the subgraph $\mathcal{G}$.

Two subgraphs, or a node and a subgraph, or two nodes are said to be *totally disjoint* if they do not embrace a common node; otherwise, they are said to *meet*, and they meet *with* that commonly embraced node (possibly with more than one such node).

Next, let P^α be an α-path, where $0 \leq \alpha \leq \mu$; see (2.6) with ρ replaced by α. P^α has been explicitly defined for $\alpha = 0, 1$ and will be recursively defined for $\alpha > 1$ once we complete the arguments of this section. In the meantime, we can take the definitions of this paragraph as formal ideas having intuitive significance implied by their meanings for $\alpha = 0, 1$. (The precise meanings of these intuitive ideas are established by recursion.) If P^α terminates on one side at a δ-node x_t^δ $(0 \le \delta \le \alpha)$, then P^α traverses a $(\delta-1)$-tip $t^{\delta-1}$ (e.g., an elementary tip if $\delta = 0$) that is a member of x_t^δ. On the other hand, if P^α extends infinitely on one side, it will traverse an α-tip t^α on that side. In either case, we call $t^{\delta-1}$ or t^α a *terminal tip* of P^α. Furthermore, let x^ρ be a ρ-node that embraces that terminal tip (thus, $\rho > \delta - 1$ or $\rho > \alpha$); then, we say that P^α is *terminally incident to* or *reaches* x^ρ *with* or *through* that terminal tip. If P^α does terminate at x_t^δ and if x^ρ embraces x_t^δ, then we say that P^α is *terminally incident to* x^ρ *with* x_t^δ and that P^α *terminates at* or *meets* x^ρ *with* x_t^δ. In this case, we also say that P^α *meets* all the nodes embraced by x^ρ. Finally, we say that P^α is *terminally incident to* x^ρ *but otherwise is totally disjoint from* x^ρ if P^α reaches x^ρ with a terminal tip and if x^ρ does not embrace any other tip traversed by P^α. At times we will assign an orientation to a path—perhaps implicitly—and then will speak of that path as *starting at* or *stopping at* any node to which is terminally incident.

Here too, "incidence" and "meeting" are different concepts for one-ended and endless α-paths when $0 \le \alpha < \mu$. Such paths are not subgraphs of $\mathcal{G}^\mu$ because they do not contain the $(\alpha+1)$-node to which they are incident. Thus, they do not meet those $(\alpha+1)$-nodes.

We are now ready to define a "μ-path" for any natural number $\mu \ge 1$. A *trivial* μ*-path* is a singleton $\{x^\mu\}$, where x^μ is a μ-node. A *nontrivial* μ*-path* P^μ is an alternating sequence of three or more elements:

$$P^\mu = \langle \ldots, x_m^\mu, P_m^{\alpha_m}, x_{m+1}^\mu, P_{m+1}^{\alpha_{m+1}}, \ldots \rangle \tag{2.11}$$

containing μ-nodes x_m^μ, α_m-paths $P_m^{\alpha_m}$, and possibly terminal nodes of ranks no larger than μ, where $0 \le \alpha_m < \mu$ with the α_m allowed to differ for different m, where the indices $\ldots, m, \ldots$ are restricted to a consecutive set of integers and where the conditions stated in the next three paragraphs (a), (b), (c) are satisfied:

(a) Every $P_m^{\alpha_m}$ is a nontrivial α_m-path. Every x_m^μ is a μ-node except possibly when the sequence (2.11) terminates on the left and/or on the right. In the latter case, each terminal element is a δ-node x_t^δ $(0 \le \delta \le \mu)$, and its adjacent path is an α-path $(\alpha < \mu)$.

(b) Every $P_m^{\alpha_m}$ is terminally incident to the two nodes adjacent to it in (2.11) but is otherwise totally disjoint from those nodes. Moreover, $P_m^{\alpha_m}$ either reaches each of its adjacent nodes with a terminal $(\mu-1)$-tip or it meets that adjacent node with a terminal δ-node where $\delta \le \mu - 1$. If that adjacent node is a terminal δ-node x_t^δ of (2.11), then $P_m^{\alpha_m}$ reaches x_t^δ with a $(\delta-1)$-tip, and, if in addition $\delta < \mu$, then x_t^δ is also a terminal node of $P_m^{\alpha_m}$.

(c) Every two elements in (2.11) that are not adjacent therein are totally disjoint.

Another fact [51, Lemma 2.2-2] of some importance is the following.

Lemma 2.4-2. *Let P^μ $(\mu \ge 1)$ be a nontrivial μ-path given by (2.11). If x_{m+1}^μ is not a terminal node, then at least one of its adjacent paths $P_m^{\alpha_m}$ and $P_{m+1}^{\alpha_{m+1}}$ reaches x_{m+1}^μ with a $(\mu-1)$-tip. Consequently, at least one of the ranks α_m and α_{m+1} equals $\mu-1$.*

P^μ is called *two-ended,* or *one-ended,* or *endless* when (2.11) is respectively a finite, or one-way infinite, or two-way infinite sequence.

It follows from all this that the μ-path P^μ will be a subgraph of $\mathcal{G}^\mu$. On the other hand, for $\rho < \mu$, a ρ-path P^ρ will be a subgraph of the ρ-graph of $\mathcal{G}^\mu$.[4]

A *μ-loop* is defined exactly as is a nontrivial two-ended μ-path except that one of its terminal nodes is required to embrace its other terminal node.

A ρ-path or a ρ-loop $(0 \leq \rho \leq \mu)$, all of whose branches are embraced by a subgraph, are said to be *in* that subgraph.

We can assign one of the two possible *orientations* to P^μ in accordance with increasing or decreasing indices m in (2.11), and we also speak of a *tracing* of P^μ as a progression along the branches of P^μ in accordance with one of those orientations. The same can be done for a loop.

Here is another fact; its simple proof is virtually the same as that of [51, Lemma 2.2-4].

Lemma 2.4-3. *Every two-ended ρ-path and every ρ-loop in $\mathcal{G}^\mu$ $(\rho \leq \mu)$ embraces every tip the ρ-path or ρ-loop traverses.*

We can round off this stage of our recursive definitions by defining equivalent classes of one-ended μ-paths and thus μ-tips just as equivalent classes of one-ended $(\mu - 1)$-tips were defined at the beginning of this section. From this we can define $(\mu + 1)$-nodes and then $(\mu + 1)$-graphs. But, there is no need for this since our recursive argument for defining transfinite graphs of all natural-number ranks is now complete.

2.5 $\vec{\omega}$-Graphs

In order to define transfinite graphs of ranks ω and larger, we first define a special kind of graph $\mathcal{G}^{\vec{\omega}}$ of rank $\vec{\omega}$, called an "$\vec{\omega}$-graph." As defined in Sec. 1.2, $\vec{\omega}$ is the first arrow rank; it is less than the rank ω but greater than all the natural-number ranks. It will be possible to define an $\vec{\omega}$-graph so long as the constructions of the node sets $\mathcal{X}^0, \mathcal{X}^1, \ldots$ of progressively higher ranks can be continued throughout all the natural-number ranks μ. In other words, given the branch set $\mathcal{B}$, the infinite set $\{\mathcal{G}^\mu : \mu \in \mathbb{N}\}$ of μ-graphs generated recursively by defining μ-nodes must be such that each $\mathcal{G}^\mu$ has one-ended μ-paths (and thereby an infinity of μ-nodes), for otherwise there would be no μ-tips with which to define $\mathcal{G}^{\mu+1}$. So, let us assume that, for each natural number μ, nonempty sets $\mathcal{X}^\mu$ have been constructed.

We now introduce a new kind of node, an *$\vec{\omega}$-node*:

$$x^{\vec{\omega}} = \langle x_0^{\mu_0}, x_1^{\mu_1}, x_2^{\mu_2}, \ldots \rangle \tag{2.12}$$

is an infinite sequence of μ_k-nodes $x_k^{\mu_k}$ $(k = 0, 1, 2, \ldots)$ having the following properties.

---

[4]A more concise way of specifying P^ρ when $\rho < \mu$ can be obtained from its corresponding reduced graph; see Sec. 2.7 in this regard.

(a) Each rank μ_k is a natural number.
(b) Each node $x_k^{\mu_k}$ is the exceptional element of the next node $x_{k+1}^{\mu_{k+1}}$.
(c) $x_0^{\mu_0}$ does not have an exceptional element.

Thus, $\mu_0 < \mu_1 < \mu_2 < \cdots$. The set of all $\vec{\omega}$-nodes will be denoted by $\mathcal{X}^{\vec{\omega}}$. As usual, we say that an $\vec{\omega}$-node $x^{\vec{\omega}}$ *embraces* itself, all its nodes, and all the tips in all its nodes. Moreover, all those embraced elements are said to be *shorted together* or simply *shorted* by $x^{\vec{\omega}}$. An $\vec{\omega}$-node cannot be "open" in the sense of a singleton node. It follows from this definition that two different $\vec{\omega}$-nodes are disjoint; in fact, as an easy consequence of Lemma 2.4-1, we have

Lemma 2.5-1. *If two $\vec{\omega}$-nodes embrace a common node, then the two $\vec{\omega}$-nodes are the same node. Moreover, if an $\vec{\omega}$-node $x^{\vec{\omega}}$ and a μ-node x^μ ($\mu < \vec{\omega}$) embrace a common node, then $x^{\vec{\omega}}$ embraces x^μ.*

Our definition of an $\vec{\omega}$-node differs substantially from that of a μ-node x^μ with a natural-number rank μ in two ways. First, $x^{\vec{\omega}}$ does not contain any tips of any ranks, whereas x^μ always contains μ-tips. (But, $x^{\vec{\omega}}$ does embrace all the $(\mu_k - 1)$-tips contained in every $x_k^{\mu_k}$. The *degree* of $x^{\vec{\omega}}$ is the cardinality of the set of all its embraced tips, as before.) Secondly, $x^{\vec{\omega}}$ embraces an infinity of nodes, whereas x^μ embraces only finitely many nodes — perhaps none at all.

Here is the definition of an *$\vec{\omega}$-graph.* It is the infinite set of sets:

$$\mathcal{G}^{\vec{\omega}} = \{\mathcal{B}, \mathcal{X}^0, \mathcal{X}^1, \ldots, \mathcal{X}^{\vec{\omega}}\}, \tag{2.13}$$

where $\mathcal{B}$ is a set of branches and $\mathcal{X}^\nu$ ($\nu = 0, 1, \ldots, \vec{\omega}$) is a set of ν-nodes constructed as stated above. For each natural number μ, the node set $\mathcal{X}^\mu$ is nonempty. However, $\mathcal{X}^{\vec{\omega}}$ may be empty. (See [51, pages 38-41] for examples of $\vec{\omega}$-graphs.) For each $\mu = 0, 1, 2, \ldots$, $\{\mathcal{B}, \mathcal{X}^0, \ldots, \mathcal{X}^\mu\}$ is called the *μ-graph of $\mathcal{G}^{\vec{\omega}}$*.

A *subgraph* $\mathcal{G}_s$ of an $\vec{\omega}$-graph $\mathcal{G}^{\vec{\omega}}$ induced by a nonempty subset $\mathcal{B}_s$ of the branch set $\mathcal{B}$ is defined much as before. With μ being a natural number, a μ-node of $\mathcal{G}^{\vec{\omega}}$ is a μ-node of $\mathcal{G}_s$ if and only if it has a $(\mu - 1)$-tip with a representative all of whose branches are in $\mathcal{B}_s$. Also, an $\vec{\omega}$-node $x^{\vec{\omega}}$ of $\mathcal{G}^{\vec{\omega}}$ is an $\vec{\omega}$-node of $\mathcal{G}_s$ if and only if an infinity of its members are nodes having the stated property. The rank of the resulting subgraph $\mathcal{G}_s$ may be either $\vec{\omega}$ or a natural number μ. In the former case, $\mathcal{X}^{\vec{\omega}}$ may be empty.

Here too, a subgraph $\mathcal{G}_s$ of $\mathcal{G}^{\vec{\omega}}$ is not in general a graph, but we can generate a graph by reducing nodes. That is, we remove an embraced tip of a node of $\mathcal{G}_s$ when every representative of that tip has a branch not in $\mathcal{B}_s$. The result is a *reduced node.* Then, $\mathcal{B}_s$ along with the sets of reduced nodes at the various ranks comprise the *reduced graph $\mathcal{G}_r$*.

The terminology that was defined for the subgraphs and paths of a μ-graph carries over to $\vec{\omega}$-graphs virtually word-for-word. Thus, just as before, we define *maximal nodes*, the *partitioning* of a graph or subgraph by a set of subgraphs, the *traversing* of a tip by a subgraph or path, *incidence* for tips, nodes, branches, and subgraphs, *total disjointness* for subgraphs and/or nodes, and the *meeting* of subgraphs and/or nodes.

However, we now encounter a rather different kind of one-ended path. A *one-ended $\vec{\omega}$-path* $P^{\vec{\omega}}$ is a one-way infinite sequence

$$P^{\vec{\omega}} = \langle x_0^{\mu_0}, P_0^{\alpha_0}, x_1^{\mu_1}, P_1^{\alpha_1}, x_2^{\mu_2}, P_2^{\alpha_2}, \ldots \rangle \tag{2.14}$$

where the μ_m are natural numbers satisfying $\mu_0 < \mu_1 < \mu_2 < \ldots$, the α_m are natural numbers satisfying $0 \leq \alpha_m \leq \mu_{m+1} - 1$ for every m, each $x_m^{\mu_m}$ is a (not necessarily maximal) μ_m-node, each $P_m^{\alpha_m}$ is a nontrivial α_m-path, and the following conditions are satisfied.

(a) $P_0^{\alpha_0}$ meets $x_0^{\mu_0}$ with $x_0^{\mu_0}$ (that is, $P_0^{\alpha_0}$ terminates on the left with $x_0^{\mu_0}$).
(b) Each $P_m^{\alpha_m}$ is terminally incident to the two nodes adjacent to it in (2.14) but is otherwise totally disjoint from those nodes. Moreover, $P_m^{\alpha_m}$ either reaches $x_m^{\mu_m}$ (resp. $x_{m+1}^{\mu_{m+1}}$) through a terminal $(\mu_m - 1)$-tip (resp. a terminal $(\mu_{m+1} - 1)$-tip) or meets it with a terminal δ-node where $\delta \leq \mu_m - 1$ (resp. $\delta \leq \mu_{m+1} - 1$).
(c) Every two elements in (2.14) that are not adjacent therein are totally disjoint.

Here too, $P^{\vec{\omega}}$ is a subgraph of $\mathcal{G}^{\vec{\omega}}$ induced by the branches embraced by $P^{\vec{\omega}}$.

Also, a version of Lemma 2.4-2 holds for $\vec{\omega}$-paths; namely, for each $m > 0$, at least one of the paths $P_m^{\alpha_m}$ and $P_{m+1}^{\alpha_{m+1}}$ meets $x_{m+1}^{\mu_{m+1}}$ with a $(\mu_{m+1} - 1)$-tip.

There is no such thing as a two-ended $\vec{\omega}$-path or a trivial $\vec{\omega}$-path, but an *endless $\vec{\omega}$-path* is meaningful. Such a path arises when two one-ended $\vec{\omega}$-paths start at the same initial node $x_0^{\mu_0}$ but are otherwise totally disjoint. It can be written as a two-way infinite sequence:

$$\langle \ldots, x_{-2}^{\mu_{-2}}, P_{-2}^{\alpha_{-2}}, x_{-1}^{\mu_{-1}}, P_{-1}^{\alpha_{-1}}, x_0^{\mu_0}, P_0^{\alpha_0}, x_1^{\mu_1}, P_1^{\alpha_1}, x_2^{\mu_2}, \ldots \rangle \tag{2.15}$$

where $x_0^{\mu_0}$ and the terms to the left (resp. right) of $x_0^{\mu_0}$ comprise a one-ended $\vec{\omega}$-path, but those to the left are written in reverse order along with changes in its subscripts. Thus, for example, $\cdots > \mu_{-2} > \mu_{-1} > \mu_0 < \mu_1 < \mu_2 < \cdots$ and $0 \leq \alpha_{-m} \leq \mu_{-m} - 1$ for $-m < 0$.

An *orientation* can be assigned to a one-ended or endless $\vec{\omega}$-path, and it can be *traced* in the same way as for μ-paths.

Also, there is no such thing as an $\vec{\omega}$-loop, but there are loops of higher ranks as we shall see.

There is another difference between μ-paths and $\vec{\omega}$-paths that is worth mentioning. A μ-path can be written in only one way because the nodes displayed explicitly in its sequence (2.11) are all of a single rank μ. This is not so for $\vec{\omega}$-paths. For example, the first four terms of (2.14) can be encompassed within the path:

$$\langle x_0^{\mu_0}, P_0^{\alpha_0}, x_1^{\mu_1}, P_1^{\alpha_1}, y^{\rho_2} \rangle = \langle x_0^{\mu_0}, Q_0^{\alpha}, y^{\rho_2} \rangle$$

where y^{ρ_2} is the node that contains the terminal tip on the right of $P_1^{\alpha_1}$; y^{ρ_2} is embraced by $x_2^{\mu_2}$. Thus, (2.14) can be rewritten as

$$P^{\vec{\omega}} = \langle x_0^{\mu_0}, Q_0^{\alpha}, x_2^{\mu_2}, P_2^{\alpha_2}, \ldots \rangle.$$

Nonetheless, both versions of $P^{\vec{\omega}}$ have the same branch set with the same total ordering as determined by a tracing of $P^{\vec{\omega}}$; in this sense we can consider both versions as being the "same" $\vec{\omega}$-path.

The extremities of an $\vec{\omega}$-graph are its "$\vec{\omega}$-tips," which are equivalence classes of one-ended $\vec{\omega}$-paths. Now, however, we have to alter our definition of "equivalence" because one-ended $\vec{\omega}$-paths do not have unique representations as sequences, in contrast to one-ended μ-paths. We can use the fact that the branches embraced by a given one-ended $\vec{\omega}$-path $P^{\vec{\omega}}$ form a totally ordered set, the ordering being defined by a tracing of $P^{\vec{\omega}}$ starting at its terminal node. We shall say that two one-ended $\vec{\omega}$-paths $P_1^{\vec{\omega}}$ and $P_2^{\vec{\omega}}$ are *equivalent* if there is a node x_1 embraced by $P_1^{\vec{\omega}}$ and a node x_2 embraced by $P_2^{\vec{\omega}}$ such that the two totally ordered sets of branches following x_1 in $P_1^{\vec{\omega}}$ and following x_2 in $P_2^{\vec{\omega}}$ are identical. Thus, $P_1^{\vec{\omega}}$ and $P_2^{\vec{\omega}}$ are equivalent if and only if there is a third one-ended $\vec{\omega}$-path $P_3^{\vec{\omega}}$ that is "in" both $P_1^{\vec{\omega}}$ and $P_2^{\vec{\omega}}$ in the sense that all the branches of $P_3^{\vec{\omega}}$ are embraced by both $P_1^{\vec{\omega}}$ and $P_2^{\vec{\omega}}$.

This is truly an equivalence relationship, and it partitions the set of all one-ended $\vec{\omega}$-paths in $\mathcal{G}^{\vec{\omega}}$ into equivalence classes. Each such class will be called an *$\vec{\omega}$-tip*. A *representative* of a given $\vec{\omega}$-tip is any one of the paths in that equivalence class. Thus, for example, the endless $\vec{\omega}$-path (2.15) has exactly two $\vec{\omega}$-tips, and a representative of one tip or the other tip is the one-ended $\vec{\omega}$-path starting at $x_0^{\mu_0}$ and extending to the left or right. Each $\vec{\omega}$-tip of an $\vec{\omega}$-path $P^{\vec{\omega}}$ is called a *terminal tip* of $P^{\vec{\omega}}$. Also, if $P^{\vec{\omega}}$ is one-ended and therefore starts at a δ-node x_t^δ $(0 \leq \delta < \vec{\omega})$, the $(\delta - 1)$-tip with which $P^{\vec{\omega}}$ reaches x_t^δ is the other *terminal tip* of $P^{\vec{\omega}}$.

2.6 ω-Graphs

Our construction of an ω-graph is much the same as that of a μ-graph, except that now $\vec{\omega}$-tips replace $(\mu - 1)$-tips. Let $\mathcal{T}^{\vec{\omega}}$ denote the set of all $\vec{\omega}$-tips in the $\vec{\omega}$-graph $\mathcal{G}^{\vec{\omega}}$ at hand. We are assuming that $\mathcal{T}^{\vec{\omega}}$ is nonempty.[5] Partition $\mathcal{T}^{\vec{\omega}}$ into subsets $\mathcal{T}_\tau^{\vec{\omega}}$ in any fashion. Thus, $\mathcal{T}^{\vec{\omega}} = \cup_\tau \mathcal{T}_\tau^{\vec{\omega}}$, where each $\mathcal{T}_\tau^{\vec{\omega}}$ is nonempty and $\mathcal{T}_{\tau_1}^{\vec{\omega}} \cap \mathcal{T}_{\tau_2}^{\vec{\omega}} = \emptyset$ if $\tau_1 \neq \tau_2$. Also, for each τ, let $\mathcal{X}_\tau^{\vec{\omega}}$ be either the empty set or a singleton whose only member x_t^α is either an $\vec{\omega}$-node or a μ-node, μ being a natural number as always.

For each τ, the set

$$x_\tau^\omega = \mathcal{T}_\tau^{\vec{\omega}} \cup \mathcal{X}_\tau^{\vec{\omega}} \tag{2.16}$$

is called an *ω-node* so long as the following condition is satisfied: Whenever $\mathcal{X}_\tau^{\vec{\omega}}$ is nonempty, its single member x_τ^α $(0 \leq \alpha \leq \vec{\omega})$ is not a member of any other β-node $(\alpha < \beta \leq \omega)$.

When x_τ^α exists, we call it the *exceptional element* of x_τ^ω. As usual, we say that x_τ^ω *embraces* itself, all its elements, and all the elements embraced by its exceptional element if the latter exists. When x_τ^ω is a nonsingleton, all the elements embraced by

[5]In general, $\mathcal{T}^{\vec{\omega}}$ may be empty; see [51, Example 2.3-2].

x_τ^ω are said to be *shorted together by* x_τ^ω. However, when x_τ^ω is a singleton, its sole $\vec{\omega}$-tip is called *open*.

As an analog to Lemma 2.4-1, we have

Lemma 2.6-1. *If an* α*-node* $(0 \le \alpha \le \omega)$ *and an* ω*-node embrace a common node, then the* ω*-node embraces the* α*-node; in addition, if* $\alpha = \omega$*, then the* α*-node is identical to the* ω*-node.*

An ω*-graph* $\mathcal{G}^\omega$ is an infinite set of sets:

$$\mathcal{G}^\omega = \{\mathcal{B}, \mathcal{X}^0, \mathcal{X}^1, \ldots, \mathcal{X}^{\vec{\omega}}, \mathcal{X}^\omega\} \tag{2.17}$$

where again $\mathcal{B}$ is a set of branches and each $\mathcal{X}^\rho$ $(\rho = 0, 1, \ldots, \vec{\omega}, \omega)$ is a set of ρ-nodes constructed as stated above. (For every ρ other than $\rho = \vec{\omega}$, $\mathcal{X}^\rho$ is nonempty; $\mathcal{X}^{\vec{\omega}}$ may be empty.) For each $\rho = 0, 1, \ldots, \vec{\omega}$, $\{\mathcal{B}, \mathcal{X}^0, \ldots, \mathcal{X}^\rho\}$ is called the ρ*-graph of* $\mathcal{G}^\omega$.

Here too, a *subgraph* $\mathcal{G}_s$ of an ω-graph $\mathcal{G}^\omega$ induced by a subset $\mathcal{B}_s$ of $\mathcal{B}$ is defined as before. The nodes of $\mathcal{G}_s$ with natural-number ranks or with rank $\vec{\omega}$ are specified as they were for a subgraph of an $\vec{\omega}$-graph. Also, an ω-node of $\mathcal{G}^\omega$ is an ω-node of $\mathcal{G}_s$ if and only if it has an $\vec{\omega}$-tip with a representative all of whose branches are in $\mathcal{B}_s$. The rank of the resulting subgraph $\mathcal{G}_s$ may be either μ, $\vec{\omega}$, or ω. Furthermore, by removing from every node every tip that is not traversed by $\mathcal{G}_s$, we get the *reduced graph* induced by $\mathcal{B}_s$.

Moreover, the following terms are defined for ω-graphs in the same way as they were for μ-graphs (see Sec. 2.4): the *degree* of a node, a *maximal node*, the *partitioning* of a graph or subgraph by a set of subgraphs, the *traversing* of a tip by a subgraph or path, *incidence* for tips, nodes, branches, and subgraphs, *total disjointness* for subgraphs and/or nodes, and the *meeting* of subgraphs and/or nodes.

If x^ω is an ω-node that embraces a terminal tip of an α-path P^α $(0 \le \alpha \le \vec{\omega})$, then we say as before that P^α is *terminally incident to* or *reaches* x^ω *with* or *through* that terminal tip; if in addition P^α terminates at a δ-node x_t^δ $(0 \le \delta < \vec{\omega})$ and if x^ω embraces x_t^δ, then we also say that P^α is *terminally incident to* x^ω *with* x_t^δ and that P^α *meets* x^ω *with* x_t^δ. Finally, we say that P^α is *terminally incident to* x^ω *but otherwise totally disjoint from* x^ω if P^α reaches x^ω with a terminal tip and if x^ω does not embrace any other tip traversed by P^α.

A *trivial* ω*-path* is a singleton $\{x^\omega\}$, where x^ω is an ω-node. As for a "nontrivial ω-path," consider the alternating sequence of three or more elements:

$$P^\omega = \langle \ldots, x_m^\omega, P_m^{\alpha_m}, x_{m+1}^\omega, P_{m+1}^{\alpha_{m+1}}, \ldots \rangle \tag{2.18}$$

containing ω-nodes x_m^ω and α_m-paths $P_m^{\alpha_m}$, where m is restricted to a consecutive set of integers and where $0 \le \alpha_m \le \vec{\omega}$; the α_m may differ for different m. P^ω is called a *nontrivial* ω*-path* if the following three conditions are satisfied:

(a) Each $P_m^{\alpha_m}$ is a nontrivial α_m-path. Every x_m^ω is an ω-node except possibly when the sequence (2.18) terminates on the left and/or on the right. In the latter case, each terminal element is a δ-node x_t^δ, where either $\delta = \omega$ or $0 \le \delta < \vec{\omega}$, and its adjacent path is either an $\vec{\omega}$-path or an α-path $(\alpha \le \vec{\omega})$, respectively.

(b) Each $P_m^{\alpha_m}$ is terminally incident to the two nodes adjacent to it in (2.18) but is otherwise totally disjoint from those nodes. Moreover, $P_m^{\alpha_m}$ either reaches each of its adjacent nodes with a terminal $\vec{\omega}$-tip or meets it with a terminal δ-node where $\delta < \vec{\omega}$. If that adjacent node is a terminal node x_t^δ ($\delta \leq \omega$) of (2.18), then $P_m^{\alpha_m}$ reaches x_t^δ with a $(\delta - 1)$-tip; if in addition $\delta < \vec{\omega}$, then x_t^δ is also a terminal node of $P_m^{\alpha_m}$.

(c) Every two elements in (2.18) that are not adjacent therein are totally disjoint.

Here, too, an ω-path P^ω is a subgraph of $\mathcal{G}^\omega$ induced by the branches embraced by P^ω. Also, an analog of Lemma 2.4-2 holds; namely, if x_{m+1}^ω is not a terminal node of (2.18), then at least one of its adjacent paths $P_m^{\alpha_m}$ and $P_{m+1}^{\alpha_{m+1}}$ reaches x_{m+1}^ω with an $\vec{\omega}$-tip and must therefore be of rank $\vec{\omega}$.

P^ω is called *two-ended, one-ended,* or *endless* when (2.18) is respectively a finite, one-way infinite, or two-way infinite sequence. An ω*-loop* is defined exactly like a nontrivial two-ended ω-path except that one of its terminal nodes is required to embrace the other. A ρ-path ($0 \leq \rho \leq \omega$) or ρ-loop ($0 \leq \rho < \vec{\omega}$ or $\rho = \omega$), all of whose branches are embraced by a subgraph, is said to be *in* that subgraph. An analog of Lemma 2.4-3 holds here as well; namely, every two-ended ω-path and every ω-loop embraces every tip that that path or loop traverses.

Finally, as before, we can define the ω*-tips* of $\mathcal{G}^\omega$ as equivalence classes of one-ended ω-paths, two such paths being equivalent if their one-way infinite sequences differ on no more than finitely many elements of those sequences. This would prepare us for the definitions of $(\omega + 1)$-nodes and $(\omega + 1)$-graphs.

2.7 A Concise Characterization of Transfinite Paths and Loops

We have defined paths and loops several times while considering in turn higher ranks of transfiniteness. Those definitions specified the structure of those paths and loops. Moreover, their wording was much the same, except for the $\vec{\omega}$-paths—their structure is rather different. However, by introducing the idea of the "degree of a node with respect to a reduced graph," we can subsume all those definitions, including that for an $\vec{\omega}$-path, into a single characterization of paths and loops.

Let $\mathcal{G}^\nu$ be a ν-graph ($0 \leq \nu \leq \omega$), and let $\mathcal{R}$ be the reduced graph of a subgraph $\mathcal{G}_s$ of $\mathcal{G}^\nu$. A node of $\mathcal{R}$ that is not embraced by a node of $\mathcal{R}$ of higher rank will be called an $\mathcal{R}$*-maximal* node. For any $\mathcal{R}$-maximal node x of $\mathcal{R}$, let $d_{x|\mathcal{R}}$ denote the $\mathcal{R}$*-degree* of x, that is, $d_{x|\mathcal{R}}$ is the cardinality of the set of tips embraced by x. (Recall that, by the definition of a reduced graph, every tip embraced by x is traversed by $\mathcal{R}$.) Let us apply these ideas to the case where $\mathcal{G}_s$ is a path or a loop.

The nodes specified in the definition of a path or loop given in any prior section may embrace tips that are not traversed by the path or loop. For our characterization, we will strip from those nodes all the tips that are not traversed by the path or loop. On the other hand, a one-ended ρ-path P^ρ in the ν-graph $\mathcal{G}^\nu$, where $\rho < \nu$, will be incident to a $(\rho + 1)$-node $x^{\rho+1}$ containing the ρ-tip of P^ρ. Again for our characterization, we will append to P^ρ the singleton $(\rho + 1)$-node $x^{\rho+1}$ that contains that

ρ-tip but not any other tip embraced by $x^{\rho+1}$. Similarly, we will append two such singleton $(\rho+1)$-nodes to an endless ρ-path when $\rho < \nu$. All this will convert the one-ended or endless ρ-path into a two-ended $(\rho+1)$-path. However, if $\rho = \nu$, $\mathcal{G}^\nu$ will not contain any $(\nu+1)$-nodes, and the ν-path will remain one-ended or endless. All this can be accomplished by replacing the original path or loop P by the reduced graph $\mathcal{R}$ induced by the branches of the path or loop. We will call that $\mathcal{R}$ the *reduced path* or *reduced loop* for P. $\mathcal{R}$ will be called *nontrivial* if it contains at least two $\mathcal{R}$-maximal nodes.[6] Also, the *rank* of $\mathcal{R}$ is the largest of the ranks of the nodes of $\mathcal{R}$ if such a largest rank exists; otherwise, $\mathcal{R}$'s rank is $\vec{\omega}$. Finally, $\mathcal{R}$ is called *connected*[7] if for every two maximal nodes in $\mathcal{R}$ there is a path in $\mathcal{R}$ terminating at those nodes. $\mathcal{R}$ is connected if and only if $\mathcal{G}_s$ is connected.

Theorem 2.7-1. *Let $\mathcal{G}_s$ be a subgraph of the ν-graph $\mathcal{G}^\nu$ ($0 \le \nu \le \omega$, possibly $\nu = \vec{\omega}$), and let $\mathcal{R}$ be the reduced graph corresponding to $\mathcal{G}_s$.*

(i) $\mathcal{G}_s$ either is a loop of any rank or is an endless ν-path in $\mathcal{G}^\nu$ if and only if $\mathcal{R}$ is connected and $d_{x|\mathcal{R}} = 2$ for every $\mathcal{R}$-maximal node x of $\mathcal{R}$.
(ii) $\mathcal{G}_s$ is a one-ended ν-path if and only if $\mathcal{R}$ is connected and $d_{x|\mathcal{R}} = 2$ for all $\mathcal{R}$-maximal nodes x except for exactly one $\mathcal{R}$-maximal node y having $d_{y|\mathcal{R}} = 1$.
(iii) $\mathcal{G}_s$ is a two-ended path of any rank in $\mathcal{G}^\nu$ if and only if $\mathcal{R}$ is connected and $d_{x|\mathcal{R}} = 2$ for all $\mathcal{R}$-maximal nodes x except for exactly two $\mathcal{R}$-maximal nodes, y and z, having $d_{y|\mathcal{R}} = d_{z|\mathcal{R}} = 1$.

Proof. It follows obviously from the definitions of a path and loop in $\mathcal{G}^\nu$ that the "only if" parts of (i), (ii), and (iii) hold. For the "if" parts, we need only show that $\mathcal{R}$ is a loop or path of the kind stated because then $\mathcal{G}_s$ will be the same kind of loop or path.

Note that by the definition of a reduced graph, a ρ-tip of $\mathcal{R}$ of rank less than ν will be incident in $\mathcal{R}$ to a $(\rho+1)$-node of $\mathcal{R}$. Thus, $\mathcal{R}$ can be a one-ended or endless path in $\mathcal{R}$ only if its rank is ν.

Now, consider the "if" part of (i). $\mathcal{R}$ is now assumed to be connected. Choose any $\mathcal{R}$-maximal node y in $\mathcal{R}$; y has exactly two incident tips. Start tracing $\mathcal{R}$ along one of those tips. Because every $\mathcal{R}$-maximal node has exactly two incident tips. there is only one way of tracing through any subsequent node. Let us consider some conceivabilities: (1) The tracing twice meets a node x other than y. This will imply that $d_{x|\mathcal{R}} > 2$, in contradiction to our assumption; hence, this case does not occur. (2) The tracing meets y for a second time. In this case, a loop has been traced. That loop must have passed through all the nodes of $\mathcal{R}$ because otherwise $\mathcal{R}$ would not be connected—another consequence of the condition that $d_{x|\mathcal{R}} = 2$ for all $\mathcal{R}$-maximal nodes x of $\mathcal{R}$. This is one of the two possibilities cited in (i). (3) The only remaining conceivability is that the tracing proceeds indefinitely without repeating any node. Thus, a one-ended path is traced, and its rank must be ν. But then, we can also start tracing from y along the other tip incident to y. Again, another one-ended path of

[6] Thus, a trivial $\mathcal{R}$ would simply be a self-loop.

[7] Connectedness is discussed in Sec. 3.1.

rank ν must be traced, and the two tracings must not meet at a node x other than y, for otherwise $d_{x|\mathcal{R}} = 2$ would be violated. Thus, the two tracings together comprise an endless ν-path. That $\mathcal{R}$ is connected again implies that the endless ν-path passes through all the nodes of $\mathcal{R}$. The "if" part of (i) has hereby been established.

The "if" parts of (ii) and (iii) are proven by much the same arguments. So, let us simply summarize these cases:

(ii) This time start at the node y for which $d_{y|\mathcal{R}} = 1$. The tracing will pass along a one-ended ν-path through all the nodes of $\mathcal{R}$.

(iii) This time start at one of the two nodes, y and z. The tracing will have to be along a two-ended path of any rank no larger than ν that terminates at y and z. That two-ended path will pass through all the nodes of $\mathcal{R}$. ♣

Let us summarize how the form of the nontrivial reduced path or reduced loop $\mathcal{R}$ relates to its rank and to the $\mathcal{R}$-degrees of its nodes. $\mathcal{R}$ is connected, and $d_{x|\mathcal{R}} = 2$ for all nonterminal $\mathcal{R}$-maximal nodes of $\mathcal{R}$. If there are exactly two $\mathcal{R}$-maximal nodes in $\mathcal{R}$ for which $d_{x|\mathcal{R}} = 1$, then $\mathcal{R}$ is a two-ended path; $\mathcal{R}$ can have any rank less than or equal to ν other than $\vec{\omega}$, and the nodes of largest rank in $\mathcal{R}$ are now finite in number. If there is exactly one $\mathcal{R}$-maximal node in $\mathcal{R}$ for which $d_{x|\mathcal{R}} = 1$, then $\mathcal{R}$ is a one-ended path; the rank of $\mathcal{R}$ must now be ν (possibly $\nu = \vec{\omega}$). If there is no $\mathcal{R}$-maximal node in $\mathcal{R}$ for which $d_{x|\mathcal{R}}$ equals 1, then three cases arise:

(a) If there are infinitely many nodes of largest rank in $\mathcal{R}$, then $\mathcal{R}$ is an endless ν-path with $\nu \neq \vec{\omega}$.
(b) If there is no largest rank for the nodes in $\mathcal{R}$, then $\mathcal{R}$ is an endless $\vec{\omega}$-path.
(c) If there is at least one but no more than finitely many nodes of largest rank in $\mathcal{R}$, then $\mathcal{R}$ is a loop whose rank can be less than or equal to ν. (Remember that there is no such thing as a $\vec{\omega}$-loop.)

2.8 Graphs of Higher Ranks

We can mimic the constructions of Secs. 2.3 and 2.4 to obtain transfinite graphs of ranks $\omega + \mu$, where μ is any positive natural number. Then, imitating what was done in Sec. 2.5, we get a $\vec{\omega} \cdot 2$-graph, which in turn leads to an $\omega \cdot 2$-graph, and so on to still higher ranks of transfiniteness.

How far can all this be carried? We would like to say throughout all the countable ordinals. Indeed, why not employ transfinite recursion to do so all at once? The possible stumbling block is that the transfinite-recursion argument at a ν-graph $\mathcal{G}^\nu$ or at a $\vec{\nu}$-graph $\mathcal{G}^{\vec{\nu}}$ requires that the analogs of all the conditions used in Secs. 2.4 and 2.5 for the constructions of a μ-graph and an $\vec{\omega}$-graph should hold for the graphs of all ranks less than ν and $\vec{\nu}$. It is not apparent that such a situation can hold for any arbitrary countable ordinal ν.

We will not explore this extension to higher ranks and will assume throughout the rest of this book that the rank ν of $\mathcal{G}^\nu$ will be no larger than ω.

2.9 Why Not Restrict "Extremities" to "Ends"?

A much-studied structure of conventionally infinite graphs is that of an "end." Ends have become a common way of representing the infinite extremities of such graphs. Examples of early works that discuss this idea are [25], [27], [32], and [33], and research on it has continued.

An "end" can be defined as follows: Let $\mathcal{G}^0 = \{\mathcal{B}, \mathcal{X}^0\}$ be a 0-graph having at least one 0-tip (that is, at least one one-ended 0-path). With $\mathcal{B}$ being the set of branches of $\mathcal{G}^0$, let $\mathcal{E}$ be any finite subset of $\mathcal{B}$, and let $\mathcal{G}_s(\mathcal{B} \setminus \mathcal{E})$ denote the subgraph of $\mathcal{G}^0$ induced by all the branches of $\mathcal{B}$ that are not in $\mathcal{E}$. An *end* of $\mathcal{G}^0$ is a mapping e that takes every finite subset $\mathcal{E}$ of $\mathcal{B}$ into an infinite component $e(\mathcal{E})$ of $\mathcal{G}_s(\mathcal{B} \setminus \mathcal{E})$ in accordance with the following rule: If $\mathcal{E}_1$ and $\mathcal{E}_2$ are finite subsets of $\mathcal{B}$ with $\mathcal{E}_1 \subset \mathcal{E}_2$, then $e(\mathcal{E}_2)$ is a subgraph of $e(\mathcal{E}_1)$.

An equivalent way of defining an end is based on certain equivalence classes of one-ended 0-paths in $\mathcal{G}^0$. Let P^0 and Z^0 be two one-ended 0-paths in $\mathcal{G}^0$ oriented toward their 0-tips. Let us say that P^0 and Z^0 *meet persistently* if, given any node x in P^0 and any node y in Z^0, there is a node z beyond x in P^0 and beyond y in Z^0 at which P^0 and Z^0 meet. Another way of saying this is that every one-ended subpath of P^0 meets every one-ended subpath of Z^0. Then, two one-ended 0-paths P^0 and Q^0 in $\mathcal{G}^0$ will be called *end-equivalent* if there is a third one-ended 0-path Z^0 that meets both P^0 and Q^0 persistently. This partitions the set of all one-ended 0-paths in $\mathcal{G}^0$ into equivalence classes, and it can be shown that each such equivalence class coincides with an end of $\mathcal{G}^0$ as defined in the preceding paragraph.

Furthermore, it can be shown that the ends of $\mathcal{G}^0$ partition the set of 0-tips of $\mathcal{G}^0$, with each 0-tip belonging to a particular end but with each end having possibly many 0-tips.

Altogether, the ends of $\mathcal{G}^0$ can be interpreted as representing particular infinite extremities of $\mathcal{G}^0$. So, why not use this well-used construct as the places where connections at infinite extremities of $\mathcal{G}^0$ are to be made? The answer is that "ends" are in general too crude to allow all the various such connections we may want to make. For example, a one-way infinite ladder has only one end, and yet we may wish to connect, say, a branch to two 0-tips within that single end. (Fig. 6.7(a) in Chapter 6 shows such a connection.)

To be sure, 0-tips provide a perhaps overly fine multitude of infinite extremities of $\mathcal{G}^0$, but, on the other hand, ends are too coarse for certain purposes.

Actually, one may use either or both such representations of infinite extremities depending upon the purposes at hand. Indeed, the idea of an end can be generalized transfinitely, as is done in [54, Sec. 3.6], and it provides some potent means of examining various phenomena relating to transfinite graphs, such as random walks on them [51, Chapter 7], [54, Chapter 8] or electrical current flows through infinite extremities [54, Secs. 5.3 to 5.5]. But, the fact remains that ends do not provide enough structure to encompass other phenomena, such as electrical behavior in the presence of connections at different tips within a single end [50, Chapters 3 to 5], [51, Chapter 5], [54, Chapter 5].

In short, ends are fine for certain purposes, but tips are needed for other purposes.

3

Connectedness, Trees, and Hypergraphs

The first section, Sec. 3.1, is the more important one for this chapter. It defines the various ranks of connectedness for transfinite graphs, presents the critical Condition 3.1-2 that insures that connectedness is transitive, and shows how at each rank this partitions a transfinite graph into nonoverlapping subgraphs, called "sections." Sec. 3.2 explains how a transfinite tree can be easily contracted to and expanded from a conventional tree. Sec. 3.3 relates the ν-nodes and $(\nu - 1)$-sections of a ν-graph to a hypergraph and gives an example of how a result from hypergraph theory can be transferred to transfinite graph theory.

3.1 Transfinite Connectedness

Given any ν-graph $\mathcal{G}^\nu$ $(0 \le \nu \le \omega)$, let ρ be any rank such that $0 \le \rho \le \nu$. Here, either or both of ν and ρ may equal $\vec{\omega}$. Also, let x and y be two totally disjoint nodes, not necessarily of the same rank. x and y are said to be *ρ-connected* if there is a two-ended α-path P^α, with $\alpha \le \rho$ when ρ is an ordinal and with $\alpha < \rho$ when $\rho = \vec{\omega}$, such that P^α meets both x and y. As a special case, we also say that two nodes are *ρ-connected* for every $\rho \le \nu$ if one node embraces the other. Two branches are said to be *ρ-connected* if a 0-node of one branch and a 0-node of the other branch are ρ-connected or are the same 0-node; if this is so for all pairs of branches in $\mathcal{G}^\nu$, then $\mathcal{G}^\nu$ itself is called *ρ-connected.* A branch is *ρ-connected* to a node x if a 0-node y^0 of the branch is ρ-connected to x or x embraces y^0. The definition of ρ-connectedness is extended to tips by saying that a tip is *ρ-connected* to some entity if a node embracing that tip is ρ-connected to that entity. Note that ρ-connectedness implies λ-connectedness whenever $\rho < \lambda \le \nu$.

A *ρ-section* $\mathcal{S}^\rho$ is a subgraph of the ρ-graph of $\mathcal{G}^\nu$ induced by a maximal set of branches that are pairwise ρ-connected.[1] A *component* of $\mathcal{G}^\nu$ is a ν-section. As an

[1]This is a somewhat sharper definition of a ρ-section than that given in [51, page 49] because there a ρ-section was merely required to be a subgraph of $\mathcal{G}^\nu$, whereas now a ρ-section is required to be a subgraph of the ρ-graph of $\mathcal{G}^\nu$. The only difference between

additional special case, a (-1)-*section* is taken to be a single branch; in this case alone it is not a subgraph.

Connectedness is a binary relationship on the set of all branches, nodes, and tips. It is obviously reflexive and symmetric, but it need not be transitive. As a consequence of this nontransitivity, it can happen that different ρ-sections of a ν-graph $\mathcal{G}^\nu$ *overlap* in the sense that they share branches, as the next example shows. However, 0-connectedness is always transitive; it is at the higher ranks of ρ-connectedness ($\rho > 0$) that nontransitivity can occur.

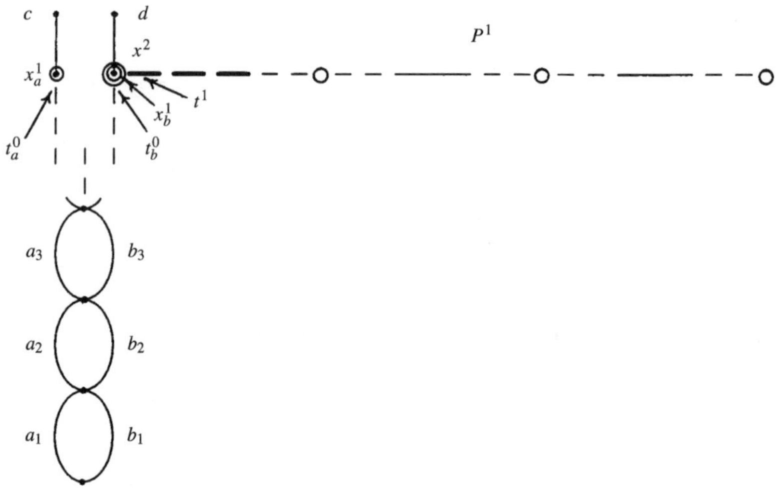

Fig. 3.1. A 2-graph. The heavy dots denote 0-nodes, the line segments between heavy dots denote branches, the small circles denote 1-nodes, and the larger circle denotes a 2-node x^2.

Example 3.1-1. Here is an example showing some of the complications that can arise when connectedness is nontransitive. (For other examples, see [51, Examples 3.1-5 and 3.1-6] and [54, Example 3.1-1].) Consider the 2-graph shown in Fig. 3.1. There, t_a^0 (resp. t_b^0) is the 0-tip of the one-ended 0-path consisting of the branches a_k (resp. b_k), where $k = 1, 2, \ldots$. The 1-node x_a^1 (resp. x_b^1) embraces t_a^0 (resp. t_b^0) and an elementary tip of the branch c (resp. branch d). The 2-node x^2 embraces x_b^1 (and x_b^1's embraced elements of course) and also the 1-tip t^1 of the one-ended 1-path P^1. There are also uncountably many 1-nodes (not shown), all taken to be singletons, each of which contains a 0-tip resulting from a selection of the a_k and b_k branches,

the definition of a ρ-section $\mathcal{S}^\rho$ given in [51] and the one used here is that in [51] nodes of ranks higher than ρ are allowed to be in $\mathcal{S}^\rho$, whereas no such node can be in $\mathcal{S}^\rho$ according to our present definition; our present definition avoids some unnecessary complications. Let us also note that there is a sharper idea, called a "subsection"; it plays an important role in [51], but we will have no need for it in this book.

infinitely many of each. There is no path of any rank connecting the branches c and d because any attempt to trace such a path will perforce repeat some of the 0-nodes between the a_k and b_k branches.

The sections of this 2-graph are as follows:

- A 0-section induced by all the a_k and b_k ($k = 1, 2, \ldots$), another 0-section induced by c, another 0-section induced by d, and finally infinitely many 0-sections within P^1. These 0-sections do not "overlap," meaning that they do not share branches.
- A 1-section induced by c and all the a_k and b_k, another 1-section induced by d and all the a_k and b_k, and another 1-section induced by the branches of P^1 (i.e., P^1 is a 1-section by itself). The first two 1-sections overlap.
- A 2-section induced by c and all the a_k and b_k, and finally a 2-section induced by d, all the a_k and b_k, and all the branches of P^1. These two 2-sections overlap.

♣

However, the transitivity of ρ-connectedness can be assured by restricting the graph $\mathcal{G}^\nu$ appropriately. Two criteria, each of which suffices for such transitivity, are [51, Condition 3.2-1 and Condition 3.5-1]. The first criterion is rather lengthy and of limited use. The second one is quite simple to state and not overly restrictive, but the proof of it given in [51] is quite long and complicated.[2] This latter case can be simplified by assuming that every node of $\mathcal{G}^\nu$ is *pristine*, that is, each node does not embrace a node of lower rank—and consequently is not embraced by a node of higher rank either.[3] Under this condition, the second criterion [51, Condition 3.5-1] works again and with a much shorter proof. It turns out that that latter proof can be modified to apply to nonpristine graphs as well, yielding thereby a more elegant argument for the transitivity of ρ-connectedness. This is the subject of the present section.

We need some definitions. Two nonelementary tips (not necessarily of the same rank) are said to be *disconnectable* if one can find two representatives, one for each tip, that are totally disjoint. (Any two tips, at least one of which is elementary, are simply taken to be *disconnectable*.) As the negation of "disconnectable," we say that two tips are *nondisconnectable* if every representative of one of them meets every representative of the other tip infinitely often, that is, at infinitely many nodes. We can state this somewhat differently, as follows. Two tips, t_1^γ and t_2^δ, are called nondisconnectable if P_1^γ and P_2^δ meet at least once whenever P_1^γ is a representative of t_1^γ and P_2^δ is a representative of t_2^δ.

Here is the sufficient condition that will ensure the transitivity of connectedness.

---

[2] Please note the correction for [51, pages 63–69] stated in the Errata for that book given at the URL http://www.ee.sunysb.edu/~zeman. Namely, every elementary set of a sequence of any rank is required to be node-distinct. This was implied by the fact that the argument was based on paths, but it should have been explicitly stated.

[3] See [54, Sec. 3.1] in this regard.

Condition 3.1-2. *If two tips of ranks less than ν (and not necessarily of the same rank) are nondisconnectable, then either they are shorted together or at least one of them is open.*

Before turning to our principle result, we need to introduce some more specificity into our terminology. The maximal nodes partition all the nodes of $\mathcal{G}^\nu$; indeed, two nodes are in the same set of the partition if they are embraced by the same maximal node.[4] Thus, corresponding to any path, there is a unique minimal set of maximal nodes that together embrace all the nodes of the path, and that set of maximal nodes becomes totally ordered in accordance with an orientation assigned to the path. Moreover, two maximal nodes are perforce totally disjoint. In order to avoid explicating repeatedly which nodes embrace which nodes, it is convenient to deal only with the maximal nodes of our graph $\mathcal{G}^\nu$.

Let us expand upon this point because it is the key idea that allows us to replace the long proof in [51, pages 57–73] by a shorter, modified version of the proof in [54, pages 31–35]. Given any path P^ρ of any rank ρ ($\rho \neq \vec{\omega}$) in $\mathcal{G}^\nu$, let $\mathcal{X}$ be the set of (possibly nonmaximal) nodes encountered in the recursive construction of P^ρ; that is, $\mathcal{X}$ consists of the ρ-nodes in the sequential representation of P^ρ along with the nodes of lower ranks in the sequential representations of the paths between those ρ-nodes, and also along with the nodes of still lower ranks in the sequential representations of the paths of lower ranks in those latter paths, and so on down to the 0-nodes of the 0-paths embraced by P^ρ. For $\rho = \vec{\omega}$, a similar decomposition applied to the nodes and paths in (2.14) or (2.15) yields the set $\mathcal{X}$. Each node x^α of $\mathcal{X}$ need not be maximal, but there is a unique maximal node x^β in $\mathcal{G}^\nu$ embracing x^α. Let $\mathcal{X}_m$ denote the set of those maximal nodes. So, given P^ρ and thereby $\mathcal{X}$ and $\mathcal{X}_m$, each node of $\mathcal{X}_m$ corresponds to a unique set of nodes in $\mathcal{X}$ such that, for every two nodes in that latter set, one embraces the other. The union of those sets is $\mathcal{X}$. In this way, there is a bijection between $\mathcal{X}_m$ and the collection of the said sets, which in fact comprise a partition of $\mathcal{X}$. Moreover, when P^ρ has an orientation, that orientation induces a total ordering of $\mathcal{X}_m$. Thus, when dealing with the nodes of P^ρ, we can fix our attention on the nodes of $\mathcal{X}_m$. This we shall do. Any node x^β of $\mathcal{X}_m$ will be called a *maximal node for* P^ρ to distinguish it from the corresponding subset of nodes in $\mathcal{X}$. In general, we may have $\beta > \rho$, but not necessarily always. Also, we can transfer our terminology for the nodes of a path, such as "incident to a node," "meets a node," or "terminates at a node," to those maximal nodes. For example, a path is *incident to* a node $x \in \mathcal{X}_m$ if one of the path's terminal tips is embraced by x; this will be so if the path has a terminal node embraced by x. In the latter case, we say that the path *meets* x and also *terminates at* x. Thus, two paths *meet at* $x \in \mathcal{X}_m$ if x embraces a node y of one path and a node z of the second path such that y embraces or is embraced by z.

The proof of our main result (Theorem 3.1-4) requires another idea, namely, "path cuts." Let P^ρ be a ρ-path with an orientation. Let $\mathcal{Y}$ be the set of all the

[4] Similarly, the nodes of any subgraph are partitioned by the maximal nodes that embrace nodes embraced by the subgraph.

branches in P^ρ, all the (not necessarily maximal) nodes of all ranks in P^ρ, and all the nodes of $\mathcal{X}_m$ for P^ρ. The orientation of P^ρ totally orders $\mathcal{Y}$. With y_1 and y_2 being two members of $\mathcal{Y}$, we say that y_1 is *before* y_2 and that y_2 is *after* y_1 if in a tracing of P^ρ in the direction of its orientation y_1 is met before y_2 is met. A *path cut* $\{\mathcal{B}_1, \mathcal{B}_2\}$ for P^ρ is a partitioning of the set of branches of P^ρ into two subsets, $\mathcal{B}_1$ and $\mathcal{B}_2$, (perforce nonempty by definition of a partition) such that every branch of $\mathcal{B}_1$ is before every branch of $\mathcal{B}_2$. Another way of stating this is as follows. The partition $\{\mathcal{B}_1, \mathcal{B}_2\}$ of the branch set of P^ρ comprises a path cut for P^ρ if and only if, for each branch $b \in \mathcal{B}_1$, every branch of P^ρ before b is also a member of $\mathcal{B}_1$.

Lemma 3.1-3. *For each path cut $\{\mathcal{B}_1, \mathcal{B}_2\}$ for P^ρ, there is a unique maximal node x^γ for P^ρ (i.e., $x^\gamma \in \mathcal{X}_m$) such that every branch $b_1 \in \mathcal{B}_1$ is before x^γ and every branch $b_2 \in \mathcal{B}_2$ is after x^γ.*

Note. We will say that the path cut *occurs at* the maximal node x^γ. As was stated before, a node is called *maximal* if it is not embraced by a node of higher rank. On the other hand, a maximal node for P^ρ may embrace many other nodes. It follows that all the nodes of P^ρ other than the nodes of P^ρ embraced by x^γ are also partitioned into two sets, the nodes of one set being before x^γ and the nodes of the other set being after x^γ.

Proof. This lemma is obvious if the rank ρ of P^ρ is 0. For higher ranks, we argue inductively. If ρ is a positive natural number μ, we assume this lemma is true for every rank $\alpha < \mu$. Let P^μ be oriented in the direction of increasing indices m in the expression (2.11) for P^μ. If the path cut occurs within a $(\mu - 1)$-path $P_m^{\mu-1}$ of P^μ, then the branches of $P_m^{\mu-1}$ are appropriately partitioned at some maximal γ-node x^γ for $P_m^{\mu-1}$, and this in turn places every other $(\mu - 1)$-path in P^μ either before or after the node x^γ. Thus, all the branches in P^μ are partitioned according to $\{\mathcal{B}_1, \mathcal{B}_2\}$. The only other possibility is that the branch set of P^μ is partitioned according to $\{\mathcal{B}_1, \mathcal{B}_2\}$ at some maximal node x^γ that embraces a μ-node of P^μ.

For ρ equal to $\vec{\omega}$ or ω, this argument works in the same way when applied to (2.14), (2.15), or (2.18). ♣

Here now is our main result.

Theorem 3.1-4. *Let $\mathcal{G}^\nu$ $(0 \le \nu \le \omega)$ be a ν-graph for which Condition 3.1-2 is satisfied. Let x_a, x_b, and x_c be three different maximal nodes in $\mathcal{G}^\nu$ such that, if any one of them is a singleton, its sole tip is disconnectable from every tip embraced by the other two nodes. If x_a and x_b are ρ-connected and if x_b and x_c are ρ-connected $(0 \le \rho \le \nu)$, then x_a and x_c are ρ-connected.*

Proof. Note that, if any one of x_a, x_b, and x_c is a nonsingleton, then each of its embraced tips (perforce, nonopen) must be disconnectable from every tip embraced by the other two nodes; indeed, otherwise, the tips of that node would be shorted to the tips of another one of those three nodes, according to Condition 3.1-2 and our hypothesis, and thus x_a, x_b, and x_c could not be three different maximal nodes.

Let P_{ab}^{α} ($\alpha \leq \rho$) be a two-ended α-path that terminates at the maximal nodes x_a and x_b and is oriented from x_a to x_b, and let P_{bc}^{β} ($\beta \leq \rho$) be a two-ended β-path that terminates at the maximal nodes x_b and x_c and is oriented from x_b to x_c. Let P_{ba}^{α} be P_{ab}^{α} but with the reverse orientation. P_{ba}^{α} cannot have infinitely many α-nodes because it is two-ended.

Let $\{x_i\}_{i\in I}$ be the set of maximal nodes at which P_{ba}^{α} and P_{bc}^{β} meet, and let $\mathcal{X}_1$ be that set of nodes with the order induced by the orientation of P_{ba}^{α}. If $\mathcal{X}_1$ has a last node x_l, then a tracing along P_{ab}^{α} from x_a to x_l followed by a tracing along P_{bc}^{β} from x_l to x_c yields a path of rank no larger than ρ that connects x_a and x_c. Thus, x_a and x_c are ρ-connected in this case. This will certainly be so when $\{x_i\}_{i\in I}$ is a finite set.

So, assume $\mathcal{X}_1$ is an infinite, ordered set (ordered as stated). We shall show that $\mathcal{X}_1$ has a last node. Let Q_1 be the path induced by those branches of P_{ba}^{α} that lie between nodes of $\mathcal{X}_1$ (i.e., as P_{ba}^{α} is traced from x_b onward, such a branch is traced after some node of $\mathcal{X}_1$ and before another node of $\mathcal{X}_1$, those nodes depending upon the choice of the branch). Let $\mathcal{B}_1$ be the set of those branches. We can take it that P_{ba}^{α} extends beyond the nodes of $\mathcal{X}_1$, for otherwise $\mathcal{X}_1$ would have x_a as its last node, and x_a and x_c would be ρ-connected. Therefore, we also have a nonempty set $\mathcal{B}_2$ consisting of those branches in P_{ba}^{α} that are not in $\mathcal{B}_1$. $\{\mathcal{B}_1, \mathcal{B}_2\}$ is a path cut for P_{ba}^{α}. Consequently, by Lemma 3.1-3, there is a unique maximal node $x_1^{\gamma_1}$ at which that path cut occurs. Thus, Q_1 terminates at $x_1^{\gamma_1}$. Let $t_1^{\rho_1}$ be the ρ_1-tip through which Q_1 reaches $x_1^{\gamma_1}$. Every representative of $t_1^{\rho_1}$ contains infinitely many nodes of $\mathcal{X}_1$ (otherwise, $\mathcal{X}_1$ would have a last node).

Now, consider P_{bc}^{β}. We can take it that there is a maximal node x_d in P_{bc}^{β} different from x_c such that the subpath of P_{bc}^{β} between x_d and x_c is totally disjoint from P_{ba}^{α}, for otherwise x_a and x_c would have to be the same node according to Condition 3.1-2 and our hypothesis again. We can partition the branches of P_{bc}^{β} into two sets, $\mathcal{B}_3$ and $\mathcal{B}_4$, as follows. Each branch of the first set $\mathcal{B}_3$ is such that it lies before (according to the orientation of P_{bc}^{β}) at least one node in $\mathcal{X}_1$ of each representative of $t_1^{\rho_1}$, this being so for all such representatives.[5] The second set $\mathcal{B}_4$ consists of all the branches of P_{bc}^{β} that are not in $\mathcal{B}_3$. No branch of $\mathcal{B}_4$ can precede a branch of $\mathcal{B}_3$. Thus, we have a path cut $\{\mathcal{B}_3, \mathcal{B}_4\}$ for P_{bc}^{β} and thereby (according to Lemma 3.1-3) a unique maximal node $x_2^{\gamma_2}$ lying after the branches of $\mathcal{B}_3$ and before the branches of $\mathcal{B}_4$. Let Q_2 be the path induced by $\mathcal{B}_3$. It reaches $x_2^{\gamma_2}$ through some tip $t_2^{\rho_2}$. Furthermore, each representative of $t_2^{\rho_2}$ must meet each representative of $t_1^{\rho_1}$ at least once because they meet at at least one node of $\mathcal{X}_1$. Thus, $t_1^{\rho_1}$ and $t_2^{\rho_2}$ are nondisconnectable. Moreover, neither of those tips can be open (i.e., be in a singleton node) because the paths P_{ba}^{α} and P_{bc}^{β} pass

[5]Let us note here a correction for [54, page 34, line 19 up]: Replace the sentence on that line by the following sentence. "Let $\mathcal{N}_2$ be the set of those nodes in $\{n_i\}_{i\in I}$ that lie before (according to the orientation of P_{bc}^{β} now) at least one node of $\{n_i\}_{i\in I}$ in each representative of $t_1^{\rho_1}$, this being so for all such representatives."

through and beyond their respective nodes $x_1^{\gamma_1}$ and $x_2^{\gamma_2}$. So, by Condition 3.1-2, $x_1^{\gamma_1}$ and $x_2^{\gamma_2}$ must be the same node because the tips $t_1^{\rho_1}$ and $t_2^{\rho_2}$ are shorted together.

This means that $\mathcal{X}_1$ has a last node, namely, $x_1^{\gamma_1} = x_2^{\gamma_2}$. It follows now that x_a and x_c are ρ-connected. ♣

The last proof has established the following two results.

Corollary 3.1-5. *Under the hypothesis of Theorem 3.1-4, let P_{ab}^{α} be a two-ended α-path connecting nodes x_a and x_b, and let P_{bc}^{β} be a two-ended β-path connecting nodes x_b and x_c. Then, there is a two-ended γ-path ($\gamma \leq \max\{\alpha, \beta\}$) connecting x_a and x_c that lies in $P_{ab}^{\alpha} \cup P_{bc}^{\beta}$.*

Corollary 3.1-6. *Let $\mathcal{G}^{\nu}$ $(0 \leq \nu \leq \omega)$ be a ν-graph for which Condition 3.1-2 is satisfied. Let $\{x_i\}_{i \in I}$ be the set of maximal nodes at which two two-ended paths in $\mathcal{G}^{\nu}$ meet. Assume that, if any terminal node of either path is a singleton, its sole tip is disconnectable from every tip embraced by the two terminal nodes of the other path. Assign to $\{x_i\}_{i \in I}$ the total ordering induced by an orientation of one of these two paths. Then, $\{x_i\}_{i \in I}$ has both a first node and a last node.*

Here are some more definitions we shall need. A *bordering node*[6] of a ρ-section $\mathcal{S}^{\rho}$ is a λ-node x^{λ} $(\lambda > \rho)$ that is incident to $\mathcal{S}^{\rho}$; or equivalently it is a λ-node $(\lambda > \rho)$ that embraces an α-tip t^{α} $(\alpha \leq \rho)$ having a representative in $\mathcal{S}^{\rho}$. (Then, all the representatives of t^{α} will be in $\mathcal{S}^{\rho}$.) As a special case, we may have $\alpha = -1$ so that t^{α} is the elementary tip of a branch.

A *boundary node* x^{λ} of $\mathcal{S}^{\rho}$ is a bordering node of $\mathcal{S}^{\rho}$ that embraces a tip of any rank whose every representative meets at least one branch not in $\mathcal{S}^{\rho}$; in other words, that tip is not traversed by $\mathcal{S}^{\rho}$.

Since a ρ-section $\mathcal{S}^{\rho}$ is a ρ-graph, all the nodes in $\mathcal{S}^{\rho}$ have ranks no larger than ρ. An *internal*[7] *node* (resp. *noninternal node*) of $\mathcal{S}^{\rho}$ is a node of $\mathcal{S}^{\rho}$ that is not (resp. is) embraced in $\mathcal{G}^{\nu}$ by a node of rank greater than ρ, that is, by a bordering node of $\mathcal{S}^{\rho}$.

Example 3.1-7. In the 2-graph of Fig. 3.2, there are two 0-sections, one of which is the one-ended 0-path P^0 and the second of which consists of all the branches and 0-nodes not in P^0. The 1-node y_0^1 is a boundary node for both of these 0-sections. The other 1-nodes $y_1^1, y_2^1, y_3^1, \ldots$ are bordering nodes but not boundary nodes of that second 0-section. The singleton 2-node z^2 is not a bordering node of either 0-section because it does not embrace a 0-tip of either 0-section. The maximal (resp. nonmaximal) 0-nodes are internal (resp. noninternal) 0-nodes of their respective 0-sections.

---

[6]Here too, our definitions of bordering nodes and boundary nodes are somewhat different from those of [51, pages 49 and 81].

[7]Later on when we make use of nonstandard analysis, the adjective "internal" will have an entirely different meaning. See Appendices A12 and A20.

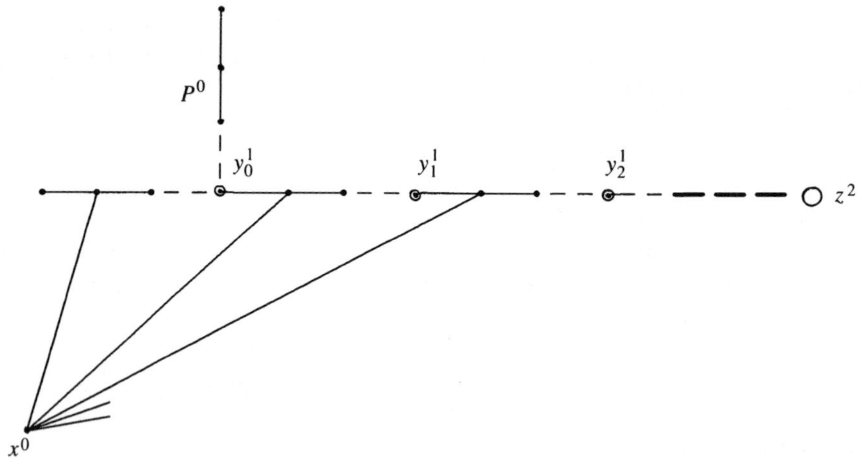

Fig. 3.2. A 2-graph. As usual, the heavy dots represent 0-nodes, the small circles are 1-nodes, and the large circle on the right is a singleton 2-node z^2. In this 2-graph there is a one-ended 1-path consisting of the 1-nodes $y_0^1, y_1^1, y_2^1, \ldots$ interspersed with one-ended 0-paths. Also, there is a 0-node x^0 of infinite degree that is adjacent to the second 0-nodes of all the one-ended 0-paths in that 1-path. In addition, there is a one-ended 0-path P^0 that reaches the 1-node y_0^1 and is otherwise totally disjoint from the rest of the graph.

Furthermore, this 2-graph has exactly one 1-section, and z^2 is a bordering node for it. All the 0-nodes and 1-nodes are internal nodes of that 1-section. ♣

An easy consequence of the definition of a boundary node is the following.

Lemma 3.1-8. *If x is a boundary node of a ρ-section $\mathcal{S}^\rho$ in a ν-graph satisfying Condition 3.1-2, then x satisfies at least one of the following two conditions:*

(i) All the tips embraced by x and traversed by $\mathcal{S}^\rho$ have ranks no less than ρ.
(ii) All the tips embraced by x and not traversed by $\mathcal{S}^\rho$ have ranks no less than ρ.

Moreover, every path that passes through x and reaches a branch in $\mathcal{S}^\rho$ and a branch not in $\mathcal{S}^\rho$ has a rank greater than ρ. Furthermore, if $\rho = \nu - 1$, then all, except possibly one, of the $(\nu-1)$-sections incident to the boundary node x^ν are so incident through tips of rank $\nu - 1$; the exceptional section, if it exists, will be incident to x^ν through tips of ranks no larger than $\nu - 1$.

Proof. If the rank conditions of (i) and (ii) are both violated, then we can conclude that there is a tip in $\mathcal{S}^\rho$ and a tip not in $\mathcal{S}^\rho$, both embraced by x, whose ranks are less than ρ. But then, those tips are shorted together by x, and their representatives will all be in $\mathcal{S}^\rho$, a contradiction. The remaining assertions follow immediately. ♣

Since some of the foregoing definitions differ from those used in [51], let us illustrate them and other definitions as well with two more simple 2-graphs.

Example 3.1-9. The 2-graph $\mathcal{G}_a^2$ shown in Fig 3.3(a) consists of a one-ended 1-path P^1,

$$P^1 = \langle x^0, P_0^0, x_1^1, P_1^0, x_2^1, P_2^0, \ldots \rangle$$

that reaches a 2-node x^2 through its 1-tip. Each 1-node x_k^1 $(k = 1, 2, 3, \ldots)$ of P^1 contains the 0-tip of its preceding one-ended 0-path P_{k-1}^0 and the terminal 0-node of its succeeding one-ended 0-path P_k^0. There are countably many 0-tips, one for each P_{k-1}^0, and exactly one 1-tip.

This 2-graph has countably many 0-sections, each being a one-ended 0-path. Each 1-node x_k^1 is a boundary node for the 0-sections P_{k-1}^0 and P_k^0. The 1-graph of this 2-graph is the one-ended path P^1, which is also a 1-section of $\mathcal{G}_a^2$. The 2-node x^2 is a nonboundary bordering node of the 1-section P^1. ♣

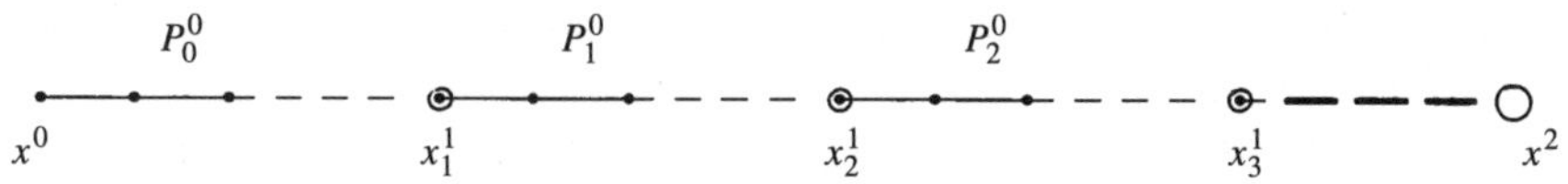

Fig. 3.3a. The 2-graph $\mathcal{G}_a^2$ discussed in Example 3.1-9. The heavy dots denote 0-nodes, the lines between such dots denote branches, the small circles denote 1-nodes, and the larger circle on the right denotes a 2-node x^2.

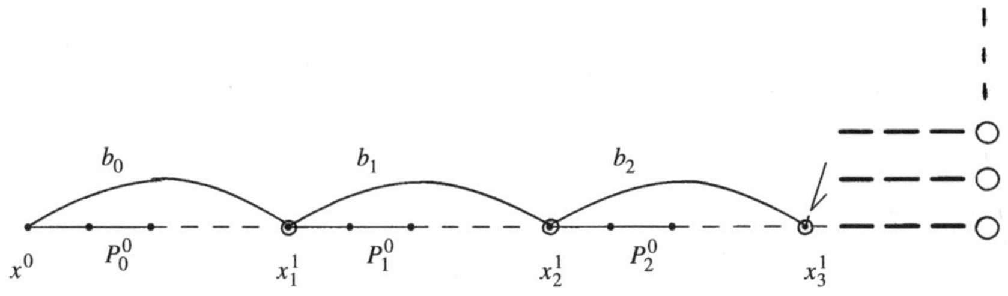

Fig. 3.3b. The 2-graph $\mathcal{G}_b^2$ discussed in Example 3.1-10. This is obtained from the 2-graph of Fig. 3.3(a) by appending the branches b_k $(k = 0, 1, 2, \ldots)$ as shown. The large circles on the right are all singleton 2-nodes.

Example 3.1-10. Let us alter the 2-graph of the preceding example by appending a branch b_k $(k = 0, 1, 2, \ldots)$ between the initial terminal 0-node of each 0-path P_k^0 and the 1-node x_{k+1}^1 that P_k^0 reaches. This shown in Fig. 3.3(b). This 2-graph $\mathcal{G}_b^2$ also has countably many 0-tips, one for each P_k^0 and another one for the one-ended 0-path P_b^0 consisting of the branches b_k. Let us assume that the 0-tip for P_b^0 is open and therefore contained in a singleton 1-node x_b^1 (not shown in Fig. 3.3(b)). Now, however, there are uncountably many 1-tips, each having a representative 1-path obtained by choosing either b_k or P_k^0 for each $k = 0, 1, 2, \ldots$. We choose to make each 1-tip open; thus, there are uncountably many singleton 2-nodes, one for each 1-tip.

The 0-graph of this 2-graph $\mathcal{G}_b^2$ is a single 0-section $\mathcal{S}^0$ induced by all the branches. The 1-nodes x_k^1 ($k = 1, 2, 3, \ldots$) and x_b^1 are now nonboundary bordering nodes of $\mathcal{S}^0$, and the 2-nodes are also all nonboundary bordering nodes of $\mathcal{S}^0$.

The 1-graph of $\mathcal{G}_b^2$ also consists of a single 1-section $\mathcal{S}^1$, again induced by all the branches. Now, however, the 1-nodes x_k^1 are all internal nodes of $\mathcal{S}^1$. The 2-nodes are all bordering nodes of $\mathcal{S}^1$, and there are no boundary nodes for $\mathcal{S}^1$. Were we to append a one-ended 1-path (totally disjoint from $\mathcal{G}_b^2$) with its 1-tip in one of the 2-nodes, that 2-node would become a boundary node for the two 1-sections then occurring. ♣

Theorem 3.1-4 asserts that, under its hypothesis, ρ-connectedness ($0 \leq \rho \leq \nu$) is transitive and therefore an equivalence relationship between branches. Thus, the branch set $\mathcal{B}$ of $\mathcal{G}^\nu$ is partitioned by ρ-connectedness. This implies that ρ-sections cannot overlap in the sense of a branch being in two or more ρ-sections. This means that ρ-sections *partition* $\mathcal{G}^\nu$ in the sense that the branch set $\mathcal{B}$ of $\mathcal{G}^\nu$ is partitioned into the branch sets of the ρ-sections in $\mathcal{G}^\nu$. Moreover, we can drop the requirement concerning singleton nodes in the hypothesis of Theorem 3.1-4 because we are now dealing with connectedness between branches (that is, between their 0-nodes) and because any elementary tip of any branch is automatically disconnectable from all other tips of any ranks. We are thus led to the following result.

Corollary 3.1-11. *Let $\mathcal{G}^\nu$ be a ν-graph that satisfies Condition 3.1-2. Then, the ρ-sections of $\mathcal{G}^\nu$ comprise a partition of $\mathcal{G}^\nu$. Also, for each $\gamma < \rho$, each ρ-section of $\mathcal{G}^\nu$ is partitioned by the γ-sections contained in that ρ-section.*

Finally, let us state another lemma that we shall need later on.

Lemma 3.1-12. *Let $\mathcal{G}^\nu$ be a ν-connected ν-graph satisfying Condition 3.1-2. Let $\mathcal{S}^\rho$ be a ρ-section in $\mathcal{G}^\nu$ having (at least) two nonsingleton bordering nodes x and y. Then, there is (at least) one two-ended path in $\mathcal{S}^\rho$ (i.e., its branches are in $\mathcal{S}^\rho$) that terminates at x and y.*

Proof. There are two tips t_x and t_y of $\mathcal{S}^\rho$ that reach x and y respectively. Let P_x and P_y be representatives of t_x and t_y in $\mathcal{S}^\rho$, and let u and v be nodes of P_x and P_y respectively. By the connectedness of $\mathcal{S}^\rho$, there is a two-ended path Q in $\mathcal{S}^\rho$ terminating at u and v. We can choose u and v to be nonsingletons too. Then, by Corollary 3.1-5, $P_x \cup Q \cup P_y$ contains a two-ended path R terminating at x and y. Since all the branches of P_x, Q, and P_y lie in $\mathcal{S}^\rho$, the same is true of R. ♣

3.2 Transfinite Trees

A *ν-tree* is a ν-connected ν-graph having no loops. Every tip of a ν-tree T^ν is disconnectable from every other tip of T^ν, for otherwise T^ν would contain a loop. Thus, Condition 3.1-2 is vacuously satisfied by T^ν. It also follows that every branch is a *bridge* (synonymously, an *isthmus*)—meaning that the removal of any branch leaves the remaining graph with exactly two components.

Now, consider a path in T^ν having the following property: The maximal nodes reached by the path are all of degree 2, except when the path has a terminal node, in which case that terminal node is embraced by a maximal node of degree 1 or more than 2. (Remember that the degree of any node is the cardinality of the set of its embraced tips.) We will call such a path a *maximal series path* in T^ν. Such a path may extend infinitely on either side, but, if it does, it will reach a ν-tip on that side.

Given any ν-tree ($\nu > 0$), we can replace each of its maximal series paths by a single branch. This replaces the ν-tree by a 0-tree (i.e., a conventional tree), each of whose nodes is of degree 1 or more than 2. Conversely, any 0-tree can be expanded into a ν-tree by replacing each of its maximal series 0-paths by a transfinite path that preserves the two-endedness, one-endedness, or endlessness of that 0-path. Since any transfinite tree can be so easily contracted to or expanded from a 0-tree in this way, not much can be said about a transfinite tree that is significantly different from the properties of conventional trees. For instance, the various characterizations of a conventional finite tree (see, for example, [45, Theorem 9A]) can be restated appropriately for transfinite trees having only finitely many maximal nodes of degrees 1 or more than 2. Similarly, the properties of a transfinite tree having infinitely many such nodes can be readily obtained from the properties of infinite 0-trees.

So, we drop this topic and turn to a more interesting kind of ν-tree having a certain tree-like structure only with regard to its $(\nu - 1)$-sections.

3.3 Hypergraphs from ν-Graphs

A *hypergraph* $\mathcal{H}$ is a set $\mathcal{X}$ along with a "family" $\mathcal{E}$ of subsets of $\mathcal{X}$ [6]. By a *family* we mean a collection of elements wherein elements may repeat in the collection;[8] this contrasts with a set wherein elements of the set are required to be pairwise distinct. $\mathcal{H}$ is called *finite* if $\mathcal{X}$ is a finite set and the cardinality of $\mathcal{E}$ is finite too when each of its elements are counted as many times as that element appears in $\mathcal{E}$.

We will relate a ν-graph $\mathcal{G}^\nu$ to a unique hypergraph $\mathcal{H}$. We continue to assume that the ν-graph $\mathcal{G}^\nu$ satisfies Condition 3.1-2. Let us also assume throughout this section that $\mathcal{G}^\nu$ is ν-connected and has at least two $(\nu - 1)$-sections. Thus, $\mathcal{G}^\nu$ has at least one boundary ν-node. We shall now exploit the theory of hypergraphs [6]. Corresponding to $\mathcal{G}^\nu$ there is a hypergraph $\mathcal{H}$ whose vertices are the boundary ν-nodes of $\mathcal{G}^\nu$ and whose edges are the $(\nu - 1)$-sections of $\mathcal{G}^\nu$. In particular, corresponding

[8]Equivalently, we may define $\mathcal{H}$ as the triplet of a set $\mathcal{X}$, a set $\mathcal{E}$, and a mapping (not necessarily injective) of $\mathcal{E}$ into the power set of $\mathcal{X}$.

to each $(\nu-1)$-section $\mathcal{S}^{\nu-1}$ there is an edge of $\mathcal{H}$ whose vertices correspond to the boundary ν-nodes incident to $\mathcal{S}^{\nu-1}$.[9] Because of the ν-connectedness of $\mathcal{G}^\nu$ and the presence of at least two $(\nu-1)$-sections, no edge is empty, and the union of the edges is the set of vertices—that is, $\mathcal{H}$ is truly a hypergraph. Consequently, many of the properties of hypergraphs can be transferred to ν-graphs so far as the interconnections between its ν-nodes and its $(\nu-1)$-sections are concerned. However, most of hypergraph theory is restricted to finite hypergraphs, and thus we should restrict our ν-graphs accordingly in order to apply certain ideas concerning hypergraphs to transfinite graphs.

As an example of such an application, consider a finite hypergraph $\mathcal{H}$ without "cycles." A *cycle* C in $\mathcal{H}$ is a closed alternating sequence of vertices v_k and edges e_k:

$$C = \langle v_0, e_0, v_1, e_1, \ldots, v_K, e_K, v_0 \rangle,$$

where $K \geq 1$; $v_0, \ldots, v_K$ are all distinct vertices of $\mathcal{H}$; $e_0, \ldots, e_K$ are all distinct edges of $\mathcal{H}$; and $v_k, v_{k+1} \in e_k$ for $k = 0, \ldots, K$ with $v_{K+1} = v_0$. A cycle can be related to a ν-loop in $\mathcal{G}^\nu$. However, not every ν-loop L^ν in $\mathcal{G}^\nu$ corresponds to a cycle in $\mathcal{H}$. Indeed, if all the branches of L^ν are in a single $(\nu-1)$-section, there is no cycle in $\mathcal{H}$ corresponding to L^ν. On the other hand, if L^ν passes through two or more $(\nu-1)$-sections (this we assume henceforth), there will be at least one and perhaps several cycles corresponding to portions of L^ν. One of them can be found as follows.

A tracing of L^ν generates an alternating sequence of ν-nodes and $(\nu-1)$-sections, where two consecutive ν-nodes are incident to the intervening $(\nu-1)$-section in the sequence. In fact, since L^ν is a closed sequence, there will be a pair x_k^ν, x_m^ν with $m > k+1$ (perhaps several such pairs) of ν-nodes in L^ν such that the alternating subsequence

$$\langle x_k^\nu, \mathcal{S}_k^{\nu-1}, x_{k+1}^\nu, \mathcal{S}_{k+1}^{\nu-1}, \ldots, x_{m-1}^\nu, \mathcal{S}_{m-1}^{\nu-1}, x_m^\nu \rangle \tag{3.1}$$

has the following properties: The $\mathcal{S}_k^{\nu-1}, \ldots, \mathcal{S}_{m-1}^{\nu-1}$ are all distinct $(\nu-1)$-sections; the $x_k^\nu, \ldots, x_{m-1}^\nu$ are all distinct boundary ν-nodes with $\mathcal{S}_l^{\nu-1}$ incident to x_l^ν and x_{l+1}^ν for $l = k, \ldots, m-1$; finally, x_m^ν is incident to $\mathcal{S}_k^{\nu-1}$. If $x_m^\nu = x_k^\nu$, then (3.1) corresponds to a cycle in $\mathcal{H}$. If $x_m^\nu \neq x_k^\nu$, then

$$\langle x_m^\nu, \mathcal{S}_k^{\nu-1}, x_{k+1}^\nu, \mathcal{S}_{k+1}^{\nu-1}, \ldots, x_{m-1}^\nu, \mathcal{S}_{m-1}^{\nu-1}, x_m^\nu \rangle \tag{3.2}$$

corresponds to a cycle in $\mathcal{H}$. (We invoke Lemma 3.1-12 at this point to find a path in $\mathcal{S}_k^{\nu-1}$ that meets x_m^ν and x_{k+1}^ν.) Any ν-loop that passes through two or more $(\nu-1)$-sections will be called a *cycle* in $\mathcal{G}^\nu$.

For our next purpose, a fact to note is that $\mathcal{G}^\nu$ has no cycle if and only if $\mathcal{H}$ has no cycle. In this case, we shall say that both $\mathcal{G}^\nu$ and $\mathcal{H}$ are *cycle-free*. An obvious result of our definitions is that $\mathcal{G}^\nu$ is cycle-free if and only if, given any two boundary

[9] We do not include the nonboundary bordering ν-nodes as vertices in $\mathcal{H}$ because they do not contribute to the connections between the $(\nu-1)$-sections and in this sense are superfluous.

ν-nodes and any ν-path P^ν terminating at them, the alternating sequence of boundary ν-nodes and distinct $(\nu - 1)$-sections generated by a tracing of P^ν is unique. (By requiring that the $(\nu - 1)$-sections be distinct, we are ignoring those boundary ν-nodes met by P^ν through tips lying in the same $(\nu - 1)$-section.)

Another result [6, page 392] concerning a connected, finite hypergraph $\mathcal{H}$ is that $\mathcal{H}$ is cycle-free if and only if

$$\sum_{i=1}^{E}(|e_i| - 1) = B - 1 \tag{3.3}$$

where e_i denotes the ith edge, $|e_i|$ is the number of vertices in e_i, E is the number of edges in $\mathcal{H}$, and B is the total number of vertices.

We will convert this result to make it applicable to certain ν-graphs. Another definition is needed. An *end* (resp. *non-end*) $(\nu - 1)$-section of $\mathcal{G}^\nu$ is a $(\nu - 1)$-section having exactly one incident boundary ν-node (resp. having two or more incident boundary ν-nodes).

Lemma 3.3-1. *Assume that the ν-graph $\mathcal{G}^\nu$ is cycle-free, ν-connected, satisfies Condition 3.1-2, has at least two $(\nu - 1)$-sections, and has only finitely many boundary ν-nodes. Then, $\mathcal{G}^\nu$ has only finitely many non-end $(\nu - 1)$-sections and at least two end $(\nu - 1)$-sections.*

Proof. If there are no non-end $(\nu - 1)$-sections, the conclusions are trivially satisfied. So, assume otherwise, and choose any non-end $(\nu - 1)$-section. Label it and all its boundary ν-nodes by "1." There will be at least two 1-labeled boundary ν-nodes. Label by "2" all the non-end $(\nu - 1)$-sections that share boundary ν-nodes with that 1-labeled $(\nu - 1)$-section (if such exist), and label their unlabeled boundary ν-nodes by "2," as well. Each 2-labeled section shares exactly one boundary ν-node with the 1-labeled $(\nu - 1)$-section and does not share any 2-labeled ν-node with any other 2-labeled $(\nu - 1)$-section, for otherwise $\mathcal{G}^\nu$ would contain a loop and therefore would not be cycle-free. It follows that the number of labeled ν-nodes is no less than the number of labeled $(\nu - 1)$-sections. Next, label by "3" all the non-end $(\nu - 1)$-sections that share boundary ν-nodes with the 2-labeled sections (if such exist), and label their unlabeled boundary ν-nodes by "3," as well. Again, each 3-labeled $(\nu - 1)$-section shares exactly one boundary ν-node with exactly one 2-labeled $(\nu - 1)$-section and does not share any 3-labeled ν-node with any other 3-labeled $(\nu - 1)$-section, for otherwise $\mathcal{G}^\nu$ would not be cycle-free. Here, too, it follows that the number of labeled ν-nodes is no less than the number of labeled $(\nu - 1)$-sections. Continue this way. At each step, the number of labeled ν-nodes will be no less than the number of labeled $(\nu - 1)$-sections. Since there are only finitely many boundary ν-nodes and since these are the ν-nodes that have been labeled, $\mathcal{G}^\nu$ can have only finitely many non-end $(\nu - 1)$-sections.

For the second conclusion, note that this labeling procedure terminates when the only unlabeled $(\nu - 1)$-sections are all end $(\nu - 1)$-sections. Since the labeling process started off with at least two 1-labeled boundary ν-nodes, there must be at least two end $(\nu - 1)$-sections. ♣

As a variation of our previously described correspondence between $\mathcal{G}^\nu$ and $\mathcal{H}$, we now proceed to replace $\mathcal{H}$ by another hypergraph $\mathcal{H}'$ as follows. The boundary ν-nodes of $\mathcal{G}^\nu$ are taken to correspond bijectively with the vertices of a hypergraph $\mathcal{H}'$, as before. Also, the non-end $(\nu-1)$-sections of $\mathcal{G}^\nu$ are taken to correspond injectively with the edges of the hypergraph $\mathcal{H}'$. However, there may be infinitely many end $(\nu-1)$-sections in $\mathcal{G}^\nu$. For each boundary ν-node x^ν, we represent all the end $(\nu-1)$-nodes incident to x^ν by a single edge in $\mathcal{H}'$; that edge will have only one vertex, the vertex corresponding to x^ν. Altogether, these are the only edges of $\mathcal{H}'$. $\mathcal{H}'$ is now taken to be the finite hypergraph corresponding to $\mathcal{G}^\nu$. So, we can invoke Lemma 3.3-1 and (3.3) with $\mathcal{H}$ replaced by $\mathcal{H}'$ to obtain the following converted result for $\mathcal{G}^\nu$.

Theorem 3.3-2. *Let $\mathcal{G}^\nu$ be a ν-connected ν-graph satisfying Condition 3.1-2, having at least two $(\nu-1)$-sections, and having only finitely many boundary ν-nodes. $\mathcal{G}^\nu$ is cycle-free if and only if*

$$J = \sum_{j=1}^{J} |e_j| - B + 1 \tag{3.4}$$

where now e_j denotes the jth non-end $(\nu-1)$-section $\mathcal{S}_j^{\nu-1}$ in $\mathcal{G}^\nu$, $|e_j|$ is the number of boundary ν-nodes incident to $\mathcal{S}_j^{\nu-1}$, J is the number of non-end $(\nu-1)$-sections in $\mathcal{G}^\nu$, and B is the number of boundary ν-nodes in $\mathcal{G}^\nu$.

Proof. Since $\mathcal{H}$ is cycle-free if and only if $\mathcal{H}'$ is cycle-free, we have that $\mathcal{G}^\nu$ is cycle-free if and only if $\mathcal{H}'$ is cycle-free. Note also that, by Lemma 3.3-1, $\mathcal{G}^\nu$ has only finitely many non-end $(\nu-1)$-sections. Thus $\mathcal{H}'$ is a connected finite hypergraph. We have that $\mathcal{H}'$ is cycle-free if and only if (3.3) holds. Note also that, for any end $(\nu-1)$-section $\mathcal{S}_k^{\nu-1}$, $|e_k| - 1 = 0$. So, we can ignore the end $(\nu-1)$-sections in our converted formula. Our theorem now follows from the correspondence between $\mathcal{G}^\nu$ and $\mathcal{H}'$ and a rearrangement of (3.3). ♣

Finally, let us extend to cycle-free transfinite graphs the fact that every node in a conventional tree is a cut-node. A maximal node x (of any rank) in $\mathcal{G}^\nu$ will be called a *cut-node* if there exists two maximal nodes y and z in $\mathcal{G}^\nu$ other than x such that every path connecting y and z passes through x. Thus, a cut-node has to be a nonsingleton. Another way of defining a cut-node is in terms of its "removal." We *remove* a nonsingleton node y by replacing y by two or more singleton nodes, each of which contains exactly one of the embraced tips of y and with every embraced tip so assigned. $\mathcal{G}^\nu - y$ will denote the graph obtained by removing y from $\mathcal{G}^\nu$. Then, x is called a *cut-node* if it is a nonsingleton and if $\mathcal{G}^\nu - x$ has two or more components. (Remember that $\mathcal{G}^\nu$ is assumed to be ν-connected.) Clearly, these two definitions are equivalent.

Theorem 3.3-3. *Let $\mathcal{G}^\nu$ be a ν-connected ν-graph satisfying Condition 3.1-2 and having at least two $(\nu-1)$-sections. $\mathcal{G}^\nu$ is cycle-free if and only if every boundary ν-node in $\mathcal{G}^\nu$ is a cut-node.*

Proof. *If:* Suppose there is a cycle in $\mathcal{G}^\nu$. It can be written as in (3.2). By Lemma 3.1-12, this implies that there is a ν-loop passing through all the boundary ν-nodes in (3.2). But, this means that there is a node y in $\mathcal{S}_k^{\nu-1}$ and a node z in $\mathcal{S}_{k+1}^{\nu-1}$ connected by a path passing through x_{k+1}^ν and by another path not passing through x_{k+1}^ν. Thus, we have found a boundary ν-node, namely, x_{k+1}^ν that is not a cut-node. This contrapositive argument establishes the "if" assertion.

Only if: Suppose a boundary ν-node x^ν is not a cut-node. Then, for every pair of maximal nodes, there is a path connecting them that does not pass through x^ν. Now, x^ν is incident to at least two $(\nu - 1)$-sections, say, $\mathcal{S}_1^{\nu-1}$ and $\mathcal{S}_2^{\nu-1}$. Let y_1 and y_2 be two internal nodes of $\mathcal{S}_1^{\nu-1}$ and $\mathcal{S}_2^{\nu-1}$ respectively. There is a path P (resp. Q) with branches in $\mathcal{S}_1^{\nu-1}$ (resp. $\mathcal{S}_2^{\nu-1}$) terminating at x^ν and y_1 (resp. y_2). Also, there is a path R terminating at y_1 and y_2 that does not pass through x^ν. By Corollary 3.1-6, upon tracing P from y_1 to x^ν, we will find a last node node z_1 different from x^ν at which P and R meet. Also, by tracing Q from x^ν to y_2, we will find a first node z_2 different from x^ν at which Q and R meet. Then, a tracing from z_1 along P to x^ν, then from x^ν along Q to z_2, and then from z_2 along R to z_1 follows a loop L. L passes through at least two $(\nu - 1)$-sections, namely, $\mathcal{S}_1^{\nu-1}$ and $\mathcal{S}_2^{\nu-1}$ and is of rank ν since it passes through x^ν (invoke Lemma 3.1-8); that is, L is a cycle, in contradiction to the hypothesis that $\mathcal{G}^\nu$ is cycle-free. Consequently, x^ν must be a cut-node. ♣

4

Ordinal Distances in Transfinite Graphs

The idea of distances in connected finite graphs has been quite fruitful, with much research directed toward both theory and applications. See, for example, [8], [9], [14] and the references therein. Such distances are given by a metric that assigns to each pair of nodes the minimum number of branches for all the paths connecting those two nodes. Thus, the metric takes its values in the set $\aleph_0$ if natural numbers. The objective in this chapter is to extend that metric to connected transfinite graphs and then to establish several transfinite generalizations of distance-related facts about finite graphs. This requires that our distance metric now take its values in the set $\aleph_1$ of all countable ordinals.

The distance metric used herein is quite different from the distance metrics used in [13], [54]. Those metrics take their values in the nonnegative real line, comprise an infinite hierarchy—one metric for each rank of transfiniteness, require a variety of restrictions such as local finiteness, and do not reduce to the branch-count metric for finite graphs. Our present branch-count metric can be applied far more generally and is defined for all nonsingleton nodes and some singleton nodes as well without most of the restrictions needed by the other metrics. However, it does require that there be no self-loops and that the transfinite graph be connected; these two conditions are tacitly assumed throughout this chapter.

Let us emphasize why, we feel, that for this chapter it is inappropriate to use a real-valued metric that makes infinite extremities of a conventionally infinite graph look as though they are only finitely distant from any node of the graph. If branch counts are to determine distances between nodes in a conventionally infinite graph, then no node is closer to any infinite extremity than any other node. Indeed, all nodes should be viewed as equally distant from any infinite extremity, and that distance should be ω, the first transfinite ordinal. This property cannot be avoided if branch counts are to prevail. Similarly, for transfinite graphs, the role played by ω is taken by larger limit ordinals.

We assume throughout this chapter that the ν-graph $\mathcal{G}^\nu$ we are dealing with is ν-connected and has no self-loops.

4.1 Natural Sums of Ordinals

We will use a variety of results from the theory of ordinal and cardinal numbers. A good exposition on this subject is [1]. Much more concisely, [51, Appendix A] summarizes all the facts concerning ordinal and cardinal numbers that we shall need, except that the "normal expansions" of transfinite ordinals and their "natural sums" [1, pages 354–355] are not discussed in [51, Appendix A]. So, let us now state the pertinent facts about the latter two ideas—restricting ourselves to the kinds of ordinals relating to branch-count distances in transfinite graphs.

As we shall see below, for a given transfinite graph $\mathcal{G}^\nu$ ($\nu \leq \omega$) every such ordinal distance α is bounded by $\omega^{\omega+1}\cdot 2$. Such an ordinal has a unique representation, called its *normal expansion*:

$$\alpha = \omega^{k_1}\cdot j_1 + \omega^{k_2}\cdot j_2 + \cdots + \omega^{k_m}\cdot j_m \qquad (4.1)$$

where m and j_i ($i = 1, \ldots, m$) are natural numbers with $m \geq 1$ and $j_i > 0$ for such i and where $\omega + 1 \geq k_1 > k_2 > \cdots > k_m \geq 0$. Thus, for $m = 1$ and $k_m = 0$, we have a finite distance $\alpha = \omega^0\cdot j_1 = j_1$. Because of the way the addition of ordinals is defined [1, page 327], it is appropriate to write the normal expansion in the order of decreasing k_i as shown in (4.1) because ordinal addition is not commutative. Later on, we will be adding ordinal lengths of paths, in which case those lengths should be arranged into their normal expansions.

Given the normal expansions of two or more ordinals it is now a simple matter to add them. For each i just add the corresponding coefficients j_i of their natural expansions. Such an addition is called the *natural sum* of ordinals. For example, the natural sum of ω^3, $\omega^2\cdot 2 + 4$, and $\omega^3\cdot 2 + \omega^2\cdot 3 + \omega\cdot 2 + 1$ is $\omega^3\cdot 3 + \omega^2\cdot 5 + \omega\cdot 2 + 5$. Another illustration of this is given in the forthcoming Example 4.2-1.

4.2 Lengths of Paths

In this section we define the *length* $|P|$ of a nontrivial path P. The length of any trivial path is 0.

0-Paths:

Let P^0 be a nontrivial 0-path (see Equation (2.3)). If P^0 is one-ended, its length $|P^0|$ is defined to be ω. If P^0 is endless, we set $|P^0| = \omega\cdot 2$. If P^0 is two-ended, we set $|P^0| = \tau_0$, where τ_0 is the number of branches in P^0. We might motivate these definitions by noting that we are using ω to denote the infinity of branches in a one-ended 0-path and using $\omega\cdot 2$ to represent the fact that an endless 0-path is the union of two one-ended paths. Equivalently, we can identify ω with each 0-tip traversed; a one-ended 0-path has one 0-tip, and an endless 0-path has two 0-tips—hence the length $\omega\cdot 2$.

1-Paths:

Let P^1 be a nontrivial 1-path as in (2.5). By definition, we have the following lengths. If P^1 is one-ended, $|P^1| = \omega^2$. If P^1 is endless, $|P^1| = \omega^2 \cdot 2$. When P^1 is two-ended, we set $|P^1| = \sum_m |P_m^0|$, where the sum is over the finitely many 0-paths in (2.5). This gives the normal expansion of the ordinal

$$|P^1| \;=\; \omega \cdot \tau_1 \;+\; \tau_0, \tag{4.2}$$

where τ_1 is the number of 0-tips P^1 traverses, and τ_0 is the number of branches in all the 0-paths in (2.5) that are two-ended.

μ-Paths:

Now, let P^μ be a nontrivial μ-path of positive natural-number rank μ; see Equation (2.11). If P^μ is one-ended, set $|P^\mu| = \omega^{\mu+1}$. If P^μ is endless, set $|P^\mu| = \omega^{\mu+1} \cdot 2$. Moreover, if P^μ is two-ended, set $|P^\mu| = \sum_m |P_m^{\alpha_m}|$, where again we have by definition the natural sum over the ordinal lengths $|P_m^{\alpha_m}|$ of the paths $P_m^{\alpha_m}$ in (2.11). Recursively, this gives

$$|P^\mu| \;=\; \omega^\mu \cdot \tau_\mu \;+\; \omega^{\mu-1} \cdot \tau_{\mu-1} \;+\; \cdots \;+\; \omega \cdot \tau_1 \;+\; \tau_0, \tag{4.3}$$

where $\tau_\mu, \tau_{\mu-1}, \ldots, \tau_0$ are natural numbers. τ_μ is the number of $(\mu - 1)$-tips among all the one-ended and endless $(\mu - 1)$-paths (i.e., when $\alpha_m = \mu - 1$) appearing in (2.11); τ_μ is not 0 because P^μ is of rank μ and must therefore have at least one $(\mu - 1)$-tip (see Lemma 2.4-2). For $k = \mu - 1, \mu - 2, \ldots, 1$, we set τ_k equal to the number of $k - 1$-tips generated by these recursive definitions. Finally, τ_0 is one-half the number of elementary tips generated recursively by these definitions; thus, τ_0 is a number of branches because each branch has exactly two elementary tips. Any τ_k $(0 \le k < \mu)$ can be 0.

Example 4.2-1. Let P^3 be the two-ended 3-path:

$$P^3 \;=\; \langle x_1^2, P_1^2, x_2^3, P_2^2, x_3^3, P_3^2, x_4^3 \rangle \,.$$

Here, P_1^2 is assumed to be a one-ended 2-path terminating on the left with x_1^2 and reaching x_2^3 through a 2-tip. Hence, $|P_1^2| = \omega^3$. We take P_2^2 to be the two-ended 2-path

$$P_2^2 \;=\; \langle y_1^2, Q_1^1, y_2^2, Q_2^0, y_3^2 \rangle,$$

where y_1^2 and y_3^2 are members of x_2^3 and x_3^3 respectively, Q_1^1 is an endless 1-path reaching the 2-nodes y_1^2 and y_2^2 with 1-tips, and Q_2^0 is a finite 0-path with four branches, whose terminal 0-nodes are members of y_2^2 and y_3^2. Hence, $|Q_1^1| = \omega^2 \cdot 2$ and $|Q_2^0| = 4$. Finally, we take P_3^2 to be an endless 2-path reaching x_3^3 and x_4^3 through 2-tips. Hence, $|P_3^2| = \omega^3 \cdot 2$.

Altogether then, with a rearrangement of the following ordinal sum to get a normal expansion, we may write

$$\begin{aligned} |P^3| &= |P_1^2| + |P_2^2| + |P_3^2| \\ &= \omega^3 + |Q_1^1| + |Q_2^0| + \omega^3 \cdot 2 \\ &= \omega^3 + \omega^2 \cdot 2 + 4 + \omega^3 \cdot 2 \\ &= \omega^3 \cdot 3 + \omega^2 \cdot 2 + 4 \end{aligned}$$

♣

$\vec{\omega}$-paths:

$\vec{\omega}$-paths occur within paths of ranks ω (and higher), but they are never two-ended or trivial. The length of an $\vec{\omega}$-path $P^{\vec{\omega}}$ is defined to be $|P^{\vec{\omega}}| = \omega^{\omega}$ when $P^{\vec{\omega}}$ is one-ended, and $|P^{\vec{\omega}}| = \omega^{\omega} \cdot 2$ when $P^{\vec{\omega}}$ is endless.

ω-Paths:

For a nontrivial ω-path P^{ω} as indicated in (2.18), we assign the following lengths. When P^{ω} is one-ended, $|P^{\omega}| = \omega^{\omega+1}$. When P^{ω} is endless, $|P^{\omega}| = \omega^{\omega+1} \cdot 2$. When P^{ω} is two-ended, we set

$$|P^{\omega}| = \sum_m |P_m^{\alpha_m}| = \omega^{\omega} \cdot \tau_{\omega} + \sum_{k=0}^{\infty} \omega^k \cdot \tau_k \tag{4.4}$$

with the natural summation being understood. Here, τ_{ω} is the number of $\vec{\omega}$-tips among all the one-ended and endless $\vec{\omega}$-paths appearing as elements $P_m^{\alpha_m}$ in (2.18) (i.e., when $\alpha_m = \vec{\omega}$); τ_{ω} is not 0. On the other hand, the τ_k are determined recursively, as they are in (4.3). There are only finitely many nonzero terms in the summation within (4.4) because there are only finitely many paths $P_m^{\alpha_m}$ in a two-ended ω-path and each $|P_m^{\alpha_m}|$ is a finite sum as in (4.3).

An immediate result of all these definitions is the following.

Lemma 4.2-2. *If Q^{β} is a subpath of a γ-path P^{γ} $(0 \leq \beta \leq \gamma)$, then $|Q^{\beta}| \leq |P^{\gamma}|$.*

As was mentioned in Sec. 4.1, it is easy to add ordinals when they are in normal-expansion form—simply add their corresponding coefficients, that is, use the natural sum. Thus, the length of the union of two paths that are totally disjoint except for incidence at a terminal node (a "series connection") is obtained by adding their lengths in normal-expansion form.

4.3 Metrizable Sets of Nodes

In a connected finite graph, for every two nodes there is at least one path terminating at them. This is not in general true for transfinite graphs.

Example 4.3-1. The 1-graph of Fig. 4.1 provides an example. In that graph, x_a^1 (resp. x_b^1) is a nonsingleton 1-node containing the 0-tip t_a^0 (resp. t_b^0) for the one-ended path of a_k branches (resp. b_k branches) and also embracing an elementary tip of branch d (resp. e). There are, in addition, uncountably many 0-tips for paths that alternate infinitely often between the a_k and b_k branches by passing through c_k branches; those tips are contained in singleton 1-nodes, one for each. x^1_{abc} denotes one such singleton 1-node; the others are not shown. Note that there is no path connecting x^1_{abc} to x^1_a (or to any other 1-node) because any tracing between x^1_{abc} and x^1_a must repeat 0-nodes; that is, t^0_a and the single 0-tip in x^1_{abc} are nondisconnectable. Thus, our definition (given in the next section) of the distance between two nodes as the minimum path length for all paths connecting those nodes cannot be applied to x^1_{abc} and x^1_a. We seek some means of applying that distance concept to at least some pairs of nodes. ♣

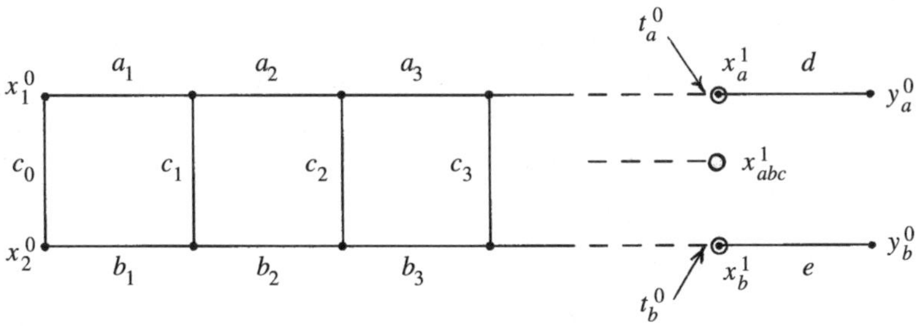

Fig. 4.1. A 1-graph consisting of a one-way infinite ladder along with two branches, d and e, connected to infinite extremities of the ladder. x^1_a and x^1_b are the only nonsingleton 1-nodes. All the other 1-nodes are singletons. There are uncountably many of the latter, and x^1_{abc} is one of them.

To this end, we impose Condition 3.1-2 on the transfinite graph $\mathcal{G}^\nu$ and assume that $\mathcal{G}^\nu$ is ν-connected. These conditions are understood to hold henceforth throughout this chapter. The 1-graph of Fig. 4.1 satisfies these conditions.

The following results ensue: Since $\mathcal{G}^\nu$ is ν-connected, for any two branches there is a two-ended path P^ρ of some rank ρ ($\rho \leq \nu$) that meets those two branches. However, as was noted in Example 4.3-1, there may be two nodes not having any path that meets them. Now, as will be established by Lemma 4.3-2 below, if $\mathcal{G}^\nu$ satisfies Condition 3.1-2, then, for any two nonsingleton nodes, there will be at least one two-ended path terminating at them. As a result, we will be able to define distances between nonsingleton nodes. Furthermore, some singleton nodes may be amenable to such distance measurements, as well. This will occur when the single tip in any singleton node is disconnectable from all the embraced tips in all the nonsingleton nodes and from the tips in the other chosen singleton nodes, too. To test this, we need merely append a new branch b to a singleton node x^α by adding an elementary tip of b to x^α as an embraced elementary tip to get a nonsingleton node $\hat{x}^\alpha$, with the other elementary tip of b left open (i.e., b is added as an end branch)—and then check to see if Condition 3.1-2 is maintained. More generally, with $\mathcal{G}^\nu$ being ν-connected and satisfying Condition 3.1-2, let $\mathcal{M}$ be a set consisting of all the nonsingleton maximal nodes in $\mathcal{G}^\nu$ and possibly other singleton maximal nodes having the property that, if end branches are appended to those singleton nodes simultaneously, Condition 3.1-2 is still satisfied by the resulting network. Any such set $\mathcal{M}$ will be call a *metrizable set* of nodes.

Lemma 4.3-2. *Let $\mathcal{M}$ be a metrizable set of nodes in $\mathcal{G}^\nu$. Then, for any two nodes of $\mathcal{M}$, there exists a two-ended path terminating at those nodes.*

Proof. Let x_a^α and x_b^β be two different nodes in $\mathcal{M}$. Since they are maximal, they must be totally disjoint. Then, by Condition 3.1-2, any tip in x_a^α is disconnectable from every tip in x_b^β; indeed, if they were nondisconnectable, they would have to be shorted together, making x_a^α and x_b^β the same node. Thus, we can choose a representative path P_a for that tip in x_a^α that is totally disjoint from a representative path P_b for a tip in x_b^β. By the definition of ν-connectedness, there will be a path P_{ab} connecting a branch of P_a and a branch of P_b. By Corollary 3.1-5, there is in the subgraph $P_a \cup P_{ab} \cup P_b$ induced by the branches of those three paths a two-ended path terminating at x_a^α and x_b^β. ♣

Example 4.3-3. All the 0-nodes in Fig. 4.1 along with the 1-nodes x_a^1 and x_b^1 comprise a metrizable set.

As another illustration, remove in Fig. 4.1 branches d and e along with the 0-nodes y_a^0 and y_b^0, thereby making x_a^1 and x_b^1 singleton 1-nodes. Then, the remaining 0-nodes along with x_{abc}^1 comprise another metrizable set. However, those remaining 0-nodes along with x_a^1, x_b^1, and x_{abc}^1 do not comprise a metrizable set. ♣

4.4 Distances between Nodes

Our objective now is to define ordinal distances between nodes whereby the metric axioms are satisfied. Let $\mathcal{M}$ be a metrizable set of nodes in $\mathcal{G}^\nu$. We define the *distance function* $d\colon \mathcal{M} \times \mathcal{M} \rightsquigarrow \aleph_1$ as follows: If x_a^α and x_b^β are different nodes in $\mathcal{M}$, we set

$$\begin{aligned} d(x_a^\alpha, x_b^\beta) \;=\; & \min\{|P_{ab}|\colon P_{ab} \\ & \text{is a two-ended path terminating at } x_a^\alpha \text{ and } x_b^\beta\}. \end{aligned} \tag{4.5}$$

If $x_a^\alpha = x_b^\beta$, we set $d(x_a^\alpha, x_b^\beta) = 0$. By our constructions in Sec. 4.2, $|P_{ab}|$ is a countable ordinal no larger than $\omega^\omega \cdot k$, where k is a natural number. Moreover, any set of ordinals is well-ordered and thus has a least member. Therefore, the minimum indicated in (4.5) exists, and is a countable ordinal.

Obviously, $d(x_a^\alpha, x_b^\beta) > 0$ if $x_a^\alpha \neq x_b^\beta$. Moreover, $d(x_a^\alpha, x_b^\beta) = d(x_b^\beta, x_a^\alpha)$. It remains to prove the triangle inequality; namely, if x_a^α, x_b^β, and x_c^γ are any three (maximal) nodes in $\mathcal{M}$, then

$$d(x_a^\alpha, x_b^\beta) \leq d(x_a^\alpha, x_c^\gamma) + d(x_c^\gamma, x_b^\beta). \tag{4.6}$$

This is easily done by using Corollary 3.1-5 and Lemma 4.2-2 and taking minimums of sets of ordinals.

Proposition 4.4-1. *d satisfies the metric axioms.*

Clearly, d reduces to the standard (branch-count) distance function when $\mathcal{G}^\nu$ is replaced by a finite graph. We have hereby extended transfinitely the branch-count distance function to any metrizable set of nodes in $\mathcal{G}^\nu$.

Example. 4.4-2. For the 1-graph of Fig. 4.1 and with $\mathcal{M}$ consisting of all the 0-nodes along with x_a^1 and x_b^1, we have $d(x_1^0, x_2^0) = 1$, $d(x_1^0, x_a^1) = d(x_1^0, x_b^1) = \omega$, $d(x_1^0, y_a^0) = d(x_1^0, y_b^0) = \omega + 1$, and $d(y_a^0, y_b^0) = \omega \cdot 2 + 2$. ♣

Because the minimum in (4.5) is achieved, we can sharpen Lemma 4.3-2 as follows.

Lemma 4.4-3. *Given any two nodes x and y in $\mathcal{M}$, there exists a path Q_{xy} terminating at x and y for which $|Q_{xy}| = d(x, y)$.*

There may be more than one such path. We call each of them an *x-to-y geodesic*.

4.5 Eccentricities and Related Ideas

As we have seen, the distance between any two nodes of $\mathcal{M}$ is a countable ordinal. However, given any $x \in \mathcal{M}$, the set $\{d(x, y) : y \in \mathcal{M}\}$ may have no maximum ordinal. For example, this is the case for a one-ended 0-path P^0 where x is any fixed node of P^0 and y ranges through all the 0-nodes of P^0. On the other hand, for finite graphs the said maximum ordinal exists and is the "eccentricity" of x. We will be able to define an "eccentricity" for every node of $\mathcal{M}$ if we allow eccentricities to be ranks instead of just ordinals. As was indicated in Sec. 1.2, the set $\mathcal{R}$ of ranks is obtained from the set $\aleph_1$ of countable ordinals by inserting an *arrow rank* $\vec{\rho}$ immediately before each limit-ordinal rank $\rho \in \aleph_1$. (We often refer to an arrow rank simply as an "arrow.")

To obtain eccentricities as both arrow ranks and ordinal ranks, note first of all that the lengths of all paths in a ν-graph $\mathcal{G}^\nu$ are bounded by $\omega^{\nu+1} \cdot 2$ because the longest possible paths in $\mathcal{G}^\nu$ are the endless paths of rank ν. Therefore, all distances in $\mathcal{G}^\nu$ are also bounded above by $\omega^{\nu+1} \cdot 2$. Hence, the definition of arrow eccentricities as equivalence classes of persistently increasing sequences of ordinal distances uniformly bounded above can be used (see Sec. 1.2).

The *eccentricity* $e(x)$ of any node $x \in \mathcal{M}$ is defined by

$$e(x) \;=\; \sup\{d(x, y) \colon y \in \mathcal{M}\}. \tag{4.7}$$

Two cases arise: First, the supremum is achieved at some node $\hat{y} \in \mathcal{M}$. In this case, $e(x)$ is an ordinal; so, we can replace "sup" by "max" in (4.7) and write $e(x) = d(x, \hat{y})$. Second, the supremum is not achieved at any node in $\mathcal{M}$. In this case, $e(x)$ is an arrow.

The ideas of radii and diameters for finite graphs [9, page 32], [14, page 20] can also be extended transfinitely. Given $\mathcal{G}^\nu$ and $\mathcal{M}$, the *radius* $\mathrm{rad}(\mathcal{G}^\nu, \mathcal{M})$ is the least eccentricity among the nodes of $\mathcal{M}$:

$$\mathrm{rad}(\mathcal{G}^\nu, \mathcal{M}) \;=\; \min\{e(x) \colon x \in \mathcal{M}\}. \tag{4.8}$$

We also denote this simply by rad with the understanding that $\mathcal{G}^\nu$ and $\mathcal{M}$ are given. The minimum exists as a rank (either as an ordinal or as an arrow) because the set of ranks is well-ordered. Thus, there will be at least one $x \in \mathcal{M}$ with $e(x) = \mathrm{rad}$.

Furthermore, the *diameter* $\mathrm{diam}(\mathcal{G}^\nu, \mathcal{M})$ is defined by

$$\mathrm{diam}(\mathcal{G}^\nu, \mathcal{M}) \;=\; \sup_{x \in \mathcal{M}} \sup_{y \in \mathcal{M}} \{d(x, y)\} \;=\; \sup\{d(x, y) \colon x, y \in \mathcal{M}\}. \tag{4.9}$$

With $\mathcal{G}^\nu$ and $\mathcal{M}$ understood, we denote the diameter simply by diam. As we have noted before, each $d(x, y)$ is no greater than $\omega^{\nu+1} \cdot 2$. Therefore, diam exists either as an ordinal or as an arrow.

The ideas of the center and periphery of finite graphs can also be extended. The *center* of $(\mathcal{G}^\nu, \mathcal{M})$ is the set of nodes in $\mathcal{M}$ having the least eccentricity, namely, rad. The center is never empty.

The *periphery* of $(\mathcal{G}^\nu, \mathcal{M})$ is the set of nodes in $\mathcal{M}$ having the greatest eccentricity, namely, diam. If diam is an ordinal, there will be at least two nodes of $\mathcal{M}$ in the periphery. Indeed, if there did not exist at least two nodes in the periphery, then the ordinal diam could only be approached from below by distances between pairs of nodes that are less than diam; thus, the supremum in (4.9) would have to be an arrow—a contradiction. On the other hand, if diam is an arrow, the periphery can have any positive number of nodes—even just one or an infinity of them, as the following examples will show. It seems that the periphery will never be empty, but presently this is only a conjecture.

Example 4.5-1. Let $\mathcal{G}^0$ be a one-ended 0-path with $\mathcal{M}$ being the set of all its 0-nodes. (We do not yet assign a 1-node at the path's infinite extremity.) Then, every 0-node has an eccentricity of $\vec{\omega}$. Thus, rad = diam = $\vec{\omega}$, and $\mathcal{M}$ is both the center and the periphery of $(\mathcal{G}^0, \mathcal{M})$.

We have here a graphical realization of Aristotle's "potential infinity" represented by the eccentricity $\vec{\omega}$. Aristotle's opinion was that "potential infinities" can exist but that "actual infinities" cannot exist. A potential infinity arises he asserted when there is a collection of entities having sizes such that for every entity however large there exists a larger entity. An actual infinity would exist were there a potential infinity and also an entity whose size is larger than the sizes of all the other entities comprising the potential infinity, and this he felt was impossible. A description of his ideas in this regard can be found in [21, see the entry "Infinity and the Infinite"].[1] In our context of distances in transfinite graphs, one might interpret Aristotle's ideas by saying that each arrow eccentricity represents a potential infinity because certain distances only increase toward and approach an infinity, such as a limit ordinal, without that infinity being achieved as a distance. This is so for distances in the one-ended 0-path, with the result that the eccentricities of all the nodes are the arrow rank $\vec{\omega}$.

However, "actual infinities" are ubiquitous in modern mathematics—certainly since the advent of the calculus. In this example, we need merely append a 1-node to the one-ended 0-path to get an actual infinity ω for all the eccentricities. This challenges Aristotle's opinion. ♣

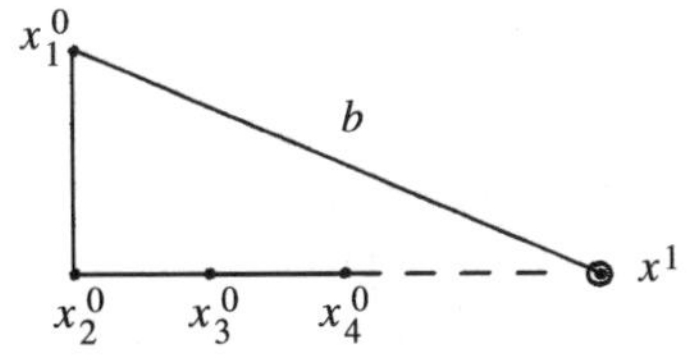

Fig. 4.2. The simplest "Aristotelian graph." The eccentricity of every node is the first arrow rank.

Example 4.5-2. Now, consider the 1-graph of Fig. 4.2. It consists of a one-ended 0-path with the maximal 0-nodes $x_1^0, x_2^0, x_3^0, \ldots$. That 0-path reaches the 1-node x^1. In addition, there is a branch b incident to x_1^0 and x^1. Because of b, the distance between any x_k^0 and x^1 is k (not ω), and the distance between x_k^0 and x_m^0 is $|k-m|$. With regard to eccentricities, this implies that $e(x^1) = \vec{\omega}$ and moreover $e(x_k^0) = \vec{\omega}$ for every k, too. Note that, in contrast to the one-ended 0-path of Example 4.5-1, the arrow eccentricities now encountered cannot be converted into ordinal eccentricities by appending another 1-node; there is no place to put it.

For this graph at least, Aristotle's opinion is vindicated, and therefore it seems appropriate to call this an "Aristotelian graph." It is the simplest of the graphs in which one or more of the eccentricities are arrows. In the present case, all the maximal nodes have the eccentricity $\vec{\omega}$ and comprise both the center and the periphery, with rad = diam = $\vec{\omega}$. ♣

Example 4.5-3. Consider now the 1-graph of Fig. 4.3. The eccentricities of the nodes are as follows: $e(x_k) = \omega\cdot 2+k$ for $k = 1, 2, 3, \ldots$; $e(y^1) = \omega\cdot 2$; $e(z_k)^0 = \vec{\omega}\cdot 2$ for $k = \ldots, -1, 0-1, \ldots$; $e(w^1) = \vec{\omega}\cdot 3$. Thus, rad $= \vec{\omega}\cdot 2$, diam $= \vec{\omega}\cdot 3$, the center is $\{z_k : k = \ldots, -1, 0, 1, \ldots\}$, and the periphery is the singleton $\{w^1\}$.

[1] See also [3] or [36] for a popular history of the roles potential and actual infinities have played in mathematics.

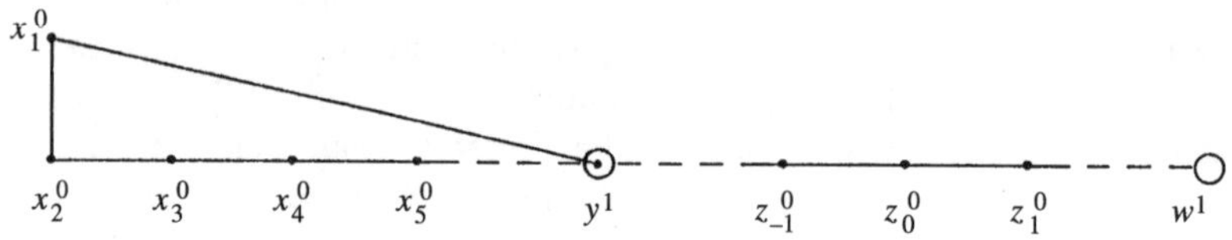

Fig. 4.3. The 1-graph of Example 4.5-3.

Here, we have two different representations of Aristotle's potential infinity, given by the eccentricities $\vec{\omega}\cdot 2$ and $\vec{\omega}\cdot 3$; each of these, too, cannot be converted into an ordinal by appending a transfinite node—there is no place to put it. Note also that the periphery has only one node—in contrast to the peripheries of finite graphs, which must have two or more nodes. ♣

Example 4.5-4. The 1-graph of Fig. 4.1 has no arrow eccentricities when we let $\mathcal{M}$ be the set of all maximal 0-nodes along with the 1-nodes x_a^1 and x_b^1. (Ignore x_{abc}^1 and all other singleton 1-nodes.) The 0-nodes to the left of the 1-nodes all have the eccentricity $\omega + 1$. Also, $e(x_a^1) = e(x_b^1) = \omega\cdot 2 + 1$, and $e(y_a^0) = e(y_b^0) = \omega\cdot 2 + 2$. Thus, rad = $\omega + 1$ and diam = $\omega\cdot 2 + 2$. The center consists of all the 0-nodes to the left of the 1-nodes, and the periphery is $\{y_a^0, y_b^0\}$. By appending more "end" branches incident to either x_a^1 or x_b^1, we can increase the number of nodes in the periphery.

Next, consider the 1-graph obtained from Fig. 4.1 by deleting the branches d and e and the 0-nodes y_a^0 and y_b^0 but appending a new branch incident to x_1^0 and x_a^1. Let $\mathcal{M}$ consist of all the nonsingleton nodes along with the singleton 1-node that x_b^1 becomes when the elementary tip of e is deleted. Then, the eccentricity of every node is ω. Thus, rad = diam = ω, and the center and periphery are the same, namely, $\mathcal{M}$. ♣

Of course, these examples can be converted into examples of eccentricities of higher ranks by replacing branches by endless paths.

4.6 Some General Results

Henceforth in this chapter, let it be understood that the metrizable set $\mathcal{M}$ has been chosen and fixed for the ν-graph $\mathcal{G}^\nu$, and that any node we refer to is in $\mathcal{M}$ unless a nonmaximal node is explicitly specified. (Remember that all the nodes in $\mathcal{M}$ are maximal and that Condition 3.1-2 is assumed to hold.)

In the next theorem, $\mathcal{S}^\rho$ is any ρ-section whose bordering nodes are incident to $\mathcal{S}^\rho$ only through ρ-tips. For example, in the 1-graph of Fig. 4.1 with $\mathcal{M}$ being the set of all maximal 0-nodes along with the 1-nodes x_a^1 and x_b^1, x_a^1 and x_b^1 are bordering nodes of the 0-section to the left of those nodes, and the condition is satisfied, that is, those 1-nodes are incident to that 0-section only through 0-tips. However, branch d

induces a 0-section by itself, and the condition is not satisfied because d reaches x_a^1 through a (-1)-tip (i.e., a tip of branch d); similarly for e and x_b^1.

Theorem 4.6-1. *Let $\mathcal{S}^\rho$ be a ρ-section in $\mathcal{G}^\nu$ $(0 \le \rho < \nu)$ all of whose bordering nodes are incident to $\mathcal{S}^\rho$ only through ρ-tips. Then, all the internal nodes of $\mathcal{S}^\rho$ have the same eccentricity.*

Proof. By virtue of our hypothesis and the ρ-connectedness of $\mathcal{S}^\rho$, for any internal node x^α $(\alpha \le \rho)$ and any bordering node z^γ $(\gamma > \rho)$ of $\mathcal{S}^\rho$ in $\mathcal{M}$, there is a representative ρ-path P^ρ for a ρ-tip embraced by z^γ and lying in $\mathcal{S}^\rho$, and there also is a two-ended path Q lying in $\mathcal{S}^\rho$ and terminating at x^α and a node of P^ρ. So, by Condition 3.1-2 and Corollary 3.1-5, there is in $P \cup Q$ a one-ended ρ-path R^ρ that terminates at x^α and reaches z^γ through a ρ-tip. Moreover, all paths that terminate at x^α, that lie in $\mathcal{S}^\rho$, and that reach z^γ must be one-ended ρ-paths. Therefore, $d(x^\alpha, z^\gamma) = \omega^{\rho+1}$. For any other node y^β $(\beta \le \rho)$ in $\mathcal{S}^\rho$, we have $d(x^\alpha, y^\beta) < \omega^{\rho+1}$ by the ρ-connectedness of $\mathcal{S}^\rho$. So, if $\mathcal{G}^\nu$ consists only of $\mathcal{S}^\rho$ and its bordering nodes (so that $\nu = \rho + 1$), we can conclude that $e(x^\alpha) = \omega^{\rho+1}$, whatever be the choice of the internal node x^α in $\mathcal{S}^\rho$ and in $\mathcal{M}$.

Next, assume that there is a node v^δ of $\mathcal{G}^\nu$ lying outside of $\mathcal{S}^\rho$ and different from all the bordering nodes of $\mathcal{S}^\rho$. By the ν-connectedness of $\mathcal{G}^\nu$, there is a path P_{xv} terminating at x^α and v^δ. Let z^γ now be the last bordering node of $\mathcal{S}^\rho$ that P_{xv} meets. Let P_{zv} be that part of P_{xv} lying outside of $\mathcal{S}^\rho$. Then, by what we have shown above, there is a one-ended ρ-path Q_{xz}^ρ that terminates at x^α, lies in $\mathcal{S}^\rho$, and reaches z^γ through a ρ-tip. Then, $R_{xv} = Q_{xz}^\rho \cup P_{zv}$ is a two-ended path that terminates at x^α and v^δ. Moreover, $|R_{xv}| \le |P_{xv}|$.

Now, let y^β $(\beta \le \rho)$ be any other internal node of $\mathcal{S}^\rho$ in $\mathcal{M}$ (i.e., different from x^α). Again, there is a one-ended ρ-path Q_{yz}^ρ satisfying the same conditions as Q_{xz}^ρ. We have $d(x^\alpha, z^\gamma) = d(y^\beta, z^\gamma) = \omega^{\rho+1}$. Let $R_{yv} = Q_{yz}^\rho \cup P_{zv}$. Thus, $|R_{xv}| = |R_{yv}|$. We have shown that, for each one-ended path R_{xv} terminating at x^α and v^δ and passing through exactly one bordering node of z^γ of $\mathcal{S}^\rho$, there is another path R_{yv} of the same length terminating at y^β and v^δ and identical to R_{xv} outside $\mathcal{S}^\rho$. It follows that $d(x^\alpha, v^\delta) = d(y^\beta, v^\delta)$ whatever be the choice of v^δ outside of $\mathcal{S}^\rho$. We can conclude that $e(x^\alpha) = e(y^\beta)$ whatever be the choices of x^α and y^β in $\mathcal{S}^\rho$ and $\mathcal{M}$. ♣

Figs. 4.1 and 4.3 provide examples for Theorem 4.6-1. In Fig. 4.1, all the 0-nodes to the left of the 1-nodes have the same eccentricity $\omega + 1$. In Fig. 4.3 the 0-nodes z_k^0 have the same eccentricity $\vec{\omega \cdot 2}$.

A standard result [14, page 21] can be extended to the transfinite case, albeit in a more complicated way. Given $\mathcal{G}^\nu$ and $\mathcal{M}$, rad may be either an ordinal or an arrow rank. If it is an arrow rank, we let rad^+ denote the limit ordinal immediately following rad.

Theorem 4.6-2.

(i) If rad is an ordinal, then rad $\le$ diam $\le$ rad$\cdot 2$.
(ii) If rad is an arrow rank, then rad $\le$ diam $\le$ rad$^+ \cdot 2$.

Proof. The proofs of (i) and (ii) are much the same. So, let us consider (ii) alone. That rad $\leq$ diam follows directly from the definitions (4.7), (4.8), and (4.9). Next, by the definition of the diameter (4.9), we can choose two sequences $\langle y_k : k \in \mathbb{N}\rangle$ and $\langle z_k : k \in \mathbb{N}\rangle$ of nodes such that the sequence $\langle d(y_k, z_k) : k \in \mathbb{N}\rangle$ approaches or achieves diam. Let x be any node in the center. By the triangle inequality,

$$d(y_k, z_k) \leq d(y_k, x) + d(x, z_k).$$

Now, $d(y_k, x) \leq \text{rad} \leq \text{rad}^+$, and similarly for $d(x, z_k)$. Therefore, $d(y_k, z_k) \leq \text{rad}^+ + \text{rad}^+ = \text{rad}^+ \cdot 2$. Hence, diam $\leq \text{rad}^+ \cdot 2$. ♣

Another standard result is that the nodes of any finite graph comprise the center of some finite connected graph [14, page 22].[2] This, too, can be extended transfinitely—in fact, in several ways, but the proofs are more complicated than that for finite graphs. Nonetheless, the scheme of the proofs remains the same. First, we need the following lemma. As always, $\nu - 1$ denotes $\vec{\omega}$ when $\nu = \omega$. Also, we shall say that a path *lies branchwise in* a $(\nu-1)$-section $\mathcal{S}^{\nu-1}$ if all the path's branches lie in $\mathcal{S}^{\nu-1}$. However, a node of that path may be of rank ν and thus not in $\mathcal{S}^{\nu-1}$.

Lemma 4.6-3. *Let $\mathcal{S}^{\nu-1}$ be a $(\nu-1)$-section of $\mathcal{G}^\nu$, where $1 \leq \nu \leq \omega$ and $\nu \neq \vec{\omega}$. Let u^ν be a ν-node incident to $\mathcal{S}^{\nu-1}$ (thus, a bordering node of $\mathcal{S}^{\nu-1}$), and let x^α $(\alpha < \nu)$ be an α-node in $\mathcal{S}^{\nu-1}$ (thus, an internal node of $\mathcal{S}^{\nu-1}$). Then, there exists a two-ended path connecting x^α and u^ν that lies branchwise in $\mathcal{S}^{\nu-1}$ and whose length is no larger than ω^ν.*

Proof. If u^ν is incident to $\mathcal{S}^{\nu-1}$ through a single branch, the conclusion is obtained through an easy adjustment of the following argument. So, assume this is not the case.

That u^ν is incident to $\mathcal{S}^{\nu-1}$ means that there is in $\mathcal{S}^{\nu-1}$ a one-ended β-path P^β with $\beta \leq \nu - 1$ whose β-tip is embraced by u^ν, that is, P^β reaches u^ν. Let $P^{\beta+1}$ be the two-ended path obtained by appending to P^β the $(\beta+1)$-node $y^{\beta+1}$ embraced by u^ν and reached by P^β. ($y^{\beta+1}$ will not be maximal if $\beta + 1 < \nu$; otherwise, $y^{\beta+1} = u^\nu$.) $P^{\beta+1}$ lies branchwise in $\mathcal{S}^{\nu-1}$. The length $|P^{\beta+1}|$ of $P^{\beta+1}$ is equal to $\omega^{\beta+1}$ because $P^{\beta+1}$ traverses only one β-tip; all other tips traversed by $P^{\beta+1}$ are of lesser rank. Let z^γ be any node (not necessarily maximal) of P^β; thus, $\gamma \leq \beta$. By the $(\nu-1)$-connectedness of $\mathcal{S}^{\nu-1}$, there is in $\mathcal{S}^{\nu-1}$ a two-ended λ-path Q^λ $(0 \leq \lambda \leq \nu - 1)$ terminating at x^α and z^γ. The tips traversed by Q^λ have ranks no greater than $\lambda - 1$, hence, no greater than $\nu - 2$. By Corollary 3.1-5, there is a two-ended path R^δ in $P^{\beta+1} \cup Q^\lambda$ terminating at x^α and $y^{\beta+1}$. All the tips traversed by R^δ are of ranks no greater than $\nu - 1$, and there is at most one traversed tip of rank $\nu - 1$. Hence, the length of R^δ satisfies $|R^\delta| \leq \omega^\nu$. R^δ is the path we seek. ♣

[2]This result extends immediately to infinite 0-graphs with infinitely many 0-nodes. We are now considering transfinite graphs of rank 1 or greater.

Given any ν-graph $\mathcal{G}^\nu$ with $1 \le \nu \le \omega$ and $\nu \neq \vec{\omega}$, let us construct a larger ν-graph $\mathcal{H}^\nu$ by appending six additional ν-nodes p_i^ν and q_i^ν $(i = 1, 2, 3)$ and also appending isolated endless $(\nu - 1)$-paths[3] that reach ν-nodes as shown in Fig. 4.4. Such paths connect p_1^ν to p_2^ν, p_2^ν to p_3^ν, p_3^ν to every ν-node in a metrizable set for $\mathcal{G}^\nu$, and similarly for p_i^ν replaced by q_i^ν. To get a metrizable set $\mathcal{M}$ of $\mathcal{H}^\nu$, we choose a metrizable set for $\mathcal{G}^\nu$ and append all the nodes of the said isolated endless $(\nu - 1)$-paths and all the p_i^ν and q_i^ν, too.

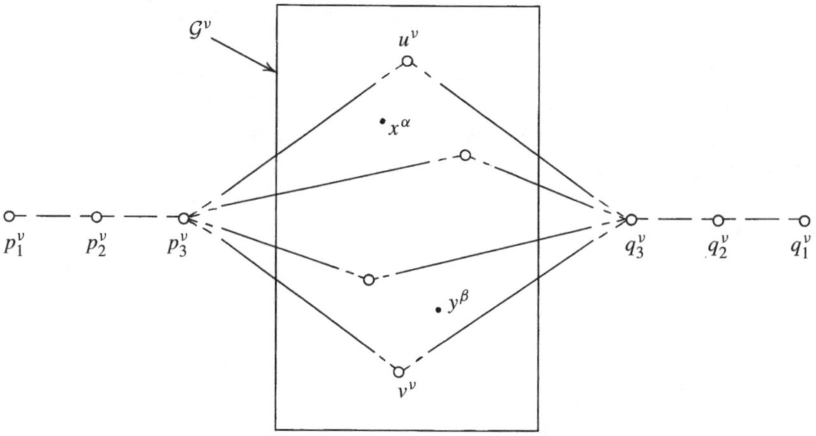

Fig. 4.4. The ν-graph $\mathcal{H}^\nu$. The lines (other than those of the rectangle) denote isolated endless $(\nu - 1)$-paths. The ν-nodes p_3^ν and q_3^ν are connected to every ν-node in a metrizable set for $\mathcal{G}^\nu$ through such isolated endless paths—as well as to p_2^ν and q_2^ν respectively.

Theorem 4.6-4. *The ν-nodes in a metrizable set for $\mathcal{G}^\nu$ $(1 \le \nu \le \omega,\ \nu \neq \vec{\omega})$ comprise the center of $\mathcal{H}^\nu$, and the periphery of $\mathcal{H}^\nu$ is $\{p_1^\nu, q_1^\nu\}$.*

Proof. We look for bounds on the eccentricities of all the nodes in $\mathcal{M}$. Let x^α and y^β be any two nodes whose ranks satisfy $0 \le \alpha, \beta < \nu$. It follows that x^α (resp. y^β) is an internal node of a $(\nu-1)$-section in $\mathcal{G}^\nu$, and that section has at least one ν-node u^ν (resp. v^ν) as a boundary node because $\mathcal{G}^\nu$ and therefore $\mathcal{H}^\nu$ are ν-connected. By the triangle inequality,

$$d(x^\alpha, y^\beta) \le d(x^\alpha, u^\nu) + d(u^\nu, p_3^\nu) + d(p_3^\nu, v^\nu) + d(v^\nu, y^\beta).$$

By Lemma 4.6-3, $d(x^\alpha, u^\nu) \le \omega^\nu$ and $d(v^\nu, y^\beta) \le \omega^\nu$. Clearly, $d(u^\nu, p_3^\nu) = d(p_3^\nu, v^\nu) = \omega^\nu \cdot 2$. Thus, $d(x^\alpha, y^\beta) \le \omega^\nu \cdot 6$. This also shows that, for any

[3]An isolated endless path embraces no tips other than the ones it traverses. Thus, to reach any other part of a graph in which the isolated path is a subgraph, one must proceed through a terminal tip of that path.

ν-node v^ν in $\mathcal{G}^\nu$, $d(x^\alpha, v^\nu) \le \omega^\nu \cdot 5$. Since $d(u^\nu, p_1^\nu) = \omega^\nu \cdot 6$, we have $d(x^\alpha, p_1^\nu) = d(x^\alpha, u^\nu) + d(u^\nu, p_1^\nu) \le \omega^\nu + \omega^\nu \cdot 6 = \omega^\nu \cdot 7$. Now, $d(x^\alpha, u^\nu) \ge 1$ because there is at least one branch in any path connecting x^α and u^ν. Thus, we also have $d(x^\alpha, p_1^\nu) \ge \omega^\nu \cdot 6 + 1$. Note also that the distance from x^α to any node of the appended endless paths is strictly less than $\omega^\nu \cdot 7$. All these results hold for p_i^ν replaced by q_i^ν. Altogether then, we can conclude the following: For any node in $\mathcal{G}^\nu$ of rank less than ν, say, x^α, the eccentricity $e(x^\alpha)$ of x^α is bounded as follows:

$$\omega^\nu \cdot 6 + 1 \le e(x^\alpha) \le \omega^\nu \cdot 7 .$$

Next, consider any two ν-nodes of $\mathcal{G}^\nu$, say, u^ν and v^ν again. By what we have already shown, $d(u^\nu, v^\nu) \le \omega^\nu \cdot 4$ and $d(u^\nu, p_1^\nu) = d(u^\nu, q_1^\nu) = \omega^\nu \cdot 6$. The distance from u^ν to any node of the appended endless $(\nu - 1)$-paths is less than $\omega^\nu \cdot 6$. Also, for any node y^β in $\mathcal{G}^\nu$ of rank less than ν, $d(u^\nu, y^\beta) \le \omega^\nu \cdot 5$. So, the largest distance between u^ν and any other node in $\mathcal{H}^\nu$ is equal to $\omega^\nu \cdot 6$; that is, $e(u^\nu) = \omega^\nu \cdot 6$.

Finally, we have $e(p_3^\nu) = e(q_3^\nu) = \omega^\nu \cdot 8$, $e(p_2^\nu) = e(q_2^\nu) = \omega^\nu \cdot 10$, and $e(p_1^\nu) = e(q_1^\nu) = \omega^\nu \cdot 12$. The eccentricities of the nodes of the appended endless paths lie between these values.

We have considered all cases. Comparing these equalities and inequalities for all the eccentricities, we can draw the conclusion of the theorem. ♣

As an immediate corollary, we have the following generalization of a result for finite graphs.

Corollary 4.6-5. *The ν-nodes of a metrizable set for $\mathcal{G}^\nu$ $(1 \le \nu \le \omega, \nu \ne \vec{\omega})$ comprise the center of some ν-graph $\mathcal{H}^\nu$.*

Variations of Corollary 4.6-5 can also be established through much the same proofs. For instance, all the nodes of one or more specified ranks in a metrizable set for $\mathcal{G}^\nu$ can be made to comprise the center of some ν-graph. This is because the $(\nu - 1)$-sections of $\mathcal{G}^\nu$ partition $\mathcal{G}^\nu$.

4.7 When the Nodes of Highest Rank Are Pristine

The nodes of highest rank in $\mathcal{G}^\nu$ $(\nu \ne \vec{\omega})$ are the ν-nodes, of course. A ν-node is said to be pristine if it does not embrace a node of lower rank. Thus, a pristine ν-node consists only of $(\nu - 1)$-tips. Henceforth in this chapter, we assume that $1 \le \nu \le \omega$, $\nu \ne \vec{\omega}$, and the following.

Condition 4.7-1. *All the ν-nodes are pristine.*

Because of this, we can view the ν-nodes as lying only at infinite extremities of the $(\nu - 1)$-sections to which they are incident because they can be reached only through $(\nu - 1)$-tips of such sections; they are, in fact, the bordering nodes of such

$(\nu - 1)$-sections. All the other nodes are of ranks less than ν and are internal nodes of $(\nu - 1)$-sections. Given any $(\nu - 1)$-section $\mathcal{S}^{\nu-1}$, the set of all internal nodes of $\mathcal{S}^{\nu-1}$ will be denoted by $i(\mathcal{S}^{\nu-1})$ and will be called the *interior* of $\mathcal{S}^{\nu-1}$.

In accordance with the definition given in Sec. 3.1, a boundary ν-node of $\mathcal{S}^{\nu-1}$ is a bordering ν-node that contains $(\nu - 1)$-tips of $\mathcal{S}^{\nu-1}$ and also $(\nu - 1)$-tips of one or more other $(\nu - 1)$-sections of $\mathcal{G}^{\nu}$. Thus, a boundary node lies at the infinite extremities of two or more $(\nu - 1)$-sections and thereby connects them. A ν-path can pass from the interior of one $(\nu-1)$-section into the interior of another $(\nu-1)$-section only by passing through a boundary node via $(\nu - 1)$-tips.

Another assumption we impose henceforth in this chapter is the following.

Condition 4.7-2. *There are only finitely many boundary ν-nodes throughout $\mathcal{G}^{\nu}$.*

Nevertheless, each $(\nu - 1)$-section may have infinitely many incident nonboundary bordering ν-nodes, and each boundary ν-node may be incident to infinitely many $(\nu - 1)$-sections.

Altogether then, we are assuming throughout the rest of this chapter that $\mathcal{G}^{\nu}$ is ν-connected, has a chosen metrizable set $\mathcal{M}$ of nodes, and satisfies Conditions 3.1-2, 4.7-1, and 4,7-2. In addition, any node we refer to is understood to be in $\mathcal{M}$ unless a nonmaximal node is specified.

Lemma 4.7-3. *If P is a two-ended ν-path, then $|P| = \omega^{\nu} \cdot k$, where k is the number of $(\nu - 1)$-tips traversed by P.*

Proof. This follows directly from the definition of the length $|P|$ and the fact that all ν-nodes are pristine. ♣

Because all ν-nodes are now pristine, we can strengthen Lemma 4.6-3 as follows.

Lemma 4.7-4.

(a) Let x^{ν} be a bordering node of a $(\nu - 1)$-section $\mathcal{S}^{\nu-1}$, and let z be an internal node of $\mathcal{S}^{\nu-1}$. Then, there exists at least one two-ended ν-path $P^{\nu}_{z,x}$ that lies branchwise in $\mathcal{S}^{\nu-1}$, terminates at z, and reaches x^{ν} through its one and only $(\nu - 1)$-tip. Moreover, the length $|P^{\nu}_{z,x}|$ of $P^{\nu}_{z,x}$ is ω^{ν}. $P^{\nu}_{z,x}$ is a z-to-x^{ν} geodesic. Finally, every two-ended ν-path lying branchwise in $\mathcal{S}^{\nu-1}$ and terminating at an internal node of $\mathcal{S}^{\nu-1}$ and at a bordering node of $\mathcal{S}^{\nu-1}$ has the length ω^{ν}.

(b) Let x^{ν} and y^{ν} be two bordering nodes of $\mathcal{S}^{\nu-1}$. Then, there exists at least one two-ended ν-path $P^{\nu}_{x,y}$ lying branchwise in $\mathcal{S}^{\nu-1}$ that reaches x^{ν} and y^{ν} through its two and only two $(\nu - 1)$-tips. Moreover, the length $|P^{\nu}_{x,y}|$ of $P^{\nu}_{x,y}$ is $\omega^{\nu} \cdot 2$. $P^{\nu}_{x,y}$ is an x^{ν}-to-y^{ν} geodesic. Finally, every two-ended ν-path lying branchwise in $\mathcal{S}^{\nu-1}$ and terminating at two bordering nodes of $\mathcal{S}^{\nu-1}$ has the length $\omega^{\nu} \cdot 2$.

Proof. The proof of part (a) is much the same as that of Lemma 4.6-3 except that now $P^\nu_{z,x}$ is incident to x^ν only through the one and only $(\nu-1)$-tip of $P^\nu_{z,x}$. The fact that all the bordering nodes of $\mathcal{S}^{\nu-1}$ are pristine is now essential to the argument.

Part (b) is proven similarly, but now we use two representative one-ended paths, one for each of x^ν and y^ν. ♣

Lemma 4.7-5. *Let x^ν be a bordering node of a $(\nu-1)$-section $\mathcal{S}^{\nu-1}$, let z be an internal node of $\mathcal{S}^{\nu-1}$, and let y be any node of $\mathcal{G}^\nu$. Then, $|d(x^\nu, y) - d(z, y)| \le \omega^\nu$.*

Proof. By Lemma 4.7-4(a), $d(z, x^\nu) = \omega^\nu$. Since, d is a metric,

$$d(x^\nu, y) \;\le\; d(x^\nu, z) + d(z, y) \;=\; \omega^\nu + d(z, y).$$

Also,

$$d(z, y) \;\le\; d(z, x^\nu) + d(x^\nu, y) \;=\; \omega^\nu + d(x^\nu, y).$$

These inequalities yield the conclusion. ♣

We will show below that, as a consequence of Conditions 4.7-1 and 4.7-2, no node of $\mathcal{G}^\nu$ can have an arrow-rank eccentricity and that all eccentricities comprise a finite set of ordinal values. But, first note that three examples of the occurrence of arrow-rank eccentricities are given in Examples 4.5-1, 4.5-2, and 4.5-3. Each of these examples violates either Condition 4.7-1 or Condition 4.7-2.

Theorem 4.7-6. *The eccentricities of all the nodes in $\mathcal{M}$ are contained within the finite set of ordinals*

$$\{\omega^\nu \cdot p \colon 1 \le p \le 2m+2\}\,. \tag{4.10}$$

Here, p and m are natural numbers, and m is the number of boundary ν-nodes.

Proof. The eccentricity of any node is at least as large as the distance between any internal node of a $(\nu-1)$-section and any bordering ν-node of that $(\nu-1)$-section. Therefore, Lemma 4.7-4(a) implies that the eccentricity of any node of $\mathcal{G}^\nu$ is at least ω^ν, whence the lower bound in (4.10). The proof of the upper bound requires more effort.

First of all, we can settle two simple cases by inspection. If $\mathcal{G}^\nu$ consists of a single $(\nu-1)$-section with exactly one bordering ν-node in $\mathcal{M}$, then all the nodes in $\mathcal{M}$ of $\mathcal{G}^\nu$ have the eccentricity ω^ν. If that one and only $(\nu-1)$-section for $\mathcal{G}^\nu$ has two or more (possibly infinitely many) bordering nodes, the internal nodes have eccentricity ω^ν, and the bordering nodes have eccentricity $\omega^\nu \cdot 2$. In both cases, the conclusion of the theorem is fulfilled with $m = 0$.

We now turn to the general case where $\mathcal{G}^\nu$ has at least one boundary ν-node and therefore at least two $(\nu-1)$-sections. $\mathcal{G}^\nu$ will have a two-ended ν-path of the form

$$P^\nu_{0,k} \;=\; \langle x_0, P^{\nu-1}_0, x^\nu_1, P^{\nu-1}_1, \ldots, x^\nu_{k-1}, P^{\nu-1}_{k-1}, x_k \rangle\,. \tag{4.11}$$

Because all ν-nodes are pristine, the x_i^ν ($i = 1, \ldots, k-1$) are nonsingleton bordering ν-nodes (possibly boundary ν-nodes), and the $P_i^{\nu-1}$ ($i = 1, \ldots, k-2$) are endless $(\nu-1)$-paths. The same is true of x_0, x_k, $P_0^{\nu-1}$, and $P_{k-1}^{\nu-1}$ if x_0 and x_k are ν-nodes, too. If x_0 (resp. x_k) is of lower rank, then it is an internal node, and $P_0^{\nu-1}$ (resp. $P_{k-1}^{\nu-1}$) is a one-ended $(\nu-1)$-path.

Let us first assume that x_0 and x_k are internal nodes in different $(\nu-1)$-sections. Let $\mathcal{S}_0^{\nu-1}$ be the $(\nu-1)$-section containing x_0. Let $x_{i_1}^\nu$ be the last ν-node in (4.11) that is incident to $\mathcal{S}_0^{\nu-1}$. $x_{i_1}^\nu$ will be a boundary ν-node because it is also incident to another $(\nu-1)$-section, say, $\mathcal{S}_1^{\nu-1}$. If need be, we can replace the subpath of (4.11) between x_0 and x_{i_1} by a ν-path $\{x_0, Q_0^{\nu-1}, x_{i_1}^\nu\}$, where $Q_0^{\nu-1}$ is a one-ended $(\nu-1)$-path and resides in $\mathcal{S}_0^{\nu-1}$, to get a shorter overall ν-path terminating at x_0 and x_k.

Now, let $\mathcal{S}_1^{\nu-1}$ be the next $(\nu-1)$-section after $\mathcal{S}_0^{\nu-1}$ through which our (possibly) reduced path proceeds. Also, let $x_{i_2}^\nu$ be the last ν-node in that path that is incident to $\mathcal{S}_1^{\nu-1}$. Then, $x_{i_2}^\nu$ will be a boundary ν-node incident to $\mathcal{S}_1^{\nu-1}$ and another $(\nu-1)$-section $\mathcal{S}_2^{\nu-1}$. If need be, we can replace the subpath between x_{i_1} and x_{i_2} by a ν-path $\{x_{i_1}^\nu, Q_1^{\nu-1}, x_{i_2}^\nu\}$, where $Q_1^{\nu-1}$ is an endless $(\nu-1)$-path residing in $\mathcal{S}_1^{\nu-1}$. This will yield a still shorter overall ν-path terminating at x_0 and x_k.

Continuing this way, we will find a boundary ν-node $x_{i_j}^\nu$ that is incident to the $(\nu-1)$-section $\mathcal{S}_j^{\nu-1}$ containing x_k. Finally, we let $Q_j^{\nu-1}$ be a one-ended $(\nu-1)$-path in $\mathcal{S}_j^{\nu-1}$ terminating at $x_{i_j}^\nu$ and x_k. Altogether, we will have the following two-ended ν-path, which is not longer than $P_{0,k}^\nu$ (actually shorter if the aforementioned replacements were made):

$$Q_{0,k}^\nu = \langle x_0, Q_0^{\nu-1}, x_{i_1}^\nu, Q_1^{\nu-1}, x_{i_2}^\nu, \ldots, x_{i_j}^\nu, Q_j^{\nu-1}, x_k\rangle . \tag{4.12}$$

Because all the ν-nodes herein are boundary nodes and pristine, the length $|Q_{0,k}|$ is obtained simply by counting the $(\nu-1)$-tips traversed by $Q_{0,k}^\nu$ and multiplying by ω^ν (see Lemma 4.7-4). We get $|Q_{0,k}^\nu| = \omega^\nu \cdot (2j)$, where $j \leq m$. Finally, we note that any geodesic path between x_0 and x_k has a length no larger than than $|Q_{0,k}^\nu| = \omega^\nu \cdot (2j)$.

Next, consider the case where x_0 is a bordering ν-node of $\mathcal{S}_0^{\nu-1}$ and x_k remains an internal node of $\mathcal{S}_j^{\nu-1}$. $P_0^{\nu-1}$ in (4.11) will be an endless $(\nu-1)$-path residing in some $(\nu-1)$-section $\mathcal{S}_0^{\nu-1}$. We let $x_{i_1}^\nu$ be the last ν-node in (4.11) incident to $\mathcal{S}_0^{\nu-1}$. Otherwise our procedure is as before, and we can now conclude that $|Q_{0,k}^\nu| = \omega^\nu \cdot (2j+1)$, because the passage from x_0 into $\mathcal{S}_0^{\nu-1}$ traverses one $(\nu-1)$-tip.

The same conclusion, namely, $|Q_{0,k}^\nu| = \omega^\nu \cdot (2j+1)$ holds if x_k is a bordering ν-node and x_0 is an internal node. Finally, if both x_0 and x_k are bordering ν-nodes, we get $|Q_{0,k}^\nu| = \omega^\nu \cdot (2j+2)$. For all cases, we can assert that the geodesic between x_0 and x_k has a length no larger than $|Q_{0,k}^\nu| = \omega^\nu \cdot (2j+2)$.

Now, the eccentricity $e(x_0)$ for x_0 is the supremum of the lengths of all geodesics starting at x_0 and terminating at all other nodes x_k. Since there are only finitely many boundary nodes and since every geodesic will have the form of (4.12), every eccentricity will be a multiple of ω^ν (there will be no arrow-rank eccentricities). Also, since $j \leq m$ where m is the number of boundary ν-nodes in $\mathcal{G}^\nu$, we can conclude that $e(x_0) \leq \omega^\nu \cdot (2m+2)$, whatever be the node x_0. ♣

The lengths of all geodesics will reside in the finite set of values (4.10). Consequently, for every node x_0 of $\mathcal{G}^\nu$ there will be at least one geodesic of maximum length starting at x_0 and terminating at some other node z of $\mathcal{G}^\nu$. Such a geodesic is called an *eccentric path* for x_0, and z is called an *eccentric node* for x_0. In general, there are many eccentric paths and eccentric nodes for a given x_0.

Corollary 4.7-7. *Let x^ν be any bordering ν-node of a $(\nu-1)$-section $\mathcal{S}^{\nu-1}$ with the eccentricity $e(x^\nu) = \omega^\nu \cdot k$, and let z be an internal node of $\mathcal{S}^{\nu-1}$ with the eccentricity $e(z) = \omega^\nu \cdot p$. Then, $|k-p| \leq 1$.*

Proof. Let $P_{z,x}$ be the two-ended ν-path obtained by appending x^ν to a one-ended $(\nu-1)$-path in $\mathcal{S}^{\nu-1}$ that reaches x^ν and terminates at the internal node z. By Lemma 4.7-4(a), $P^\nu_{z,x}$ is a z-to-x^ν geodesic, and $|P^\nu_{z,x}| = d(z, x^\nu) = \omega^\nu$. Now, let w be any node. By the triangle inequality for the metric d,

$$d(z,w) \;\leq\; d(z,x^\nu) + d(x^\nu,w) \;=\; \omega^\nu + d(x^\nu,w).$$

Next, let w be an eccentric node for z. We get $d(z,w) = e(z)$ and $e(z) \leq \omega^\nu + d(x^\nu, w)$. Moreover, $d(x^\nu, w) \leq e(x^\nu)$. Therefore,

$$e(z) \;\leq\; \omega^\nu + e(x^\nu). \tag{4.13}$$

By a similar argument with w now being an eccentric node for x^ν, we get

$$e(x^\nu) \;\leq\; \omega^\nu + e(z). \tag{4.14}$$

So, with (4.13) we have $\omega^\nu \cdot p \leq \omega^\nu + \omega^\nu \cdot k = \omega^\nu \cdot (k+1)$, or $p \leq k+1$. On the other hand, with (4.14) we have in the same way $k \leq p+1$. Whence our conclusion. ♣

Fig. 4.5. The 1-graph of Example 4.7-8.

That $k-p$ can equal 0 is verified by the next example.

Example 4.7-8. Consider the 1-graph of Fig. 4.5 consisting of a one-ended 0-path of 0-nodes w_k^0 and an endless 0-path of 0-nodes y_k^0 connected in series to two 1-nodes x^1 and z^1 as shown. The eccentricities are as follows: $e(w_k^0) = \omega \cdot 3$ for $k = 1, 2, 3, \ldots$, $e(x^1) = \omega \cdot 2$, $e(y_k^0) = \omega \cdot 2$ for $k = \ldots, -1, 0, 1, \ldots$, and $e(z^1) = \omega \cdot 3$. Thus, $e(x^1) - e(y_k^0) = 0$, as asserted. ♣

An immediate consequence of Theorem 4.7-6 and Corollary 4.7-7 is the following.

Corollary 4.7-9. *The eccentricities of all the nodes form a consecutive set of values within (4.10), with the minimum (resp. maximum) eccentricity being the radius (resp. diameter) of* $\mathcal{G}^\nu$.

By virtue of Theorem 4.6-1 and Condition 4.7-1, we have that, if an ordinal is the eccentricity of an internal node of a $(\nu-1)$-section $\mathcal{S}^{\nu-1}$, then there will be infinitely many nodes with the same eccentricity, for example, all the internal nodes of $\mathcal{S}^{\nu-1}$. Moreover, a boundary node of $\mathcal{S}^{\nu-1}$ may also have that same eccentricity; the 1-node x^1 in Example 4.7-8 illustrates this. Furthermore, it is possible for the radius of $\mathcal{G}^\nu$ to be the eccentricity of only one node in $\mathcal{G}^\nu$; this occurs when the center consists of only one node. Can another eccentricity occur for only one node? No, by virtue of Conditions 4.7-1 and 4.7-2 and Corollary 4.7-7. There must be at least two nodes for each eccentricity larger than the radius. The proof of this is virtually the same as a proof of Lesniak for finite graphs [9, page 176].

4.8 The Center Lies in a ν-Block

This is a known result for finite graphs [9, Theorem 2.2], [14, Theorem 2.9], which we now extend transfinitely. As stated before, the *center* of $\mathcal{G}^\nu$ is the set of nodes having the minimum eccentricity. To define a "ν-block," we first define (as in Sec. 3.3) the *removal* of a pristine nonsingleton ν-node x^ν to be the following procedure: x^ν is replaced by two or more singleton ν-nodes, each containing exactly one of the $(\nu-1)$-tips of x^ν and with every $(\nu-1)$-tip of x^ν being so assigned. We denote the resulting ν-graph by $\mathcal{G}^\nu - x^\nu$. Then, a subgraph $\mathcal{H}$ of $\mathcal{G}^\nu$ will be called a *ν-block* of $\mathcal{G}^\nu$ if $\mathcal{H}$ is a maximal ν-connected subgraph of $\mathcal{G}^\nu$ such that, for every ν-node x^ν, all the branches of $\mathcal{H}$ lie in the same component of $\mathcal{G}^\nu - x^\nu$.[4] A more explicit way of defining a ν-block is as follows: For any ν-node x^ν, $\mathcal{G}^\nu - x^\nu$ consists of one or more components. Choose one of those components. Repeat this for every ν-node, choosing one component for each ν-node. Then, take the intersection[5] of all those chosen components. The intersection may be empty, but, if it is not empty, it will be a ν-block of $\mathcal{G}^\nu$. Upon taking all possible intersections of components, one component from each $\mathcal{G}^\nu - x^\nu$, and then choosing the nonempty intersections, we will obtain all the ν-blocks of $\mathcal{G}^\nu$.

Furthermore, we define a *cut ν-node* to be a nonsingleton ν-node x^ν such that $\mathcal{G}^\nu - x^\nu$ has two or more components. It follows that the cut ν-nodes *separate* the ν-blocks in the sense that any path that terminates at two branches in different

[4] The definition of a ν-block can be extended by defining (as in Sec. 3.3) the removal of a nonpristine maximal node of any rank as the placement of each of its embraced tips into singleton nodes. This extension is used in Example 4.8-4.

[5] This is the subgraph induced by those branches lying in all the chosen components.

ν-blocks must pass through at least one cut ν-node. (Otherwise, the two branches would be in the same component of $\mathcal{G}^\nu - x^\nu$ for every x^ν and therefore in the same ν-block.)

Lemma 4.8-1. *The ν-blocks of $\mathcal{G}^\nu$ partition $\mathcal{G}^\nu$, and the cut ν-nodes separate the ν-blocks.*

Proof. For each x^ν, each branch will be in at least one of the components of $\mathcal{G}^\nu - x^\nu$, and therefore in at least one of the ν-blocks. On the other hand, no branch can be in two different ν-blocks because then there would be a cut ν-node that separates a branch from itself—an absurdity.

That the cut ν-nodes separate the ν-blocks has already been noted. ♣

Lemma 4.8-2. *Each $(\nu-1)$-section $\mathcal{S}^{\nu-1}$ is contained in a ν-block, and a ν-block is partitioned by its $(\nu-1)$-sections.*

Proof. Any two branches of $\mathcal{S}^{\nu-1}$ are connected through a two-ended path of rank no greater than $\nu - 1$. Such a path will not meet any ν-node because every ν-node is pristine. Thus, $\mathcal{S}^{\nu-1}$ will lie entirely within a single component of $\mathcal{G}^\nu - x^\nu$, whatever be the choice of x^ν. By the definition of a ν-block, we have the first conclusion.

The second conclusion follows from the fact that the $(\nu - 1)$-sections partition $\mathcal{G}^\nu$. ♣

By definition, all the bordering nodes of a $(\nu - 1)$-section $\mathcal{S}^{\nu-1}$ will be ν-nodes. Since a ν-block is a subgraph induced by the branches of its $(\nu - 1)$-sections, every ν-block $\mathcal{H}$ will contain the bordering ν-nodes of its $(\nu - 1)$-sections, and therefore the rank of $\mathcal{H}$ is ν. So, henceforth we denote $\mathcal{H}$ by $\mathcal{H}^\nu$. In general, a ν-node can belong to more than one $(\nu - 1)$-section and also to more than one ν-block.

Example 4.8-3. Fig. 4.6 shows a 1-graph in which the P_k^0 $(k = 1, 2, 3, 4, 5)$ are endless 0-paths and v^1, w^1, x^1, y^1, and z^1 are 1-nodes. There are two 1-blocks: One of them consists of P_1^0 along with v^1 and w^1, and the other consists of the P_k^0 $(k = 2, 3, 4, 5)$ along with w^1, x^1, y^1, and z^1. The only cut 1-node is w^1. Also, there are five 0-sections, each consisting of one endless 0-path. ♣

Example 4.8-4. The condition that the ν-nodes are pristine is needed for Lemma 4.8-2 to hold. For example, Fig. 4.7 shows a 1-graph with a nonpristine 1-node x^1, three 1-blocks,[6] and two 0-sections. Branch b_1 induces one 1-block, branch b_2 induces another 1-block, and all the branches of the one-ended 0-path P^0 induce the third 1-block. However, b_1 and b_2 together induce a single 0-section $\mathcal{S}^0$, and the branches of P^0 induce another 0-section. $\mathcal{S}^0$ lies in the union of two 1-blocks. ♣

[6]Here, we are extending the definition of a 1-block by requiring that the elementary tips of x^1 also be placed in singleton nodes when x^1 is removed.

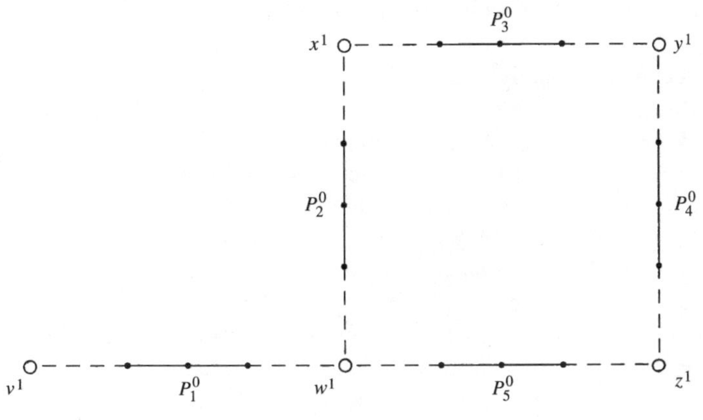

Fig. 4.6. The 1-graph of Example 4.8-3.

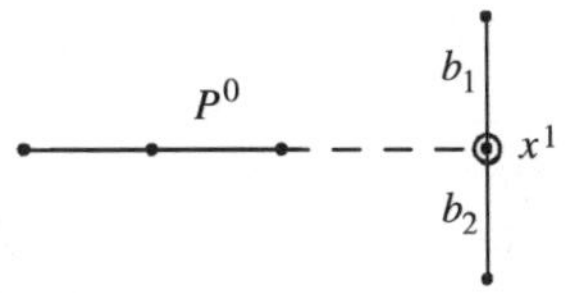

Fig. 4.7. The 1-graph of Example 4.8-4.

We are finally ready to verify the title of this section concerning the *center* of $\mathcal{G}^\nu$, which by definition is the set of nodes having the minimum eccentricity. According to Theorem 4.7-6, such nodes exist. Having set up appropriate definitions and preliminary results for the transfinite case, we can now use a proof that is much the same as that for finite graphs [9, Theorem 2.2], [14, Theorem 2.9].

Theorem 4.8-5. *The center of $\mathcal{G}^\nu$ lies in a ν-block in the sense that there is a ν-block containing all the nodes of the center.*

Proof. Suppose the center of $\mathcal{G}^\nu$ lies in two or more ν-blocks. By Lemma 4.8-1, there is a cut ν-node x^ν such that $\mathcal{G}^\nu - x^\nu$ has at least two components, say, $\mathcal{G}_1$ and $\mathcal{G}_2$, each of which contains a node of the center. Let u be an eccentric node for x^ν, and let $P_{x,u}$ be an x^ν-to-u geodesic. Thus, $|P_{x,u}| = e(x^\nu)$. $P_{x,u}$ cannot contain any node different from x^ν in at least one of $\mathcal{G}_1$ and $\mathcal{G}_2$, say, $\mathcal{G}_1$. Let w be a center node in $\mathcal{G}_1$ other than x^ν, and let $P_{w,x}$ be a w-to-x^ν geodesic. Then, $P_{w,x} \cup P_{x,u}$ is a path whose length satisfies $|P_{w,x} \cup P_{x,u}| = |P_{w,x}| + |P_{x,u}| \geq 1 + e(x)$. This shows that the eccentricity of w is greater than the minimum eccentricity, that is, w is not a center node—a contradiction that proves the theorem. ♣

4.9 The Centers of Cycle-free ν-Graphs

We now specialize our study to cycle-free ν-graphs with $\nu \geq 1$ and with at least two $(\nu-1)$-sections.[7] The cycle-free condition was defined in Sec. 3.3. In contrast to Sec. 3.3, we no longer wish to ignore the nonboundary bordering nodes in $\mathcal{M}$ because we want to consider the eccentricities of all the nodes in $\mathcal{M}$. Also, end $(\nu-1)$-sections and non-end $(\nu-1)$-sections were defined in Sec. 3.3 as well. We assume henceforth in this section that $\mathcal{G}^\nu$ has at least two $(\nu-1)$-sections—or equivalently has at least one boundary ν-node. According to Lemma 3.3-1, $\mathcal{G}^\nu$ has only finitely many non-end $(\nu-1)$-sections and at least two end $(\nu-1)$-sections.

Our next objective is to replace our cycle-free ν-graph $\mathcal{G}^\nu$ $(\nu \geq 1)$ by a conventional finite tree (i.e., a 0-connected 0-graph having no loops and only finitely many branches), a replacement that will be exploited in the proof of our final theorem. The nodal eccentricities for $\mathcal{T}^0$ will be related to the nodal eccentricities for $\mathcal{G}^\nu$ in a simple way (see Lemma 4.9-1 below).

In the following, the index k will number the boundary ν-nodes of $\mathcal{G}^\nu$ as well as certain sets of nonboundary bordering ν-nodes. By the *interior* $i(\mathcal{S}^{\nu-1})$ of a $(\nu-1)$-section $\mathcal{S}^{\nu-1}$ we mean the set of all its internal nodes. The index m will number the interiors of non-end $(\nu-1)$-sections as well as certain sets consisting of the interiors of certain end $(\nu-1)$-sections.

Consider, first of all, a non-end $(\nu-1)$-section $\mathcal{S}^{\nu-1}$. Each of its boundary ν-nodes x_k^ν is replaced by a 0-node x_k^0 having the same index number k. Also, the set of all the nonboundary bordering ν-nodes of $\mathcal{S}^{\nu-1}$ (if such exist) are replaced by a single 0-node $x_{k'}^0$. Finally, the interior $i(\mathcal{S}^{\nu-1})$ is replaced by a single 0-node y_m^0. A branch is inserted between y_m^0 and each of the x_k^0 and between y_m^0 and $x_{k'}^0$ as well. Thus, $\mathcal{S}^{\nu-1}$ is replaced by a star 0-graph. We view x_k^0 (resp. $x_{k'}^0$, resp. y_m^0) as representing x_k^ν (resp. the set of all nonboundary bordering nodes of $\mathcal{S}^{\nu-1}$, resp. the set of all the internal nodes of $\mathcal{S}^{\nu-1}$). For any $y^\gamma \in i(\mathcal{S}^{\nu-1})$, we have $d(y^\gamma, x_k^\nu) = d(y^\gamma, x_{k'}^\nu) = \omega^\nu$ (Lemma 4.7-4(a)); also, corresponding to a given $\mathcal{S}^{\nu-1}$ we have $d(y_m^0, x_k^0) = d(y_m^0, x_{k'}^0) = 1$.

Next, consider all the end $(\nu-1)$-sections that are incident to a single boundary ν-node $x_{k''}^\nu$. We represent the set of all of their interiors by a single 0-node $y_{m'}^0$, and we insert a branch between $y_{m'}^0$ and $x_{k''}^0$. If at least one of those end $(\nu-1)$-sections has a nonboundary bordering ν-node $x_{k'''}^\nu$, we represent the set of all of them for all those end $(\nu-1)$-sections incident to a single boundary ν-node by another single 0-node $x_{k'''}^0$, and we insert another branch between $y_{m'}^0$ and $x_{k'''}^0$. So, the set of all of these end $(\nu-1)$-sections incident to the chosen $x_{k''}^0$ are represented either by a single branch incident at $x_{k''}^0$ and $y_{m'}^0$ or by two branches in series incident at $x_{k''}^0$, $y_{m'}^0$, and $x_{k'''}^0$. Here, too, we have replaced the said set of end $(\nu-1)$-sections incident at a particular boundary node by a "star" 0-graph, but this time having only one branch or two branches.

---

[7] Remember that $\mathcal{G}^\nu$ is ν-connected and satisfies Conditions 3.1-2, 4.7-1, and 4.7-2.

We now connect all these "star" 0-graphs together at their end 0-nodes in the same way that the $(\nu-1)$-sections are connected together at their boundary ν-nodes. The result is a finite 0-tree $\mathcal{T}^0$. Indeed, since $\mathcal{G}^\nu$ is cycle-free, $\mathcal{T}^0$ has no loops. Also, $\mathcal{T}^0$ has only finitely many branches because $\mathcal{G}^\nu$ has only finitely many non-end $(\nu-1)$-sections and because of the way end $(\nu-1)$-sections are represented by branches in $\mathcal{T}^0$. Also, $\mathcal{T}^0$ has at least two branches because $\mathcal{G}^\nu$ has at least two $(\nu-1)$-sections.

By means of these definitions we have indirectly defined the following mapping: Corresponding to every maximal node in $\mathcal{G}^\nu$ there is a 0-node in $\mathcal{T}^0$. This mapping from the set of all maximal nodes of all ranks in $\mathcal{G}^\nu$ onto the set of all 0-nodes in $\mathcal{T}^0$ is surjective but not injective. Furthermore, for any internal node y^γ in any end $(\nu-1)$-section of $\mathcal{G}^\nu$, we have $d(y^\gamma, x^\nu_{k''}) = d(y^\gamma, x^\nu_{k'''}) = \omega^\nu$ where $x^\nu_{k''}$ is the unique boundary node and $x^\nu_{k'''}$ is any nonboundary bordering ν-node for that end $(\nu-1)$-section (if such exists). Correspondingly, $d(y^0_{m'}, x^0_{k''}) = d(y^0_{m'}, x^0_{k'''}) = 1$.

Lemma 4.9-1. *A node z of any rank in $\mathcal{G}^\nu$ has an eccentricity $e(z) = \omega^\nu \cdot p$ if and only if its corresponding 0-node z^0 in $\mathcal{T}^0$ has the eccentricity $e(z^0) = p$.*

(Here again, p is a natural number.)

Proof. An eccentric path P^ν of any node z of any rank in $\mathcal{G}^\nu$ passes alternately through $(\nu-1)$-sections and bordering ν-nodes and terminates at z and an eccentric node for z. Because all ν-nodes are pristine, the length $|P^\nu|$ is obtained by counting the $(\nu-1)$-tips traversed by P^ν and multiplying by ω^ν (Lemma 4.7-3). Furthermore, corresponding to P^ν there is a unique path Q^0 in $\mathcal{T}^0$ whose nodes x^0_k and y^0_m alternate in Q^0 and correspond to the bordering nodes x^ν_k and interiors of $(\nu-1)$-sections $\mathcal{S}^{\nu-1}_m$ traversed by P^ν. Each branch of Q^0 corresponds to one traversal of a $(\nu-1)$-tip in P^ν, and conversely. Thus, we have $|P^\nu| = \omega^\nu \cdot p$ and $|Q^0| = p$, where p is the number of branches in Q^0. Also, P^ν is an eccentric path in $\mathcal{G}^\nu$ if and only if Q^0 is an eccentric path in $\mathcal{T}^0$, whence our conclusion. ♣

Here is our principal result concerning the centers of cycle-free ν-graphs.

Theorem 4.9-2. *Let $\mathcal{G}^\nu$ be a cycle-free ν-graph having at least two $(\nu-1)$-sections. Then, the center of $\mathcal{G}^\nu$ has one of the following forms:*

(a) A single boundary ν-node x^ν.
(b) The interior $i(\mathcal{S}^{\nu-1})$ of a single non-end $(\nu-1)$-section $\mathcal{S}^{\nu-1}$.
(c) The set $i(\mathcal{S}^{\nu-1}) \cup \{x^\nu\}$, where $i(\mathcal{S}^{\nu-1})$ is the interior of a single $(\nu-1)$-section and x^ν is one of the boundary ν-nodes of that $(\nu-1)$-section $\mathcal{S}^{\nu-1}$.

Note. In the third case it is possible for $\mathcal{S}^{\nu-1}$ to be an end $(\nu-1)$-section; this occurs for instance when $\mathcal{G}^\nu$ consists of just two end $(\nu-1)$-section, one of which has a nonboundary bordering ν-node and the other does not.

Proof. By virtue of Lemma 4.9-1, we can make use of an established theorem for finite 0-trees [9, Theorem 2.1]; namely, the center of such a tree is either a single 0-node or a pair of adjacent 0-nodes. Now, our finite tree $\mathcal{T}^0$ has at least two branches, and consequently the center of $\mathcal{T}^0$ cannot contain an end 0-node. So, when the center of $\mathcal{T}^0$ is a single 0-node representing a single boundary ν-node in $\mathcal{G}^\nu$, form (a) holds. When the center of $\mathcal{T}^0$ is a single 0-node representing the interior of a $(\nu-1)$-section $\mathcal{S}^{\nu-1}$ in $\mathcal{G}^\nu$, form (b) holds; $\mathcal{S}^{\nu-1}$ cannot be an end $(\nu-1)$-section in this second case because there are at least two $(\nu-1)$-sections. Finally, when the center of $\mathcal{T}^0$ is a pair of adjacent 0-nodes, one of them will represent a boundary ν-node in $\mathcal{G}^\nu$, and the other will represent the interior of a $(\nu-1)$-section. ♣

Finally, let us observe in passing the periphery of a cycle-free ν-graph (again having at least two $(\nu-1)$-sections). As before, the periphery is the set of those nodes having the maximum eccentricity. It consists either of nonboundary bordering ν-nodes of end $(\nu-1)$-sections, or the interiors of end $(\nu-1)$-sections, or both. This is because the periphery of a finite 0-tree consists only of end 0-nodes.

5

Walk-Based Transfinite Graphs and Networks

The theory of transfinite graphs developed so far has been based on the ideas that connectedness is accomplished through paths and that the infinite extremities of the graph are specified through equivalence classes of one-ended paths. This is a natural extension of finite graphs because connectedness for finite graphs is fully characterized by paths; indeed, any walk terminating at two nodes of a finite graph contains a path doing the same. However, such is no longer the case for transfinite graphs. Indeed, path-connectedness need not be transitive as a binary relationship among transfinite nodes, and Condition 3.1-2 was imposed to ensure such transitivity. Without that condition, distances as defined by paths do not exist between certain pairs of nonsingleton nodes. This limitation is also reflected in the theory of transfinite electrical networks by the fact that node voltages need not be uniquely determined when they are defined along paths to a chosen ground node.

These troubles disappear when transfinite walks are used as the basic construct, but the resulting, more general kinds of transfinite graphs now encompass some strange structures that stress intuition based upon one's familiarity with finite graphs. Moreover, electrical network theory is similarly ensnarled. Such complications should not be surprising because infinite entities are mathematical abstractions involving a variety of counterintuitive phenomena, as for example the antinomies of infinite sets [36]. Following theory wherever it may lead, we now propose to explore walk-based transfinite graphs.

By the end of Sec. 5.5, we will have accomplished one of the objectives of this chapter, namely, to define and develop recursively transfinite graphs based upon walk-defined extremities. This opens up the possibility of establishing a variety of more general results analogous to those already proven for path-based transfinite graphs. Such will be done in Secs. 5.6 through 5.9, where we examine ordinal-valued distances in walk-based transfinite graphs. In particular, an ordinal-valued distance function can now be defined on all pairs of walk-connected nodes, in contrast to the path-based theory wherein no distance function is definable for those pairs of nodes that are not path-connected even though they are walk-connected. Some results concerning eccentricities, centers, and blocks are presented in this more general walk-based graph theory.

Furthermore, Secs. 5.10 through 5.15 are devoted to the development of an electrical network theory for networks whose graphs are walk-based. A unique current-voltage regime is established under certain conditions. The current regime is built up from current flows in closed walks—in contrast to a prior theory [51, Chap. 5] based upon flows in loops. A notable advantage of the present approach is that node voltages with respect to a given ground node are always unique whenever they exist. The present approach is more general in that it provides nontrivial current-voltage regimes for certain networks for which the prior approach would only provide trivial solutions having only zero currents and zero voltages everywhere.

We will be defining several entities and concepts in this chapter that are analogous to the tips, nodes, graphs, path-connectedness, eccentricities, centers, etc. of Chapters 3 and 4 but are based upon transfinite walks rather than on transfinite paths. So as not to introduce new terminology and some consequent unnecessary confusion, we will again use the path-based terminology but will employ the letter"w" as a prefix to indicate that our entities are now walk-based. Thus, we write "wtip," "wnode," "wgraph," "wconnectedness," "weccentricites," "wcenters," etc.[1]

Let us explicate another bit of terminology: A one-ended infinite sequence

$$\langle a_0, a_1, \ldots, a_m, \ldots \rangle$$

or a two-way infinite sequence

$$\langle \ldots, a_{-m}, \ldots, a_{-1}, a_0, a_1, \ldots, a_m, \ldots \rangle$$

will be said to possess *eventually* a certain property if there exists a natural number m_0 such that the subsequences having all indices m with $|m| > m_0$ possess that property. Similarly, two one-way infinite sequences will be said to be *eventually identical* if, with possibly an appropriate shifting of the indices of one of them, their elements are eventually the same.

In this chapter, we no longer impose Condition 3.1-2

5.1 0-Walks and 1-Wgraphs

A *nontrivial 0-walk* W^0 is an alternating sequence of 0-nodes x_m^0 and branches b_m:

$$W^0 = \langle \ldots, x_{m-1}^0, b_{m-1}, x_m^0, b_m, x_{m+1}^0, \ldots \rangle \tag{5.1}$$

where the indices m traverse a set of consecutive integers and, for each m, the branch b_m is incident to the two 0-nodes x_m^0 and x_{m+1}^0. In contrast to a path, nodes and branches may repeat in the sequence. We allow any branch to be a self-loop, in which case x_m^0 and x_{m+1}^0 are the same 0-node. If the sequence (5.1) terminates on either side, it terminates at a 0-node. The 0-walk W^0 is called *two-ended* or *finite* if

[1] One may pronounce wtip as "walk-tip" or simply say "tip" when the prefix "w" is understood—and similarly for the other "wentities."

it terminates on both sides, *one-ended* if it terminates on just one side, and *endless* if it terminates on neither side. We may assign to W^0 one of two possible *orientations* determined by either increasing or decreasing indices m. A *closed* 0-walk is a two-ended 0-walk whose terminal 0-nodes are the same.

A *trivial 0-walk* is a singleton containing just one 0-node.

A superfluity arises when we try to define the infinite extremities of a 0-graph $\mathcal{G}^0$ by using 0-walks. Consider a one-ended 0-walk W^0 in a 0-graph $\mathcal{G}^0$. As we trace W^0 starting from its initial 0-node, we may encounter a first 0-node x_a^0 that is encountered more than once. The closed 0-walk (possibly the self-loop) traced between the first and second encounters of x_a^0 can be removed from W^0 without affecting that part of W^0 occurring after the second encounter of x_a^0. To be precise, we mean by this *removal* the deletion of all the nodes and branches in (5.1) strictly between the first two encounters of x_a^0 along with the deletion of that second occurrence of x_a^0 but not the first. For the purpose of finding an infinite extremity of $\mathcal{G}^0$, we may work with the reduced walk resulting from the said deletions. In fact, we may as well remove such closed 0-walks as they are traced in W^0. But, if we keep doing so, the result of removing all such closed 0-walks in W^0 as they are encountered sequentially will simply be a 0-path—either a finite 0-path (possibly a trivial one) or a one-ended 0-path. It appears that nothing more will be gained by using one-ended 0-walks in place of one-ended 0-paths when defining the infinite extremities of a 0-graph; we may as well stick with one-ended 0-paths for this purpose. This we do.

It is only when we turn to the construction of graphs of higher ranks of transfiniteness that walks provide greater generality than do paths, as we shall see. At this point of our discussion, 0-tips and then 1-nodes and 1-graphs are defined exactly as they are in Secs. 2.2 and 2.3. To conform with the terminology and notation used subsequently, we may at times denote a 0-graph and a 1-graph as the doublet and triplet

$$\mathcal{G}_w^0 = \{\mathcal{B}, \mathcal{X}_w^0\}, \quad \mathcal{G}_w^1 = \{\mathcal{B}, \mathcal{X}_w^0, \mathcal{X}_w^1\},$$

where the subscript w has been appended to the graph and node-set symbols even though they are defined exactly as before. Similarly, we may refer to $\mathcal{G}_w^0$ as a 0-*wgraph* and to $\mathcal{G}_w^1$ as a 1-*wgraph*. Moreover, when dealing with a wgraph in general, a "wnode" (resp. "wtip") may in fact be a 0-node or 1-node (resp. a 0-tip). Also, in conformity with some subsequent constructions, we will say that a one-ended 0-walk

$$W^0 = \langle x_0^0, b_0, x_1^0, b_1, x_2^0, b_2, \ldots \rangle$$

is *extended* if its 0-nodes x_m^0 are eventually pairwise distinct. Thus, W^0 is extended if it eventually is identical to a one-ended 0-path. We say that W^0 *traverses* a 0-tip if it is extended and is eventually identical to a representative of that 0-tip. Finally, W^0 is said to *reach* a 1-node x^1 if it traverses a 0-tip embraced by x^1. In the same way, an endless 0-walk can *reach* two 1-nodes (or possibly reach the same node) by traversing two 0-tips, one toward the left and the other toward the right. On the other hand, if a 0-walk terminates at a 0-node that is embraced by a 1-node, we again say that the walk *reaches* both of those nodes and does so *through* an elementary tip of its branch that is incident to that 0-node.

5.2 1-Walks, 2-Wgraphs, and 2-Walks

Matters become more complicated when the infinite extremities of 1-graphs are considered.

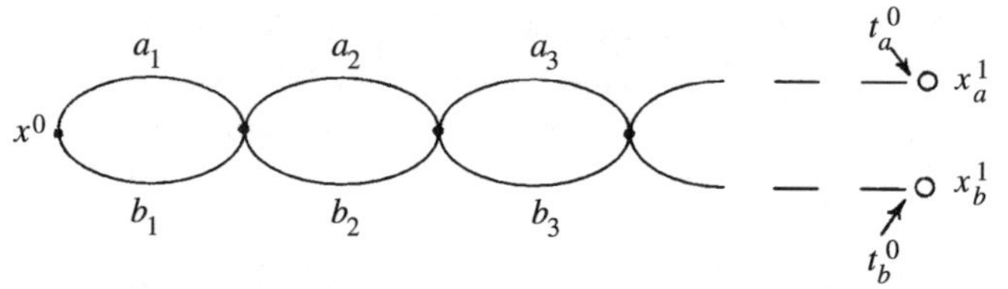

Fig. 5.1. The simplest 1-graph having no path connecting any two of its 1-nodes. (Only two of its uncountably many 1-nodes are shown.) There is, however, a walk between any two 1-nodes.

Example 5.2-1. Consider the 1-graph of Fig. 5.1. (This appeared as a subgraph in Fig. 3.1. See also Example 3.1-1.) There are many "1-walks" (defined below) connecting the 1-nodes x_a^1 and x_b^1, but no such 1-path. One such "1-walk" starts at x_a^1, passes through the a_k branches to reach the 0-node x^0, then passes through the b_k branches, and finally ends at x_b^1. We can connect infinitely many such 1-graphs "in series," as shown in Fig. 5.2, to obtain a 1-graph with an extremity not definable through paths. That extremity is reached through the "one-ended 1-walk" that starts at x_1^1, passes to x_2^1 through an endless 0-walk, then to x_3^1 through another endless 0-walk, and so on. More specifically, from x_1^1 it passes along the left side of the first 0-section to, say, the 0-node x_1^0, then along the right side of that 0-section to x_2^1, then along the left side of the second 0-section to, say, the 0-node x_2^0, then along the right side of that 0-section to x_3^1, and so forth. On the other hand, there is no one-ended 1-path that passes through $x_1^1, x_2^1, x_3^1, \ldots$. Thus, we have a new kind of extremity, something that was not considered previously in Chapter 2.[2]

Before leaving this example, let us consider a variation of it. Consider the 1-graph obtained from that of Fig. 5.2 by shorting together all the 0-nodes x_k^0 ($k = 1, 2, 3, \ldots$) along the bottom of Fig. 5.2 to obtain a 0-node x^0 of infinite degree. Then, the "one-ended 1-walk" described just before can be specified as one that keeps returning to the same 0-node x^0. This time, we will allow such a walk as a means of identifying an infinite extremity of the 1-graph. So long as the 1-nodes through which the walk passes become eventually pairwise distinct, we will take it that an infinite extremity of the 1-graph is being identified. Compare this with the situation considered in the preceding Section where a 0-walk that keeps returning to the same 0-node is reduced to a trivial 0-path by removing finite loops; that 0-walk was thus unsuitable for the specification of an extremity of the 0-graph. Analogously,

[2]There are infinitely many such extremities obtained by choosing other 0-tips through which the one-ended 1-path proceeds.

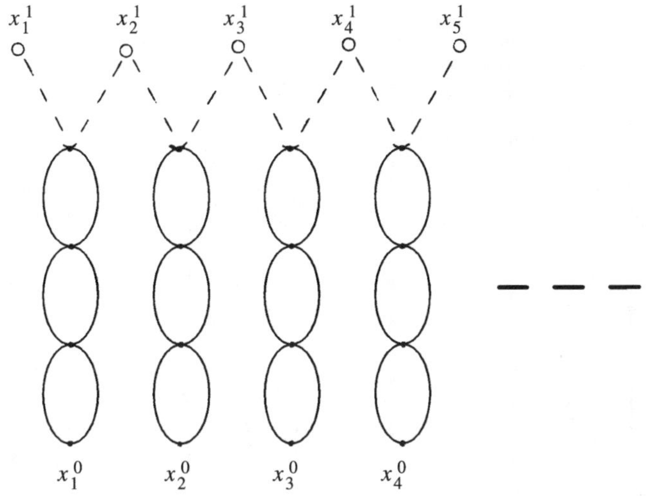

Fig. 5.2. A 1-graph having a 1-wtip but no 1-tip.

were we to have a one-ended 1-walk that keeps returning to the same 1-node, we could remove 1-loops to render that 1-walk into a trivial or two-ended 1-path, and we would then conclude that such a 1-walk is unsuitable for the identification of an extremity of a 1-wgraph. It is the fact that our 1-walk is "extended" (that is, its sequence of 1-nodes are eventually all distinct) that makes it suitable for the identification of an extremity of the 1-wgraph. ♣

Example 5.2-2. Extremities like that of Fig. 5.2 appear inherently in more familiar kinds of 1-graphs. For example, a 1-walk reaching such a walk-based extremity is indicated in Fig. 5.3 for a 1-graph consisting of two infinite checkerboard 0-graphs connected together through certain 1-nodes. x_1^1 is a singleton 1-node containing one 0-tip having as a representative a one-ended 0-path that passes horizontally toward the left and alternating up and down through single vertical branches, as shown. For $m = 2, 3, 4, \ldots$, each x_m^1 consists of two 0-tips, one for a similar 0-path toward the left and another for such a 0-path toward the right. Pairs of such consecutive 0-paths pass through infinitely many horizontal branches in opposite directions. Altogether, these infinitely many 1-nodes along with these infinitely many 0-paths form a "one-ended 1-walk" whose extremity cannot be reached through any one-ended 1-path because those pairs of consecutive 0-paths intersect infinitely often. However, each walk between two consecutive 1-nodes x_m^1 and x_{m+1}^1 is the conjunction of two one-ended 0-paths forming an endless 0-walk passing through a 0-section (one of the checkerboards) and reaching those 1-nodes through 0-tips. So, perhaps, one-ended 1-walks, which can be decomposed into endless 0-walks between eventually distinct 1-nodes, will suffice in identifying a new kind of transfinite extremity. ♣

Let us implement this idea in the following precise way: Because 0-tips and 1-nodes are now defined exactly as they are for path-based transfinite graphs and

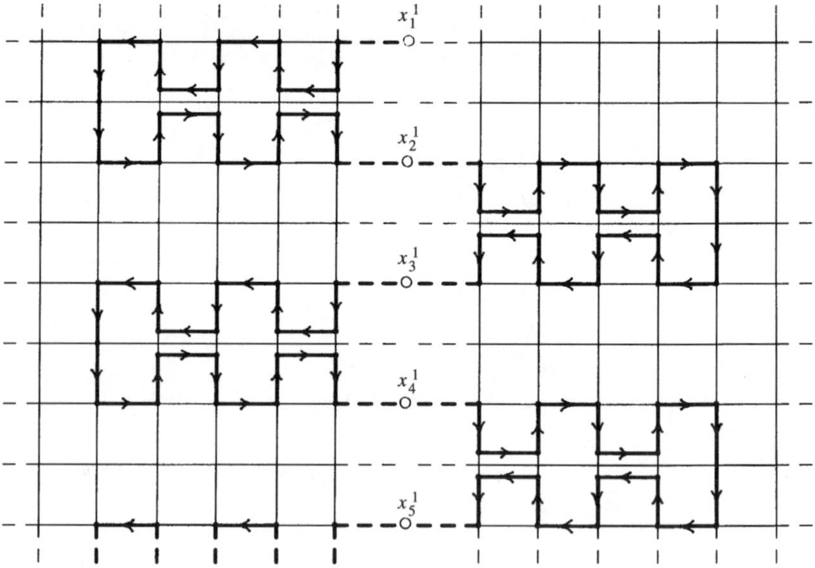

Fig. 5.3. Another 1-graph showing a 1-walk that is a representative of a 1-wtip. For the sake of clarity, some of the branch transversals are shown slightly displaced from the branch. This occurs when the branch has two oppositely directed transversals.

because 0-connectedness is transitive, every 1-graph $\mathcal{G}^1$ is partitioned into 0-sections, that is, each branch belongs to exactly one 0-section. Every maximal bordering node of any 0-section $\mathcal{S}^0$ of $\mathcal{G}^1$ is a 1-node, and every maximal boundary node is a maximal bordering node of two or more 0-sections.

A given 0-section $\mathcal{S}^0$ can be entered from one of its bordering nodes either along a branch in $\mathcal{S}^0$ (that is, along an elementary tip) or along a representative of a 0-tip of $\mathcal{S}^0$. That representative is a one-ended 0-path. So, given the bordering nodes x^1 and y^1 of $\mathcal{S}^0$ (possibly $x^1 = y^1$), there exists a 0-walk W^0 in $\mathcal{S}^0$ leaving x^1 and reaching y^1. W^0 enters $\mathcal{S}^0$ along a branch or a representative of a 0-tip, proceeds along a two-ended 0-walk, and then leaves $\mathcal{S}^0$ along a branch or representative of a 0-tip. 0-nodes and/or branches may have to repeat[3] along the walk W^0 from x^1 to y^1, but each of them will be passed through only finitely many times. We will say that W^0 is a 0-walk *through* $\mathcal{S}^0$, that W^0 *starts at* x^1 and *stops at* y^1, and that W^0 *reaches* x^1 and y^1 *through* either elementary tips or 0-tips. It can happen, of course, that a particular 0-walk through $\mathcal{S}^0$ is in fact a 0-path.

Let s (resp. t) be the elementary tip or 0-tip through which W^0 reaches x^1 (resp. y^1). In general, there will be many 0-walks through $\mathcal{S}^0$ reaching x^1 and y^1, but any two of them having the same tips s and t will be considered *equivalent*. In this way, it is only how a 0-walk W^0 enters and leaves a 0-section that will be significant; it

[3]Remember now that we are not imposing Condition 3.1-2.

will not be of any importance if finitely many of the 0-nodes and branches of W^0 are changed. We will make use of this equivalence idea later on.

A *nontrivial 1-walk* W^1 is an alternating sequence of 1-nodes and nontrivial 0-walks through 0-sections:

$$W^1 = \langle \ldots, x^1_{m-1}, W^0_{m-1}, x^1_m, W^0_m, x^1_{m+1}, \ldots \rangle \tag{5.2}$$

where, for every m, x^1_m and x^1_{m+1} are incident to the same 0-section (possibly $x^1_m = x^1_{m+1}$) and W^0_m is a 0-walk through that 0-section reaching x^1_m and x^1_{m+1}; it is also required that, for each m, at least one of W^0_{m-1} and W^0_m reaches x^1_m through a 0-tip (not an elementary tip). If the sequence in (5.2) terminates on either side, it is furthermore required that it terminate at a 0-node or 1-node. We refer to that node as a *terminal node* of W^1; in this case, the elementary tip (resp. 0-tip) with which the 0-walk in (5.2) adjacent to the terminal 0-node (resp. 1-node) reaches that terminal node will be called a *terminal* tip of W^1. (If (5.2) extends infinitely in either direction, we shall later on define another kind of "terminal wtip" for W^1.)

Note that, if W^1 terminates on either side at a 0-node, the adjacent 0-walk in the sequence (5.2) can be reduced to a 0-path by removing finite loops if it is not already a 0-path.

A 1-walk is said to *embrace* itself, all its elements, and all the elements that its elements embrace.

A *trivial* 1-walk is a singleton whose sole member is a 1-node.

A nontrivial 1-walk W^1 is called *two-ended* (resp. *one-ended*, resp. *endless*) if it terminates on both sides (resp. terminates on exactly one side, resp. terminates on neither side). If W^1 terminates at a 0-node, the adjacent 0-walk in (5.2) will be either one-ended or two-ended (not endless). However, W^1 may terminate at a 1-node that embraces a 0-node; in this case, we may also say that W^1 terminates at that 0-node.

As with 0-tips, 1-nodes, and 1-graphs, a 1-section is defined exactly as before; it is a subgraph of the 1-graph of $\mathcal{G}^\nu$ induced by a maximal set of branches that are pairwise 1-connected through 1-paths. As we have seen, 1-sections can overlap because 1-connectedness is not in general transitive. For example, consider Fig. 5.2 again; for each $m = 1, 2, 3, \ldots$, the subgraph between any x^1_m and x^1_{m+2} is a 1-section, and thus every two consecutive 1-sections overlap.

However, we can define a more general kind of connectedness, namely, "1-wconnectedness". Two branches (resp. two nodes) will be called 1-*wconnected* if there exists a two-ended 0-walk or 1-walk that terminates at those branches, that is, at a 0-node of each branch (resp. that terminates at those two nodes). We will say that two 1-walks form a *conjunction* if a terminal node of one 1-walk embraces or is embraced by a terminal node of another 1-walk. In this case, the two 1-walks taken together form another 1-walk, obviously. It follows that 1-wconnectedness is a transitive binary relationship for the branch set $\mathcal{B}$ and is in fact an equivalence relationship.

Furthermore, if every two branches in the 1-graph $\mathcal{G}^1$ are 1-wconnected, we shall say that $\mathcal{G}^1$ is 1-*wconnected*. Also, a subgraph induced by a maximal set of branches that are pairwise 1-wconnected will be called a 1-*wsection* and also a *wcomponent*.[4] The 1-graphs of Figs. 5.2 and 5.3 are 1-wconnected, and each of them is a single 1-wsection. Note that, if two branches or two nodes are 1-connected, they are also 1-wconnected because a 1-path is a special case of a 1-walk.

Our next objective is to define the new kind of extremity of a 1-graph illustrated in Examples 5.2-1 and 5.2-2. A one-ended 1-walk will be called *extended* if its 1-nodes x_m^1 are eventually pairwise distinct. Now, consider an extended one-ended 1-walk starting at the node x_0:

$$W^1 = \langle x_0, W_0^0, x_1^1, W_1^0, x_2^1, W_2^0, \ldots \rangle . \tag{5.3}$$

For each $m \geq 0$, there exist two tips s_m and t_m (elementary tips or 0-tips) with which W_m^0 reaches x_m^1 and x_{m+1}^1 respectively. Thus, we have a sequence of tips,

$$\langle s_0, t_0, s_1, t_1, s_2, t_2, \ldots \rangle, \tag{5.4}$$

corresponding to (5.3), where at least one of t_m and s_{m+1} is a 0-tip, whatever be m. When W^1 is extended, the tips in (5.4) are eventually pairwise distinct except possibly $t_m = s_{m+1}$ for infinitely many m.

Two extended one-ended 1-walks will be called *equivalent* if their tip sequences are eventually identical. This is truly an equivalence relationship for the set of all extended one-ended 1-walks, and thus that set is partitioned into equivalence classes, each of which will be called a 1-*wtip*. Specifically, with $\mathcal{W}^1$ denoting the set of all extended one-ended 1-walks, we have $\mathcal{W}^1 = \cup_{j \in J_1} \mathcal{W}_j^1$, where J_1 is an index set for the partition of $\mathcal{W}^1$ into 1-wtips $\mathcal{W}_j^1$. For any extended 1-walk W^1 in $\mathcal{W}_j^1$, we refer to $\mathcal{W}_j^1$ as the *terminal* 1-*wtip* of W^1, or simply as the 1-*wtip* of W^1. Each member of $\mathcal{W}_j^1$ is a *representative* of that 1-wtip.

Note that this equivalence relationship depends only on the tips s_m and t_m for all m sufficiently large. Also, the equivalence relationship is not disturbed by changing some branches and 0-nodes in the 0-walks W_m^0 so long as the tips s_m and t_m remain unchanged except possibly for finitely many of them. The reason for this choice of definition is that we feel intuitively that the infinite extremities of the 1-walks illustrated in Figs. 5.2 and 5.3 should not depend on how each 0-walk between consecutive 1-nodes extends into its 0-section; the same infinite extremity should ensue so long as the tip sequence (5.4) remains eventually unchanged. All this stands in contrast to the definition of 1-tips in Sec. 2.3, which depended upon the choice of the 0-path between consecutive 1-nodes. We could have developed a theory of transfinite graphs based upon tip sequences of one-ended paths, but did not do so in Chapter 2.

---

[4] Presently, a 1-wsection of $\mathcal{G}^1$ is the same as a wcomponent of $\mathcal{G}^1$, but, later on, when we consider wgraphs of higher ranks, 1-wsections will be in general different from wcomponents. In fact, 1-wsections will partition those wgraphs in the sense that each branch will lie in exactly one 1-wsection. Also, different 1-wsections may be wconnected through walks of higher ranks than 1. In any case, 1-wsections do not overlap. In contrast, a wcomponent will be a maximal subgraph whose branches are wconnected by walks of arbitrary ranks; thus, different wcomponents will not be wconnected to each other.

There is nothing sacred in our choice of definitions. All that is required (and hoped for) is that the definitions produce a self-consistent mathematical structure. We believe that such has been accomplished in both Chapter 2 and here.

A 1-walk is said to *traverse* an elementary tip (resp. 0-tip or 1-wtip) if the 1-walk embraces the branch for that elementary tip (resp. embraces the branches of a representative of that 0-tip or 1-wtip).

The next step is to define a "2-wnode." First, we arbitrarily partition the set $\mathcal{Q}^1 = \{\mathcal{W}_j^1\}_{j\in J_1}$ of 1-wtips, assuming there are 1-wtips, into subsets $\mathcal{Q}_i^1$: $\mathcal{Q}^1 = \cup_{i\in I_1}\mathcal{Q}_i^1$, where I_1 is the index set for the partition, $\mathcal{Q}_i^1 \neq \emptyset$ for all i, and $\mathcal{Q}_i^1 \cap \mathcal{Q}_k^1 = \emptyset$ if $i \neq k$. Next, for each $i \in I_1$, let $\mathcal{N}_i^1$ be either the empty set or a singleton whose only member is a 0-node or 1-node. Then, for each $i \in I_1$, we define the 2-*wnode* x_i^2 as

$$x_i^2 = \mathcal{Q}_i^1 \cup \mathcal{N}_i^1 \tag{5.5}$$

so long as the following condition is satisfied: Whenever $\mathcal{N}_i^1$ is not empty, its single α-node ($0 \leq \alpha \leq 1$) is not a member of any other 1-node or any other 2-wnode.

We say that a 2-wnode x_i^2 *embraces* itself, all its elements including the node of $\mathcal{N}_i^1$ if $\mathcal{N}_i^1$ is not empty, and all the elements embraced by that node of $\mathcal{N}_i^1$ if it exists. More concisely, we say that x_i^2 embraces itself, all its elements, and all the elements embraced by its elements. We also say that x_i^2 *shorts together* all its embraced elements. Furthermore, we say that an α-walk ($0 \leq \alpha \leq 1$) *reaches* a 2-wnode if the α-walk traverses an elementary tip or 0-tip or 1-wtip embraced by the 2-wnode.[5]

We now define a 2-*wgraph* $\mathcal{G}_w^2$ as the quadruplet

$$\mathcal{G}_w^2 = \{\mathcal{B}, \mathcal{X}_w^0, \mathcal{X}_w^1, \mathcal{X}_w^2\}, \tag{5.6}$$

where $\mathcal{B}$ is the set of branches, $\mathcal{X}_w^0$ and $\mathcal{X}_w^1$ are the sets of 0-nodes and 1-nodes respectively as defined in Chapter 2 (thus, $\mathcal{X}_w^0 = \mathcal{X}^0$ and $\mathcal{X}_w^1 = \mathcal{X}^1$), and $\mathcal{X}_w^2$ is the set of 2-wnodes.

To save words later on, we may also refer to the 0-nodes, 1-nodes, 0-graphs, and 1-graphs as 0-*wnodes*, 1-*wnodes*, 0-*wgraphs*, and 1-*wgraphs*, respectively.

Let us note at this point that a 2-graph $\mathcal{G}^2$ is a 2-wgraph if the following conditions always hold: Let t_a^1 and t_b^1 be two 1-tips of $\mathcal{G}^2$ as defined in Sec. 2.4, and assume that, if t_a^1 and t_b^1 have representative 1-paths whose sequences of tips, as indicated in (5.4), are eventually identical, then, t_a^1 and t_b^1 are shorted together (i.e., are members of the same 2-node). In this case, every 2-node is also a 2-wnode.

$\{\mathcal{B}, \mathcal{X}_w^0\}$ is the 0-*wgraph* of $\mathcal{G}_w^2$, and $\{\mathcal{B}, \mathcal{X}_w^0, \mathcal{X}_w^1\}$ is the 1-*wgraph* of $\mathcal{G}_w^2$.

We now turn to the definition of a "wsubgraph" of $\mathcal{G}_w^2$. Let $\mathcal{B}_s$ be a nonempty subset of $\mathcal{B}$. For each $\rho = 0, 1, 2$, let $\mathcal{X}_{ws}^\rho$ be the set of ρ-wnodes in $\mathcal{X}_w^\rho$, each of

[5]For the special case of a single branch connected between two 2-wtips, we take that branch and its two 0-nodes as comprising a 0-walk, in which case the 0-walk *reaches* the 2-wnode through a (-1)-tip of the branch.

which contains a $(\rho - 1)$-wtip having a representative all of whose branches are in $\mathcal{B}_s$. Then,

$$\mathcal{G}_{ws} = \{\mathcal{B}_s, \mathcal{X}_{ws}^0, \mathcal{X}_{ws}^1, \mathcal{X}_{ws}^2\}$$

is the *wsubgraph of* $\mathcal{G}_w^2\}$ *induced by* $\mathcal{B}_s$. $\mathcal{X}_{ws}^2$, and perhaps $\mathcal{X}_{ws}^1$ as well, may be empty, but it cannot happen that $\mathcal{X}_{ws}^1$ is empty while $\mathcal{X}_{ws}^2$ is nonempty.

For $\rho = 0, 1$, a ρ*-wsection* in $\mathcal{G}_w^2$ is a wsubgraph of the ρ-graph of $\mathcal{G}_w^2$ induced by a maximal set of branches that are ρ-wconnected. A 0-wsection is the same thing as a 0-section, but a 1-wsection is in general larger than a 1-section.

As was observed above, 1-wconnectedness is an equivalence relationship for the branch set $\mathcal{B}$ of $\mathcal{G}_w^2$. Consequently, the 1-wsections partition $\mathcal{G}_w^2$, in contrast to the possibly overlapping 1-sections. Similarly, the 0-wsections partition $\mathcal{G}_w^2$, and any 1-wsection is partitioned by the 0-wsections in it.

Again to conform with general terminology, we may at times refer to tips of ranks -1 or 0 as "(-1)-wtips" (alternatively, "elementary wtips") and "0-wtips", respectively. Also, a (-1)-wsection is understood to be a branch. An α-wtip $(-1 \leq \alpha \leq 1)$ is said to be *embraced by* or to be *in* or to *belong to* a β-wsection $\mathcal{S}_w^\beta$ $(\alpha \leq \beta \leq 1)$ if the branches of any one (and therefore of all) of its representative paths are all in $\mathcal{S}_w^\beta$. We also say that $\mathcal{S}_w^\beta$ *traverses* that α-wtip.

A *bordering wnode* x^β of an α-wsection $\mathcal{S}_w^\alpha$ $(-1 \leq \alpha \leq 1)$ is a wnode of rank greater than α (i.e., $\beta > \alpha$) that embraces a wtip of $\mathcal{S}_w^\alpha$. A wnode of $\mathcal{S}_w^\alpha$ that is not (resp. is) embraced by a bordering wnode of $\mathcal{S}_w^\alpha$ is called an *internal* (resp. *noninternal*) wnode of $\mathcal{S}_w^\alpha$. A *boundary wnode* of $\mathcal{S}_w^\alpha$ is a bordering wnode of $\mathcal{S}_w^\alpha$ that also embraces a wtip not belonging to $\mathcal{S}_w^\alpha$. Thus, a boundary β-wnode must be incident to two or more α-wsections, but a bordering β-wnode may be incident to only one α-wsection or to many of them.

An α-walk $(0 \leq \alpha \leq 1)$ *through* a 1-wsection $\mathcal{S}_w^1$ of $\mathcal{G}_w^2$ or simply a walk *through* $\mathcal{S}_w^1$ from a bordering wnode x^2 of $\mathcal{S}_w^1$ to a bordering wnode y^2 of $\mathcal{S}_w^1$ (possibly $x^2 = y^2$) is an α-walk whose branches are all in $\mathcal{S}_w^1$ and that reaches x^2 and y^2.

A *nontrivial 2-walk* is an alternating sequence of 2-wnodes x_m^2 and nontrivial α_m-walks $W_m^{\alpha_m}$ $(0 \leq \alpha_m \leq 1)$ through 1-wsections:

$$W^2 = \langle \ldots, x_{m-1}^2, W_{m-1}^{\alpha_{m-1}}, x_m^2, W_m^{\alpha_m}, x_{m+1}^2, \ldots \rangle \tag{5.7}$$

such that, for each m, x_m^2 and x_{m+1}^2 are bordering 2-wnodes of the 1-wsection through which $W_m^{\alpha_m}$ passes, $W_m^{\alpha_m}$ reaches x_m^2 and x_{m+1}^2, and either $W_{m-1}^{\alpha_{m-1}}$ or $W_m^{\alpha_m}$ (perhaps both) reaches x_m^2 through a 1-wtip. It is also required that, if the sequence (5.7) terminates on either side, it does so at a wnode x of rank 2 or less. Again, we say that (5.7) *terminates at* a wnode y if it terminates at a wnode x that embraces or is embraced by y. *Terminal wnodes* and *terminal wtips* are defined as before for 1-walks that terminate on either side. (If (5.7) extends infinitely in either direction, W^2 has another kind of terminal wtip, namely, a "2-wtip," which will be defined when we discuss the general case of a wgraph of rank higher than 2.)

A *trivial 2-walk* is a singleton containing a 2-wnode.

As usual, a 2-walk is either *two-ended, one-ended* or *endless* whenever (5.7) terminates on both sides, just on one side, or on neither side, respectively. A one ended 2-walk will be called *extended* if its 2-wnodes x_m^2 are eventually pairwise distinct.

Let $s_m^{\sigma_m}$ (resp. $t_m^{\tau_m}$) be the σ_m-wtip (resp. τ_m-wtip) with which $W_m^{\alpha_m}$ reaches x_m^2 (resp. x_{m+1}^2) in (5.7). Thus, $-1 \leq \sigma_m, \tau_m \leq 1$, and at least one of τ_m and σ_{m+1} equals 1 for every m. When (5.7) terminates on the left, we may write the corresponding sequence of wtips as

$$\langle s_0^{\sigma_0}, t_0^{\tau_0}, s_1^{\sigma_1}, t_1^{\tau_1}, s_2^{\sigma_2}, t_2^{\tau_2}, \ldots \rangle. \tag{5.8}$$

Wtips can repeat in this sequence, but, if (5.7) is extended, the wtips eventually do not repeat except possibly $t_m^{\tau_m} = s_{m+1}^{\sigma_{m+1}}$ for various m. Two extended one-ended 2-walks will be called *equivalent* if their wtip sequences are eventually identical. This, too, is an equivalence relationship, and it partitions the set of extended one-ended 2-walks into subsets, which we refer to as the 2-*wtips* of $\mathcal{G}_w^2$. We now take these 2-wtips to be the "infinite extremities" of $\mathcal{G}_w^2$. Any representative 2-walk of a 2-wtip is said to *traverse* that 2-wtip, and that 2-wtip is said to *belong to* or to be a *terminal 2-wtip* of that representative 2-walk.

5.3 μ-Walks and $(\mu + 1)$-Wgraphs

We are now ready to construct recursively a wgraph of any natural-number rank. We assume that ρ-wgraphs $\mathcal{G}_w^\rho$ have been constructed for all natural-number ranks ρ up to and including some natural-number rank μ. We have done so explicitly for $\mu = 2$, but remember that 0-wgraphs and 1-wgraphs are the same as 0-graphs and 1-graphs as defined in Secs. 2.2 and 2.3. We now take it that such ideas as wsections, bordering wnodes, boundary wnodes, etc. are at hand for $\mathcal{G}_w^\rho$ when $\rho \leq \mu$. They will then be defined explicitly for $\mathcal{G}_w^{\mu+1}$ in this section, thereby completing one more cycle of our recursive definitions.

Our recursive assumptions have it that, given any ranks α and β with $0 \leq \alpha < \beta \leq \mu$, α-wsections partition each β-wsection. We also have recursively the following definition: An α-walk *through* a ρ-wsection $\mathcal{S}_w^\rho$ of $\mathcal{G}_w^\mu$ (where now $0 \leq \alpha \leq \rho \leq \mu - 1$) from one bordering wnode x to another bordering wnode y of $\mathcal{S}_w^\rho$ (the ranks of x and y being greater than ρ) is an α-walk whose branches are all in $\mathcal{S}_w^\rho$ and that reaches x and y through the terminal wtips of the α-walk. (Possibly, $x = y$.) Here, too, as a special case we may have a branch in $\mathcal{S}_w^\rho$ incident to x and y, in which case that branch along with its incident 0-nodes is a 0-walk through $\mathcal{S}_w^\rho$; those 0-nodes are embraced by bordering nodes x and y of $\mathcal{S}_w^\rho$ of ranks greater than ρ, and the latter are reached through the (-1)-tips of the branch.

A *nontrivial μ-walk* W^μ is an alternating sequence of μ-wnodes x_m^μ and nontrivial α_m-walks $W_m^{\alpha_m}$ $(0 \leq \alpha_m \leq \mu - 1)$ through $(\mu - 1)$-wsections:

$$W^\mu = \langle \ldots, x_{m-1}^\mu, W_{m-1}^{\alpha_{m-1}}, x_m^\mu, W_m^{\alpha_m}, x_{m+1}^\mu, \ldots \rangle \tag{5.9}$$

where, for each m, x_m^μ and x_{m+1}^μ are bordering wnodes of the $(\mu-1)$-wsection through which $W_m^{\alpha_m}$ passes, $W_m^{\alpha_m}$ reaches x_m^μ and x_{m+1}^μ through its terminal wtips, and either $W_{m-1}^{\alpha_{m-1}}$ or $W_m^{\alpha_m}$ (perhaps both) reaches x_m^μ through a $(\mu-1)$-wtip; it is also required that, if the sequence (5.9) terminates on either side, it terminates at a wnode x of rank μ or less. We call that wnode x a *terminal wnode* of (5.9), and we call the wtip with which the adjacent walk in (5.9) reaches that terminal wnode x a *terminal wtip* of (5.9). We shall also say that (5.9) *terminates* at a wnode y if the wnode x embraces or is embraced by y. (Here, too, if (5.9) extends infinitely on either side, W^μ will have a "terminal μ-wtip" on that side; it will be defined below.)

A μ-walk is said to *embrace* itself, all its elements, and all the elements its elements embrace.

A nontrivial μ-walk is called *two-ended*, or *one-ended*, or *endless* if the sequence (5.9) terminates on both sides, or just on one side, or on neither side, respectively.

A one-ended μ-walk is called *extended* if its μ-wnodes are eventually pairwise distinct.

A *trivial μ-walk* is a singleton containing a μ-wnode.

We now assume that $\mathcal{G}_w^\mu$ has at least one extended one-ended μ-walk. Let that walk's indices be $m = 0, 1, 2, \ldots$. Corresponding to that μ-walk, we have a sequence of wtips just like (5.8):

$$\langle s_0^{\sigma_0}, t_0^{\tau_0}, s_1^{\sigma_1}, t_1^{\tau_1}, s_2^{\sigma_2}, t_2^{\tau_2}, \ldots \rangle \tag{5.10}$$

except that now $-1 \le \sigma_m, \tau_m \le \mu - 1$ and at least one of τ_m and σ_{m+1} equals $\mu - 1$ for every m. Two extended one-ended μ-walks will be considered *equivalent* if their wtip sequences are eventually identical. This, too, is an equivalence relationship, and it partitions the set $\mathcal{W}^\mu$ of all extended one-ended μ-walks into subsets $\mathcal{W}_j^\mu$, which we refer to as the *μ-wtips* of $\mathcal{G}_w^\mu$. More specifically, with J_μ denoting an index set for the partition, $\mathcal{W}^\mu = \cup_{j\in J_\mu} \mathcal{W}_j^\mu$, where $\mathcal{W}_j^\mu \ne \emptyset$ for all j and $\mathcal{W}_j^\mu \cap \mathcal{W}_k^\mu = \emptyset$ if $j \ne k$. Also, any member of a μ-wtip is called a *representative* of that μ-wtip.

A one-ended μ-walk is said to *traverse* a μ-wtip or to *have* that μ-wtip as a *terminal μ-wtip* if that μ-walk is extended and is a member of that μ-wtip. Similarly, an endless μ-walk is said to traverse two μ-wtips (possibly the same μ-wtip) if it is extended on both sides and the two one-ended μ-walks obtained by separating the endless μ-walk into two one-ended μ-walks traverse those μ-wtips.

Next, we arbitrarily partition the set $\mathcal{Q}^\mu = \{\mathcal{W}_j^\mu\}_{j\in J_\mu}$ of μ-wtips into subsets $\mathcal{Q}_i^\mu$: $\mathcal{Q}^\mu = \cup_{i\in I_\mu} \mathcal{Q}_i^\mu$, where I_μ is an index set for the partition, $\mathcal{Q}_i^\mu \ne \emptyset$ for all $i \in I_\mu$, and $\mathcal{Q}_i^\mu \cap \mathcal{Q}_q^\mu = \emptyset$ if $i \ne q$. Furthermore, for each $i \in I_\mu$, let $\mathcal{N}_i^\mu$ be either the empty set or a singleton whose sole member is an α-wnode where $0 \le \alpha \le \mu$. For each $i \in I_\mu$, we define a *$(\mu+1)$-wnode* $x_i^{\mu+1}$ by

$$x_i^{\mu+1} = \mathcal{Q}_i^\mu \cup \mathcal{N}_i^\mu \tag{5.11}$$

so long as the following condition is satisfied: Whenever $\mathcal{N}_i^\mu$ is nonempty, its single α-wnode x_i^α is not a member of another β-wnode ($\alpha < \beta \le \mu + 1$).

We define "embrace" exactly as before. In particular, the sole α-wnode of $\mathcal{N}_i^\mu$, if it exists, is called the *exceptional element* of $x_i^{\mu+1}$, and that α-wnode may contain an exceptional element (an α_1-wnode with $0 \leq \alpha_1 < \alpha$) of lower rank, which in turn may contain an exception element of still lower rank, and so on through finitely many decreasing ranks. We say that $x_i^{\mu+1}$ *embraces* itself, its exceptional element x^α if that exists, the exceptional element x^{α_1} contained in x^α if that exists, and so on down through finitely many exceptional elements. We also say that $x_i^{\mu+1}$ *embraces* its μ-wtips as well as all the wtips in all those exceptional elements. Furthermore, we say that $x_i^{\mu+1}$ *shorts together* all its embraced elements. If $x_i^{\mu+1}$ is a singleton, its sole μ-wtip is said to be *open*. Also, any ρ-wnode x^ρ $(0 \leq \rho \leq \mu + 1)$ is said to be *maximal* if x^ρ is not embraced by a wnode of higher rank.[6]

It follows[7] from these definitions that, if x^α and y^β are an α-wnode and a β-wnode respectively with $0 \leq \alpha \leq \beta$ and if x^α and y^β embrace a common wnode, then y^β embraces x^α, and moreover $x^\alpha = y^\beta$ if $\alpha = \beta$.

Next, we define the $(\mu + 1)$-*wgraph* to be the set

$$\mathcal{G}_w^{\mu+1} = \{\mathcal{B}, \mathcal{X}_w^0, \mathcal{X}_w^1, \ldots, \mathcal{X}_w^{\mu+1}\} \tag{5.12}$$

where $\mathcal{B}$ is a branch set and, for each $\rho = 0, \ldots, \mu + 1$, $\mathcal{X}_w^\rho$ is a nonempty set of ρ-wnodes built up from the wnodes and walks of lower ranks as stated. Also, for each ρ, the subset $\mathcal{G}_w^\rho = \{\mathcal{B}, \mathcal{X}_w^0, \ldots, \mathcal{X}_w^\rho\}$ is called the *ρ-wgraph of* $\mathcal{G}_w^{\mu+1}$.

Just as certain 2-graphs are 2-wgraphs (see the preceding Section), certain μ-graphs as defined in Sec. 2.4 are μ-wgraphs.

An α-walk $(0 \leq \alpha \leq \mu)$ is said to *reach* a $(\mu + 1)$-wnode if the α-walk traverses an α-wtip embraced by the $(\mu + 1)$-wnode, in which case we say that the α-walk *reaches* the $(\mu + 1)$-wnode *through* that α-wtip. As a special case, we view a branch as traversing two (-1)-tips; thus, a branch *reaches* a $(\mu + 1)$-wnode *through* a (-1)-tip if one of its (-1)-tips is embraced by the $(\mu + 1)$-wnode.

We now complete this cycle of our recursive development with a few more definitions. Let $\mathcal{B}_s$ be a nonempty subset of the branch set $\mathcal{B}$ of $\mathcal{G}_w^{\mu+1}$. For each $\rho = 0, \ldots, \mu + 1$, we let $\mathcal{X}_{ws}^\rho$ be the set of ρ-wnodes in $\mathcal{X}_w^\rho$ each of which contains a $(\rho - 1)$-wtip having a representative all of whose branches are in $\mathcal{B}_s$. (These wnodes may also contain other $(\rho - 1)$-wtips.) Then,

$$\mathcal{G}_{ws} = \{\mathcal{B}_s, \mathcal{X}_{ws}^0, \mathcal{X}_{ws}^1, \ldots, \mathcal{X}_{ws}^{\mu+1}\} \tag{5.13}$$

is called the *wsubgraph of* $\mathcal{G}_w^{\mu+1}$ *induced by* $\mathcal{B}_s$. Any one of the $\mathcal{X}_{ws}^\rho$ with $0 < \rho \leq \mu + 1$ may be empty, but there will be a maximum rank γ $(0 < \gamma \leq \mu + 1)$ for which the $\mathcal{X}_{ws}^\rho$ are nonempty for $\rho = 0, \ldots, \gamma$ and empty for $\rho = \gamma + 1, \ldots, \mu + 1$,[8] in which case the rank of $\mathcal{G}_{ws}$ is γ. In general, a wsubgraph is not a wgraph because

[6] Of course, $x_i^{\mu+1}$ is automatically maximal because there are no nodes of higher rank at this stage of our recursive development.

[7] The argument is the same as the proof of [51, Lemma 2.2-1].

[8] It can happen of course that $\gamma = \mu + 1$ so that $\mathcal{X}_{ws}^{\mu+1}$ is nonempty.

some of its wnodes may contain certain wtips having no representatives with all branches in $\mathcal{B}_s$.[9]

For each $\rho = 0, \ldots, \mu$, two branches (resp. two wnodes) in $\mathcal{G}_w^{\mu+1}$ are said to be *ρ-wconnected* if there exists a two-ended α-walk ($0 \leq \alpha \leq \rho$) that terminates at a 0-node of each branch (resp. at the two wnodes). Two walks W^α and W^β with $0 \leq \alpha, \beta \leq \mu$ are said to be *in conjunction* if a terminal wnode of W^α embraces or is embraced by a terminal wnode of W^β. The conjunction of W^α and W^β is a walk of rank $\max(\alpha, \beta)$, as is easily seen. It follows that ρ-wconnectedness is a transitive and indeed an equivalence relationship for the set $\mathcal{B}$ of branches in $\mathcal{G}_w^{\mu+1}$.

A *ρ-wsection* $\mathcal{S}_w^\rho$ ($0 \leq \rho \leq \mu$) of $\mathcal{G}_w^{\mu+1}$ is a wsubgraph of the ρ-wgraph of $\mathcal{G}_w^{\mu+1}$ induced by a maximal set of branches that are pairwise ρ-wconnected. Because ρ-wconnectedness is an equivalence relationship between branches, the ρ-wsections *partition* $\mathcal{G}_w^{\mu+1}$ (i.e., each branch is in exactly one ρ-wsection). In fact, if $0 \leq \rho < \lambda \leq \mu + 1$, any λ-wsection is partitioned by the ρ-wsections within it because ρ-wconnectedness implies λ-wconnectedness. We say that an α-wtip ($0 \leq \alpha \leq \rho$) is *traversed by* or is *embraced by* or *belongs to* or is *in* $\mathcal{S}_w^\rho$ if all the branches of any representative (and, therefore, of all representatives) of that α-wtip are in $\mathcal{S}_w^\rho$.

A *bordering wnode* of a ρ-wsection $\mathcal{S}_w^\rho$ is a wnode of rank greater than ρ that embraces a wtip belonging to $\mathcal{S}_w^\rho$. A wnode of $\mathcal{S}_w^\rho$ that is not (resp. is) embraced by a bordering wnode of $\mathcal{S}_w^\rho$ is called an *internal wnode* (resp. *noninternal wnode* of $\mathcal{S}_w^\rho$. A *boundary wnode* of $\mathcal{S}_w^\rho$ is a bordering wnode that also embraces a wtip not belonging to $\mathcal{S}_w^\rho$.

An α-walk ($0 \leq \alpha \leq \rho$) *through* a ρ-wsection $\mathcal{S}_w^\rho$ of $\mathcal{G}_w^{\mu+1}$ or simply a walk *through* $\mathcal{S}_w^\rho$ from a bordering wnode x of $\mathcal{S}_w^\rho$ to a bordering wnode y of $\mathcal{S}_w^\rho$ (possibly $x = y$) is an α-walk whose branches are all in $\mathcal{S}_w^\rho$ and that reaches x and y.

We can now define a $(\mu + 1)$-walk exactly as was a μ-walk (5.9) except that μ is replaced by $\mu + 1$. Other definitions are so-extended, too. For instance, $\mathcal{G}_w^{\mu+1}$ is said to be *(μ + 1)-wconnected* if, for every two branches, there exists a walk of rank less than or equal to $\mu + 1$ that terminates at 0-nodes of those branches. The $(\mu + 1)$-wsections of $\mathcal{G}_w^{\mu+1}$ are the *wcomponents* of $\mathcal{G}_w^{\mu+1}$. Moreover, we have the wtip sequence of any extended one-ended $(\mu + 1)$-walk, and thus the equivalence between two such $(\mu+1)$-walks defined as before. The resulting equivalence classes are the $(\mu + 1)$-wtips, which are taken to be the "infinite extremities" of $\mathcal{G}_w^{\mu+1}$.

We have completed one more cycle of our recursive development of wgraphs.

5.4 $\vec{\omega}$-Wgraphs and $\vec{\omega}$-Walks

We now assume that there is a wgraph having wnodes of all natural-number ranks. That is, the process of establishing a μ-wgraph from ρ-wgraphs ($\rho = 0, 1, \ldots, \mu - 1$) has continued unceasingly, always yielding wnodes of ever-increasing ranks. Thus, we have nonempty wnodes sets $\mathcal{X}_w^0, \mathcal{X}_w^1, \ldots, \mathcal{X}_w^\mu, \ldots$ for all natural-number ranks μ.

---

[9] As in Sec. 2.4, we can remove the latter wtips to get a wgraph, called a "reduced" wgraph.

We now define another kind of wnode, namely, one of rank $\vec{\omega}$. An *$\vec{\omega}$-wnode* $x^{\vec{\omega}}$ is an infinite sequence of μ_k-wnodes $x_k^{\mu_k}$ $(k = 0, 1, 2, \ldots)$:

$$x^{\vec{\omega}} = \langle x_0^{\mu_0}, x_1^{\mu_1}, x_2^{\mu_2}, \ldots \rangle \tag{5.14}$$

where the μ_k are increasing natural numbers: $\mu_0 < \mu_1 < \mu_2 < \cdots$, each $x_k^{\mu_k}$ is the exceptional element of $x_{k+1}^{\mu_{k+1}}$, and $x_0^{\mu_0}$ does not have an exceptional element. (In the definition of an $\vec{\omega}$-wgraph given below, it is not required that there be any $\vec{\omega}$-wnodes.) As usual, we say that $x^{\vec{\omega}}$ *embraces* itself, all its μ_k-wnodes, and all the wtips in those μ_k-wnodes. Here, too, it can be shown as an easy consequence that, if two $\vec{\omega}$-wnodes embrace a common μ-wnode of natural-number rank μ, then the two $\vec{\omega}$-wnodes are the same; also, if an $\vec{\omega}$-wnode $x^{\vec{\omega}}$ and a μ-wnode x^{μ} embrace a common wnode, then $x^{\vec{\omega}}$ embraces x^{μ}. As with μ-wnodes, we speak of $x^{\vec{\omega}}$ as *shorting together* its embraced elements.

We now define an *$\vec{\omega}$-wgraph* $\mathcal{G}_w^{\vec{\omega}}$ of rank $\vec{\omega}$ to be the infinite set of sets

$$\mathcal{G}_w^{\vec{\omega}} = \{\mathcal{B}, \mathcal{X}_w^0, \mathcal{X}_w^1, \ldots, \mathcal{X}_w^{\vec{\omega}}\} \tag{5.15}$$

where $\mathcal{B}$ is a set of branches, each $\mathcal{X}_w^{\mu}$ $(\mu = 0, 1, 2, \ldots)$ is a nonempty set of μ-wnodes, and $\mathcal{X}_w^{\vec{\omega}}$ is a (possibly empty) set of $\vec{\omega}$-wnodes. The set $\{\mathcal{B}, \mathcal{X}_w^0, \ldots, \mathcal{X}_w^{\rho}\}$, where $0 \leq \rho \leq \vec{\omega}$, is the *$\rho$-wgraph* of $\mathcal{G}_w^{\vec{\omega}}$.

Given any subset $\mathcal{B}_s$ of $\mathcal{B}$, we define the *wsubgraph of $\mathcal{G}_w^{\vec{\omega}}$ induced by $\mathcal{B}_s$* exactly as was a wsubgraph defined in the preceding Sec. 5.3 (see(5.13)) except that now there may be infinitely many nonempty wnodes sets $\mathcal{X}_{ws}^{\mu}$ $(\mu = 0, 1, 2, \ldots)$ and perhaps a nonempty $\mathcal{X}_{ws}^{\vec{\omega}}$ inserted into the right-hand side of (5.13).

Two branches (resp. two wnodes) are said to be *$\vec{\omega}$-wconnected* if there exists a two-ended walk of any natural-number rank that terminates at 0-nodes of those two branches (resp. at those two wnodes). $\vec{\omega}$-wconnectedness is an equivalence relationship for the branch set $\mathcal{B}$; indeed, it is clearly reflexive and symmetric—and also transitive because the conjunction of two walks of natural-number ranks is again a walk with a natural-number rank.

An *$\vec{\omega}$-wsection* is a wsubgraph of $\mathcal{G}_w^{\vec{\omega}}$ induced by a maximal set of branches that are $\vec{\omega}$-wconnected. Because $\vec{\omega}$-wconnectedness is an equivalence relationship for $\mathcal{B}$, the $\vec{\omega}$-wsections partition $\mathcal{G}_w^{\vec{\omega}}$. Presently, $\vec{\omega}$-wsections are simply *wcomponents* of $\mathcal{G}_w^{\vec{\omega}}$, that is, there is no walk of any rank connecting branches in two different $\vec{\omega}$-wsections, but, later on when we define ω-wgraphs, there may be such walks of rank ω.

Also, a *ρ-wsection* of $\mathcal{G}_w^{\vec{\omega}}$, where again ρ is a natural number, is defined exactly as it was in the preceding Sec. 5.3. With λ being any natural number with $\lambda > \rho$, any λ-wsection is partitioned by the ρ-wsections within it, and so, too, is $\mathcal{G}_w^{\vec{\omega}}$.

There is no such thing as a two-ended $\vec{\omega}$-walk, but we can define one-ended and endless $\vec{\omega}$-walks. In doing so, we wish to do it in such a fashion that a unique sequence of wtips, somewhat like that of (5.10), identifies a one-ended $\vec{\omega}$-walk as a representative of an infinite extremity of $\mathcal{G}_w^{\vec{\omega}}$.

A *one-ended $\vec{\omega}$-walk* is a one-way infinite alternating sequence of (not necessarily maximal) μ_m-wnodes $x_m^{\mu_m}$ and nontrivial α_m-walks $W_m^{\alpha_m}$:

$$W^{\vec{\omega}} = \langle x_0^{\mu_0}, W_0^{\alpha_0}, x_1^{\mu_1}, W_1^{\alpha_1}, x_2^{\mu_2}, \ldots, W_{m-1}^{\alpha_{m-1}}, x_m^{\mu_m}, W_m^{\alpha_m}, x_{m+1}^{\mu_{m+1}}, \ldots\rangle \quad (5.16)$$

where the μ_m comprise a strictly increasing sequence of natural numbers: $\mu_0 < \mu_1 < \mu_2 < \cdots$, and $0 \leq \alpha_m < \mu_{m+1}$ for each $m = 0, 1, 2, \ldots$; moreover, $W_m^{\alpha_m}$ reaches $x_m^{\mu_m}$ and $x_{m+1}^{\mu_{m+1}}$ through its terminal wtips, and at least one of $W_{m-1}^{\alpha_{m-1}}$ and $W_m^{\alpha_m}$ reaches $x_m^{\mu_m}$ through a $(\mu_m - 1)$-wtip.

An *endless* $\vec{\omega}$*-walk* is the *conjunction* of two one-ended $\vec{\omega}$-walks in the sense that the terminal wnode of one one-ended $\vec{\omega}$-walk embraces or is embraced by the terminal wnode of the other one.

There are many ways of representing a given one-ended $\vec{\omega}$-walk $W^{\vec{\omega}}$ as in (5.16) because the $x_m^{\mu_m}$ and $W_m^{\alpha_m}$ can be chosen in different ways. In order to get a unique sequence of wtips characterizing $W^{\vec{\omega}}$ as stated above, we proceed as follows. First of all, since $x_0^{\mu_0}$ need not be a maximal wnode, we can choose $x_0^{\mu_0}$ in (5.16) such that $W_0^{\alpha_0}$ reaches $x_0^{\mu_0}$ through a σ_0-wtip $s_0^{\sigma_0}$ where $\mu_0 = \sigma_0 + 1$. Similarly, no $x_m^{\mu_m}$ need be maximal; therefore, we can let $W_{m-1}^{\alpha_{m-1}}$ reach $x_m^{\mu_m}$ through a τ_{m-1}-wtip $t_{m-1}^{\tau_{m-1}}$ and let $W_m^{\alpha_m}$ reach $x_m^{\mu_m}$ through a σ_m-wtip $s_m^{\sigma_m}$ where $\mu_m = \max(\tau_{m-1}, \sigma_m) + 1$. Under this condition, we can continue to assume that $\{\mu_m\}_{m=0}^{\infty}$ is a strictly increasing sequence. Indeed, we can let $x_1^{\mu_1}$ be the first wnode after $x_0^{\mu_0}$ of rank μ_1 greater than μ_0 that $W^{\vec{\omega}}$ meets in accordance with the definition of (5.16). In general, for each m, we can let $x_{m+1}^{\mu_{m+1}}$ be the first wnode after $x_m^{\mu_m}$ of rank μ_{m+1} greater than μ_m that $W^{\vec{\omega}}$ meets in accordance with the definition of (5.16). Under these additional conditions, we refer to $W^{\vec{\omega}}$ as a *canonical* $\vec{\omega}$-walk. In this way, every $\vec{\omega}$-walk has a unique canonical form, and that form has a unique sequence of wtips

$$\langle s_0^{\sigma_0}, t_0^{\tau_0}, \ldots, t_{m-1}^{\tau_{m-1}}, s_m^{\sigma_m}, t_m^{\tau_m}, \ldots\rangle \quad (5.17)$$

such that, for each $m \geq 0$, $\mu_m = \max(\tau_{m-1}, \sigma_m) + 1$, as above.

The following is easily shown: Let $W_s^{\vec{\omega}}$ be a one-ended $\vec{\omega}$-walk that starts at any wnode of $W^{\vec{\omega}}$ and is identical to $W^{\vec{\omega}}$ after that wnode. The sequence of wtips obtained from a canonical representative of $W_s^{\vec{\omega}}$ will eventually be identical to (5.17). In this way, (5.17) is "eventually independent" of the choice of the initial wnode for $W^{\vec{\omega}}$.

The sequence (5.17) characterizes the way $W^{\vec{\omega}}$ leaves and enters wsections of increasing ranks. Indeed, there will be a nested sequence of wsections $\mathcal{S}_w^{\mu_m - 1}$ of increasing ranks:

$$\mathcal{S}_w^{\mu_1 - 1} \subset \mathcal{S}_w^{\mu_2 - 1} \subset \mathcal{S}_w^{\mu_3 - 1} \subset \cdots \subset \mathcal{S}_w^{\mu_m - 1} \subset \cdots \quad (5.18)$$

such that the wnode $x_m^{\mu_m}$ of the canonical $\vec{\omega}$-walk as given by (5.16) is a boundary wnode of $\mathcal{S}_w^{\mu_m - 1}$. Moreover, the truncation of the canonical walk (5.16) at $x_m^{\mu_m}$,

$$\langle x_0^{\mu_0}, W_0^{\alpha_0}, x_1^{\mu_1}, W_1^{\mu_1}, \ldots, x_m^{\mu_m}\rangle,$$

has all its branches in $\mathcal{S}_w^{\mu_m - 1}$, and $W^{\vec{\omega}}$ leaves $\mathcal{S}_w^{\mu_m - 1}$ through $x_m^{\mu_m}$ passing first through the wtip $t_{m-1}^{\tau_{m-1}}$ and then through the wtip $s_m^{\sigma_m}$. We shall refer to (5.17) as the *wtip sequence* of a canonical $\vec{\omega}$-walk, or simply as a *canonical wtip sequence*.

Let us now assume that $\mathcal{G}_w^{\vec{\omega}}$ contains at least one canonical $\vec{\omega}$-walk. Two such canonical $\vec{\omega}$-walks will be considered *equivalent* if their canonical wtip sequences are eventually identical. Thus, those two $\vec{\omega}$-walks "approach infinity" eventually along the same sequence of wnodes of strictly increasing ranks, eventually passing through those wnodes via the same wtips. This truly is an equivalence relationship for the set of canonical $\vec{\omega}$-walks in $\mathcal{G}_w^{\vec{\omega}}$, and the resulting equivalence classes will be called the *$\vec{\omega}$-wtips* of $\mathcal{G}_w^{\vec{\omega}}$. Those $\vec{\omega}$-wtips will be viewed as the "infinite extremities" of $\mathcal{G}_w^{\vec{\omega}}$.

A one-ended $\vec{\omega}$-walk is said to *traverse* an $\vec{\omega}$-wtip if the $\vec{\omega}$-walk, when written in canonical form, is a member of the $\vec{\omega}$-wtip. In the same way, an endless $\vec{\omega}$-walk can traverse two $\vec{\omega}$-wtips—or possibly the same $\vec{\omega}$-wtip.

5.5 ω-Wgraphs and ω-Walks

With μ-wgraphs for all natural numbers μ and $\vec{\omega}$-wgraphs in hand, we can define ω-wgraphs as follows. Assume that an $\vec{\omega}$-wgraph $\mathcal{G}_w^{\vec{\omega}}$ has a nonempty set $\mathcal{Q}^{\vec{\omega}}$ of $\vec{\omega}$-wtips. Partition $\mathcal{Q}^{\vec{\omega}}$ in any fashion to obtain $\mathcal{Q}^{\vec{\omega}} = \cup_{i \in I_{\vec{\omega}}} \mathcal{Q}_i^{\vec{\omega}}$, where, as before, $I_{\vec{\omega}}$ is the index set for the partition, $\mathcal{Q}_i^{\vec{\omega}}$ is nonempty for all $i \in I_{\vec{\omega}}$, and $\mathcal{Q}_i^{\vec{\omega}} \cap \mathcal{Q}_k^{\vec{\omega}} = \emptyset$ if $i \neq k$. Also, for each i, let $\mathcal{N}_i^{\vec{\omega}}$ be either the empty set or a singleton whose only member is either a μ-wnode or an $\vec{\omega}$-wnode. We also require that, if $\mathcal{N}_i^{\vec{\omega}}$ is not empty, its sole member is not the member of any other $\mathcal{N}_k^{\vec{\omega}}$ $(k \neq i)$. For each $i \in I_{\vec{\omega}}$, the set

$$x_i^{\omega} = \mathcal{Q}_i^{\vec{\omega}} \cup \mathcal{N}_i^{\vec{\omega}} \tag{5.19}$$

is called an *ω-wnode*. As usual, if $\mathcal{N}_i^{\vec{\omega}}$ is not empty, its wnode is called the *exceptional element* of x_i^{ω}. Also, as before, we say that x_i^{ω} *embraces* itself, all its elements, and all elements embraced by its exceptional element. We say that x_i^{ω} *shorts together* all its embraced elements. When x_i^{ω} is a singleton, its one and only $\vec{\omega}$-wtip is said to be *open*.

The following facts can be proven: If an α-wnode $(0 \leq \alpha \leq \vec{\omega})$ and an ω-wnode embrace a common wnode, then the ω-wnode embraces the α-wnode; on the other hand, if $\alpha = \omega$, then the α-wnode and the ω-wnode are the same ω-wnode.

A one-ended α-walk W^{α} $(-1 \leq \alpha \leq \vec{\omega})$ is said to *reach* an ω-wnode if the α-walk traverses an α-wtip embraced by the ω-wnode, in which case we say that W^{α} does so *through* that α-wtip.

Let $\mathcal{X}_w^{\omega}$ be the set of all ω-wnodes. An *ω-wgraph* $\mathcal{G}_w^{\omega}$ is defined to be

$$\mathcal{G}_w^{\omega} = \{\mathcal{B}, \mathcal{X}_w^0, \mathcal{X}_w^1, \ldots, \mathcal{X}_w^{\vec{\omega}}, \mathcal{X}_w^{\omega}\}. \tag{5.20}$$

$\mathcal{X}_w^{\vec{\omega}}$ is the set of $\vec{\omega}$-wnodes, which may be empty. All other $\mathcal{X}_w^{\nu}$ $(\nu = 0, 1, \ldots, \omega;$ $\nu \neq \vec{\omega})$ have to be nonempty in order for $\mathcal{G}_w^{\omega}$ to exist. For each ρ $(0 \leq \rho \leq \vec{\omega})$, the ρ-wgraph of $\mathcal{G}_w^{\omega}$ is $\{\mathcal{B}, \mathcal{X}_w^0, \ldots, \mathcal{X}_w^{\rho}\}$.

A wnode in $\mathcal{G}_w^{\omega}$ is called *maximal* if it is not embraced by a wnode of higher rank.

A *wsubgraph* $\mathcal{G}_{ws}$ of $\mathcal{G}_w^{\omega}$ induced by a subset $\mathcal{B}_s$ of the branch set $\mathcal{B}$ is defined exactly as before. For instance, with μ being a natural number as always, a μ-wnode

x^μ in $\mathcal{G}_w^\omega$ is also a μ-wnode in $\mathcal{G}_{ws}$ if x^μ contains a $(\mu-1)$-wtip with a representative all of whose branches are in $\mathcal{B}_s$. Similarly, an ω-wnode x^ω in $\mathcal{G}_w^\omega$ is an ω-wnode in $\mathcal{G}_{ws}$ if x^ω contains an $\vec{\omega}$-wtip with a representative all of whose branches are in $\mathcal{B}_s$. On the other hand, if infinitely many of the embraced wnodes in an $\vec{\omega}$-wnode are in $\mathcal{G}_{ws}$, then the sequence of those wnodes is an $\vec{\omega}$-wnode in $\mathcal{G}_{ws}$.

For $\mathcal{G}_w^\omega$, *μ-wconnectedness* and *$\vec{\omega}$-wconnectedness* are defined exactly as before, as are *μ-wsections* and *$\vec{\omega}$-wsections*, too. More specifically, an $\vec{\omega}$-wsection is a wsubgraph of the $\vec{\omega}$-wgraph of $\mathcal{G}^\omega$ induced by a maximal set of branches that are pairwise $\vec{\omega}$-wconnected. Wsections of a given rank *partition* every wsection of higher rank and $\mathcal{G}_w^\omega$, too. A *bordering wnode* of an $\vec{\omega}$-wsection $\mathcal{S}_w^{\vec{\omega}}$ is an ω-wnode[10] that embraces a wtip traversed by $\mathcal{S}_w^{\vec{\omega}}$. (As before, by *traversed* we mean that the wtip has a representative whose branches are all in $\mathcal{S}_w^{\vec{\omega}}$.) A wnode of $\mathcal{S}_w^{\vec{\omega}}$ that is not (resp. is) embraced by a bordering wnode of $\mathcal{S}_w^{\vec{\omega}}$ is called an *internal* (resp. *noninternal*) wnode of $\mathcal{S}_w^{\vec{\omega}}$; thus, the rank of an internal wnode of $\mathcal{S}_w^{\vec{\omega}}$ is no greater than $\vec{\omega}$. A *boundary wnode* of an $\vec{\omega}$-wsection $\mathcal{S}_w^{\vec{\omega}}$ is a bordering wnode of $\mathcal{S}_w^{\vec{\omega}}$ that also embraces a wtip not traversed by $\mathcal{S}_w^{\vec{\omega}}$.

A *nontrivial ω-walk* W^ω is an alternating sequence of ω-wnodes x_m^ω and nontrivial α_m-walks $(0 \le \alpha_m \le \vec{\omega})$ through $\vec{\omega}$-wsections:

$$W^\omega = \langle \ldots, x_{m-1}^\omega, W_{m-1}^{\alpha_{m-1}}, x_m^\omega, W_m^{\alpha_m}, x_{m+1}^\omega, W_{m+1}^{\alpha_{m+1}}, x_{m+2}^\omega, \ldots \rangle \tag{5.21}$$

such that, for each m, x_m^ω and x_{m+1}^ω are bordering wnodes of the $\vec{\omega}$-wsection through which $W_m^{\alpha_m}$ passes, $W_m^{\alpha_m}$ reaches x_m^ω and x_{m+1}^ω, and either $W_{m-1}^{\alpha_{m-1}}$ or $W_m^{\alpha_m}$ (perhaps both) reaches x_m^ω through an $\vec{\omega}$-wtip; it is also required that, if the sequence (5.21) terminates on either side, it does so at a wnode x of rank ω or less. That wnode x is called a *terminal wnode* of W^ω, and the wtip with which the adjacent walk in (5.21) reaches x is called a *terminal wtip* of W^ω. Here, too, we say that W^ω *terminates* at a wnode y if it terminates at a wnode x that embraces or is embraced by y. We define *two-ended*, *one-ended*, and *endless* ω-walks in the usual way.

A *trivial ω-walk* is a singleton whose only member is an ω-wnode.

Two branches (resp. two wnodes) are said to be *ω-wconnected* if there exists a two-ended β-walk $(0 \le \beta \le \omega, \beta \ne \vec{\omega})$ that terminates at 0-nodes of those two branches (resp. at those two wnodes). A *ρ-wsection* in $\mathcal{G}_w^\omega$, where now $0 \le \rho \le \omega$ (possibly $\rho = \vec{\omega}$), is a wsubgraph of the ρ-graph of $\mathcal{G}_w^\omega$ induced by a maximal set of branches that are ρ-wconnected. Without wgraphs of ranks higher than ω being defined, ω-wsections are simply the *wcomponents* of $\mathcal{G}_w^\omega$; that is, there is no walk of any rank terminating at the 0-nodes of two branches in different ω-wsections.

A one-ended ω-walk is called *extended* if its ω-wnodes are eventually pairwise distinct. Let the indices of that one-ended ω-walk be $m = 0, 1, 2, \ldots$. Then, corresponding to each $W_m^{\alpha_m}$ in (5.21), we have two wtips $s_m^{\sigma_m}$ and $t_m^{\tau_m}$ with which $W_m^{\alpha_m}$ reaches x_m^ω and x_{m+1}^ω, respectively. Thus, $-1 \le \sigma_m, \tau_m \le \vec{\omega}$ and $\max(\tau_m, \sigma_{m+1}) = \vec{\omega}$ for every m. Hence, corresponding to (5.21), we have a unique sequence of wtips just like (5.8) except for the revised conditions on the σ_m and τ_m:

[10] In wgraphs of ranks higher than ω, such a bordering wnode of a $\vec{\omega}$-wsection can have a rank higher than ω.

$$\langle s_0^{\sigma_0}, t_0^{\tau_0}, s_1^{\sigma_1}, t_1^{\tau_1}, s_2^{\sigma_2}, t_2^{\tau_2}, \ldots \rangle . \tag{5.22}$$

Two extended one-ended ω-walks will be called *equivalent* if their sequences of wtips, as in (5.22), are eventually identical. As before, this partitions the set of all extended one-ended ω-walks into equivalence classes. We refer to these equivalence classes as *ω-wtips* and view them as the "infinite extremities" of $\mathcal{G}_w^\omega$.

We have now completed one more cycle of our recursive development of transfinite wgraphs. We could continue as in Secs. 5.2 and 5.3 to construct wgraphs of ranks $\omega+1, \omega+2, \ldots$, then $(\vec{\omega\cdot 2})$-wgraphs, (where $\vec{\omega\cdot 2} = \omega + \vec{\omega}$) as in Sec. 5.4, then $(\omega\cdot 2)$-wgraphs as in this section, and so on through still higher ranks of wgraphs.

5.6 Walk-based Extremities

As was seen in Chapter 4, under the path-based theory of ordinal distances in transfinite graphs, not all pairs of nodes can have a distance between them because certain pairs of nodes have no paths connecting them. Examples of this are again presented below. To circumvent this difficulty, the distance function had to be restricted to subsets of the node set. Such a subset was called a "metrizable node set," and in general a transfinite graph would have many different metrizable node sets. On the other hand, in the walk-based theory presented in this chapter, a "wdistance" function can now be defined on all pairs of wnodes so long as the wgraph is wconnected, a condition we assume henceforth in this chapter. We can also state more general results concerning nodal weccentricities, wradii and wdiameters of walk-based graphs. This we do in the next three Sections. But first, let us examine some examples.

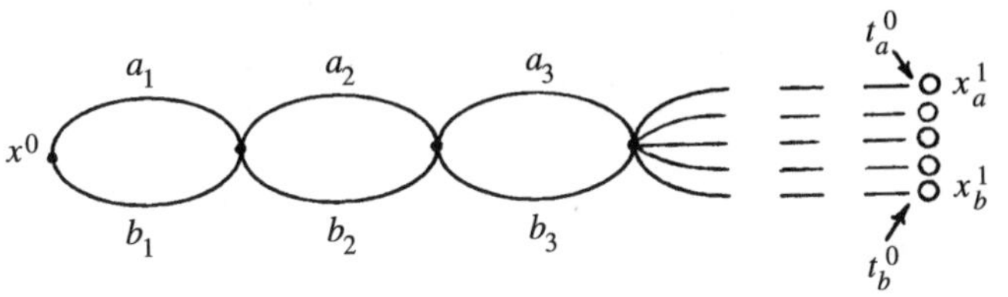

Fig. 5.4. The 1-graph of Fig. 5.1, where now more of its uncountably many 0-tips and 1-nodes are indicated. Each 1-node is a singleton containing just one 0-tip.

Example 5.6-1. Consider again the 1-graph of Fig. 5.1, which we redraw in Fig. 5.4 in order to indicate other 0-tips arising from paths that switch infinitely often between the a_k and b_k branches. There are uncountably many such paths and 0-tips; in fact, the cardinality of the set of 0-tips is the cardinality of the continuum. Indeed, consider how a one-ended path may be specified. At each 0-node encountered in a traversal toward the right, one makes a binary decision concerning the next branch

a_k or b_k to traverse; thus, each path starting at x^0 can be designated by a binary representation of a real number between 0 and 1.

As before, let x_a^1 (resp x_b^1) denote the singleton 1-node corresponding to the path proceeding only along the a_k branches (resp. b_k branches). There is a path connecting x^0 to x_a^1 and another connecting x^0 to x_b^1, but there is no path connecting x_a^1 to x_b^1 because any tracing between x_a^1 and x_b^1 must repeat 0-nodes. Thus, no distance can be assigned between x_a^1 and x_b^1 if we define distances based upon paths as Chapter 4. However, there is a walk connecting x_a^1 and x_b^1 (in fact, infinitely many of them), and a distance based upon walks can and will be assigned between x_a^1 and x_b^1 in Example 5.8-2. ♣

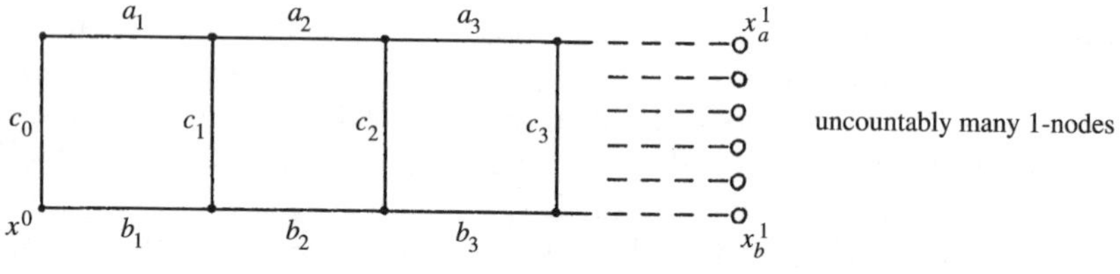

Fig. 5.5. A 1-graph containing modified forms of the 1-graph of Fig. 5.1.

Example 5.6-2. The graph of Fig. 5.4 appears in modified forms as subgraphs of many other 1-graphs, and thus those 1-graphs may also fail to have distances based upon paths for certain pairs of 1-nodes. Such is the 1-graph of Fig. 5.5, consisting of a ladder graph with uncountably many singleton 1-nodes, each containing a 0-tip. Here, too, there are uncountably many 0-paths starting at, say, the 0-node x^0 and proceeding infinitely toward the right. Let x_a^1 (resp. x_b^1) denote the 1-node containing the 0-tip of the path proceeding through all the a_k branches (resp. b_k branches). Each of the other 1-nodes corresponds to a path that passes infinitely often through some c_k branches. There is no path but there is a walk connecting any one of those latter 1-nodes to any other 1-node, and so we must resort to distances based upon walks in order to encompass all these 1-nodes in a theory of distances for transfinite graphs. Note, however, that there are paths connecting x_a^1 and x_b^1. Since a path is a special case of a walk, the distance between x_a^1 and x_b^1 as defined below will be the same as that defined in Chapter 4. ♣

Example 5.6-3. We redraw Fig. 5.2 in Fig. 5.6 to show a 2-wnode containing one of the wgraph's 1-wtip, the one that was described in Example 5.2-1. There is no path that passes through all the 1-nodes of Fig. 5.6, but there is a one-ended 1-walk that starts at the 1-node x_1^1 and passes through all the other 1-nodes $x_2^1, x_3^1, x_4^1, \ldots$ in sequence. A distance based upon walks can be defined between x^2 and any 1-node or

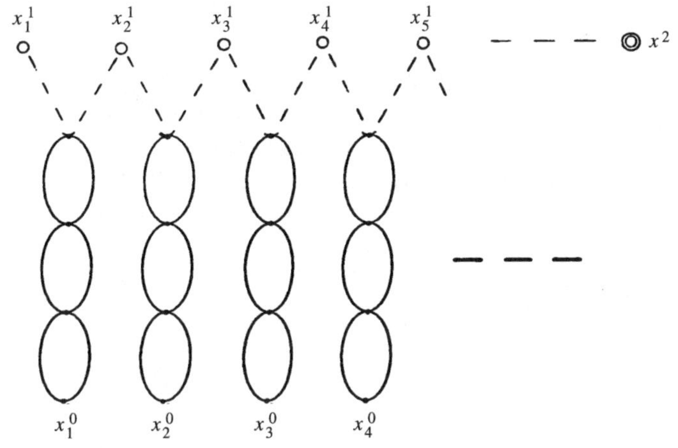

Fig. 5.6. A 2-wnode x^2 containing the 1-wtip of the 1-graph of Fig. 5.2.

0-node of Fig. 5.6, something that is impossible if we base our definition of distance upon paths. That distance will be ω^2, according to the distance definition (5.27) given below.

Now, just as infinitely many replications of the graph of Fig. 5.1 were connected in series to get the graph of Fig. 5.2, we can connect in series infinitely many two-way infinite replicates of the graph of Fig. 5.6 at their 2-wnodes. The resulting graph will have an infinite extremity of still higher rank (called a 2-wtip), which in turn can be encompassed within a transfinite wnode x^3 of rank 3 (called a 3-wnode). The distance between x^3 and any of the other μ-wnodes ($\mu = 0, 1, 2$) will be defined to be ω^3.

This construction can be continued on to still higher ranks. ♣

Example 5.6-4. Let us examine again the graph of Fig. 5.4 to note another strange property of wgraphs. We have already noted that there are uncountably many 0-tips for that graph. We can short distinct pairs of them together infinitely often to get an extended 1-walk, all of whose branches reside in that 0-graph. This will yield a 1-wtip. Moreover, only a countable set of the 0-tips is needed for this purpose. So, another disjoint set of 0-tips can be chosen to produce similarly another 1-wtip. In fact, we can do so infinitely often to get a countable set of 1-wtips, all pairwise distinct. These can be shorted together in pairs to produce an extended 2-walk and thereby a 2-wtip. Once again, only countably many of the 0-tips have been used in this construction. Therefore, we can repeat this process infinitely often to get infinitely many distinct 2-wtips, which in turn can be shorted together in pairs to get an extended 3-walk. Again only countably many of the 0-tips are used for this construction. This process can be continued to still higher ranks. In this way, wnodes of higher ranks can be so constructed at the infinite extremities of the graph of Fig. 5.4. Indeed, those ranks can reach beyond ω, and all the branches of the representative paths for the corresponding wtips will reside in the 0-graph of Fig. 5.4.

A similar construction of wnodes of higher ranks can be obtained from the wgraph of Fig. 5.3. Just create within the left-hand checkerboard 0-graph the mirror images of the 0-walks in the right-hand checkerboard graph, and then discard the latter graph. We get an extended 1-walk and thereby a 1-wtip for that left-hand graph alone. As before, we can continue this process to obtain wtips and wnodes of many higher ranks at the infinite extremities of that one checkerboard 0-graph.

All this is a notable peculiarity of wgraphs. ♣

5.7 Lengths of Walks

Given any walk W of any rank, $|W|$ will denote its length. A trivial walk of any rank is assigned the length 0. All the walks considered below are understood to be nontrivial unless triviality is explicitly stated.

0-walks: A 0-walk W^0 is simply a walk as defined in conventional graph theory. If W^0 is two-ended (i.e., finite), $|W^0|$ is the number τ_0 of branch traversals in it. In other words, it is a count of the branches in W^0 with each branch counted according to the number of times the branch appears in W^0. We do not assign a length to any arbitrary infinite 0-walk, but, if the 0-walk W^0 is one-ended and extended, we set $|W^0| = \omega$. On the other hand, if W^0 is endless and extended in both directions, we set $|W^0| = \omega\cdot 2$.

1-walks: These are defined by (5.2) and its associated conditions. We assign a length $|W^1|$ to W^1 by counting tips as follows. If W^1 is one-ended and extended, we set $|W^1| = \omega^2$. If W^1 is endless and extended in both directions, we set $|W^1| = \omega^2\cdot 2$. If, however, W^1 is two-ended, we set

$$|W^1| = \sum_m |W_m^0|, \tag{5.23}$$

where the sum is over the finitely many 0-walks W_m^0 in (5.2); thus, in this case, $|W^1| = \omega\cdot\tau_1 + \tau_0$, where τ_1 is the number of traversals of 0-tips for W^1, and τ_0 is the number of traversals of branches in all the two-ended 0-walks appearing as terms in (5.2). Thus, each traversal of any 0-tip adds 1 to the count for τ_1, and similarly for branches and τ_0. Here too, $\sum_m |W^0|$ is the natural sum of ordinals yielding a normal expansion of an ordinal [1, pages 354–355]. τ_1 is not 0 because W^1 is a nontrivial 1-walk; τ_0 may be 0.

2-walks: These are defined by (5.7) and its associated conditions. If the 2-walk W^2 is one-ended and extended, we set $|W^2| = \omega^3$. If W^2 is endless and extended in both directions, we set $|W^2| = \omega^3\cdot 2$. On the other hand, if W^2 is two-ended, we set

$$|W^2| = \sum_m |W_m^{\alpha_m}|, \tag{5.24}$$

where the summation is over the finitely many α_m-walks in (5.7). Since each α_m satisfies $0 \le \alpha_m \le 1$, each $|W_m^{\alpha_m}|$ is defined as above; once again, the summation in (5.24) denotes a normal expansion of an ordinal obtained through a natural sum of ordinals [1, pages 354–355]. So, $|W^2| = \omega^2 \cdot \tau_2 + \omega \cdot \tau_1 + \tau_0$, where the τ_k are natural numbers. τ_2 is the number of 1-wtip traversals for W^2 (counting each 1-wtip by the number of times it is traversed); τ_2 is not 0 because W^2 is two-ended and nontrivial. Also, for $k = 0, 1$, we obtain τ_k by adding ordinals in accordance with the natural summation of ordinals; τ_k can be 0.

μ-walks: We can continue recursively to define in this way the length of a μ-walk, where μ is a natural number. A μ-walk W^μ is defined by (5.9) and its associated conditions. If W^μ is one-ended and extended, we set $|W^\mu| = \omega^{\mu+1}$. If W^μ is endless and extended in both directions, we set $|W^\mu| = \omega^{\mu+1} \cdot 2$. If W^μ is two-ended, we recursively apply natural summations to get

$$|W^\mu| = \sum_m |W^{\alpha_m}| = \omega^\mu \cdot \tau_\mu + \omega^{\mu-1} \cdot \tau_{\mu-1} + \cdots + \omega \cdot \tau_1 + \tau_0 . \qquad (5.25)$$

Here too, the τ_k $(k = 0, \ldots, \mu)$ are natural numbers. Also, τ_μ is not 0, but the τ_k can be 0 if $k < \mu$.

$\vec{\omega}$-walks: As was indicated in Sec. 5.4, $\vec{\omega}$-walks always have unique canonical forms, and they are always extended. Moreover, they are either one-ended or endless—never two-ended. The length of a canonical one-ended (resp. endless) $\vec{\omega}$-walk is by definition $|W^{\vec{\omega}}| = \omega^\omega$ (resp. $|W^{\vec{\omega}}| = \omega^\omega \cdot 2$).

ω-walks: Finally, we consider an ω-walk W^ω defined by (5.21) and its associated conditions. If W^ω is one-ended (resp. endless) and extended (resp. extended in both directions), we set $|W^\omega| = \omega^{\omega+1}$ (resp. $|W^\omega| = \omega^{\omega+1} \cdot 2$). If W^ω is two-ended, we recursively apply natural summation to get the normal expansion of the ordinal length $|W^\omega|$:

$$|W^\omega| = \sum_m |W_m^{\alpha_m}| = \omega^\omega \cdot \tau_\omega + \sum_{k=0}^{\infty} \omega^k \cdot \tau_k \qquad (5.26)$$

where τ_ω and the τ_k are natural numbers. Here, too, $\tau_\omega \neq 0$. Also, only finitely many (perhaps none) of the τ_k can be nonzero because W^ω is two-ended and therefore has only finitely many $W_m^{\alpha_m}$ terms.

5.8 Wdistances between Wnodes

Henceforth, we restrict the rank of the wgraph $\mathcal{G}_w^\nu$ to be no larger than ω (i.e., $\nu \leq \omega$), and we assume that $\mathcal{G}_w^\nu$ is wconnected. Thus, given any two wnodes in $\mathcal{G}_w^\nu$ there is a two-ended γ-walk $(0 \leq \gamma \leq \nu, \gamma \neq \vec{\omega})$ that terminates at those two wnodes. Such a walk will have an ordinal length in accordance with the preceding section. Consequently, we can define the *wdistance* $d_w(x_a^\alpha, x_b^\beta)$ between any two maximal wnodes x_a^α and x_b^β in $\mathcal{G}_w^\nu$ as

$$d_w(x_a^\alpha, x_b^\beta) = \min\{|W_{a,b}|\}, \qquad (5.27)$$

where the minimum is taken over all two-ended walks $W_{a,b}$ terminating at x_a^α and x_b^β. The minimum exists because the ordinals comprise a well-ordered set. If, however, $x_a^\alpha = x_b^\beta$, we get $d_w(x_a^\alpha, x_b^\beta) = 0$ from the trivial walk at x_a^α. Thus, we have a function mapping pairs of maximal wnodes into the set of countable ordinals. If x_a^α and/or x_b^β are not maximal, we take the wdistance between them and the wdistance between the maximal wnodes embracing them as being the same. So, unless something else is explicitly stated, we will henceforth confine our attention to the maximal wnodes in $\mathcal{G}_w^\nu$.

Clearly, if $x_a^\alpha \neq x_b^\beta$, then $d_w(x_a^\alpha, x_b^\beta) > 0$; moreover, $d_w(x_a^\alpha, x_b^\beta) = d_w(x_b^\beta, x_a^\alpha)$. Furthermore, the triangle inequality

$$d_w(x_a^\alpha, x_b^\beta) \leq d_w(x_a^\alpha, x_c^\gamma) + d_w(x_c^\gamma, x_b^\beta) \tag{5.28}$$

holds for any three wnodes x_a^α, x_b^β, and x_c^γ; as always, it is understood here that we are using the "natural summation" of ordinals in order to get the right-hand side of (5.28) as the "normal expansion" of an ordinal. That (5.28) is true follows directly from the fact that the conjunction of two walks is again a walk. Thus, a walk from x_a^α to x_c^γ followed by a walk from x_c^γ to x_b^β is a walk from x_a^α to x_b^β. So, by taking minimums appropriately, we get (5.28). Thus, we have

Proposition 5.8-1. *The wdistance function d_w satisfies the metric axioms.*

Remember that an α-node and an α-wnode are the same thing for $\alpha = 0, 1$ but are different ideas for $\alpha \geq 2$.

Example 5.8-2. In the 1-graph of Fig. 5.4, the wdistance (5.27) between any 0-node and any 1-node is ω, and the wdistance between any two 1-nodes is $\omega \cdot 2$. For distances defined only by paths (see (4.5)), the distance between any 0-node and any 1-node is ω, but there is no distance defined between any two 1-nodes. ♣

Example 5.8-3. In the 1-graph of Fig. 5.5, the wdistance between any 0-node and any 1-node is ω. Also, the wdistance between any two 1-nodes is $\omega \cdot 2$; this stands in contrast to the path-based definition of distance for which the distance between those 1-nodes could only be defined for the pair x_a^1 and x_b^1. ♣

Example 5.8-4. In the 2-graph of Fig. 5.6, the wdistance $d_w(x_1^1, x_2^1)$ is $\omega \cdot 2$, according to our definition (5.27). Similarly, $d_w(x_i^1, x_k^1) = \omega \cdot (2|i-k|)$. On the other hand, for every i, $d_w(x^2, x_i^1) = \omega^2$. ♣

5.9 Weccentricities and Related Ideas

The ideas of *weccentricities* of wnodes, *wradii*, *wdiameters*, *wcenters*, and *wperipheries* for wconnected wgraphs can be defined just as their counterparts are defined for graphs, but using now the wdistance function d_w of (5.27) instead of the distance function d of (4.5). The only difference is that the restriction to a metrizable set of nodes is no longer needed. All the wnodes of a wgraph constitute a metrizable set under the distance function (5.27). So, let us present just one example illustrating this generality.

Example 5.9-1. In order to get a result distinguishable from those relating to path-based graphs, we shall use the 1-graph of Fig. 5.4 in place of a branch. Thus, in Fig. 5.7, a bold line between 1-nodes (shown as solid dots) will denote that 1-graph, where it is understood that connections to it are only made at the two 1-nodes x_a^1 and x_b^1. Also, a one-way infinite series connection of such 1-graphs will be denoted by three bold dashes. The small circles now represent 2-wnodes.

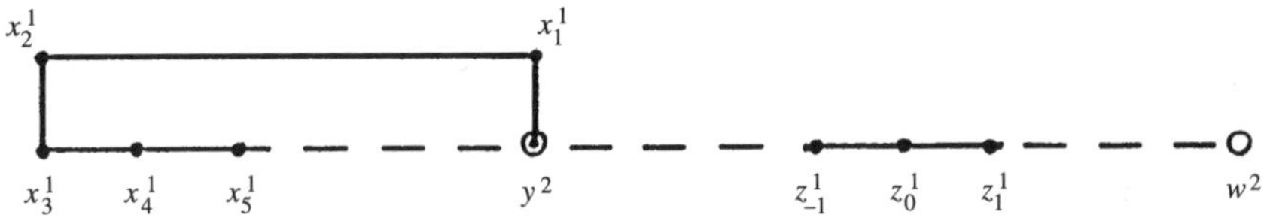

Fig. 5.7. The 2-wgraph of Example 5.9-1.

With these symbols, consider the 2-wgraph of Fig. 5.7. Let $e(v)$ denote the weccentricity of any wnode v. The weccentricities of the wnodes of Fig. 5.7 are as follows: $e(x_k^1) = \omega^2 \cdot 2 + \omega \cdot (2k)$ for $k = 1, 2, 3, \ldots$; $e(y^2) = \omega^2 \cdot 2$; $e(z_k^1) = \vec{\omega^2} \cdot 2$ for $k = \ldots, -1, 0, 1, \ldots$; and $e(w^2) = \vec{\omega^2} \cdot 3$. Consequently, the wradius of this wgraph is $\vec{\omega^2} \cdot 2$, its wdiameter is $\vec{\omega^2} \cdot 3$, its wcenter is $\{z_k^1 : k = \ldots, -1, 0, 1, \ldots\}$, and its wperiphery is $\{w^2\}$.

In contrast, there are no paths connecting these 1-nodes, and there are no 2-nodes in accordance with our path-based theory. So, the radius, diameter, center, and periphery cannot be defined. ♣

The several theorems established in Chapter 4 can be extended readily to wgraphs with virtually the same proofs. In fact, the proofs for wgraphs are simpler because of the fact that the conjunction of two walks is again a walk. Moreover, these results for wgraphs are stronger because they hold for all wnodes—not just for nodes in some metrizable set. So, let us simply state the versions of those theorems that hold for wgraphs. In the following, it is understood that every wnode discussed is a maximal wnode. The weccentricity of any nonmaximal wnode x is the same as that of the maximal wnode embracing x.

Theorem 5.9-2. *Let $\mathcal{S}_w^\rho$ be any ρ-wsection in the ν-wgraph $\mathcal{G}_w^\nu$, where $0 \le \rho < \nu \le \omega$. Assume that all the bordering wnodes of $\mathcal{S}_w^\rho$ are incident to $\mathcal{S}_w^\rho$ only through ρ-wtips. Then, all the internal wnodes of $\mathcal{S}_w^\rho$ have the same weccentricity.*

The wradius (resp. wdiameter) of a wgraph is denoted by wrad (resp. wdiam). If wrad is an arrow rank, then wrad^+ will denote the next higher limit-ordinal rank.

Theorem 5.9-3. *For any given wgraph of wradius* wrad *and wdiameter* wdiam, *the following hold:*

(i) If wrad *is an ordinal, then* $\text{wrad} \leq \text{wdiam} \leq \text{wrad}\cdot 2$.
(ii) If wrad *is an arrow rank, then* $\text{wrad} \leq \text{wdiam} \leq \text{wrad}^+ \cdot 2$.

Theorem 5.9-4. *The ν-wnodes of any wgraph $\mathcal{G}_w^\nu$ $(0 \leq \nu \leq \omega,\, \nu \neq \vec{\omega})$ comprise the wcenter of some ν-wgraph.*

In the rest of this Section, we impose the following two conditions on the ν-wgraph $\mathcal{G}_w^\nu$ at hand. (Remember that $\mathcal{G}_w^\nu$ is also assumed to be wconnected.)

Condition 5.9-5. *All the ν-wnodes are pristine (that is, no ν-wnode embraces a wnode of lower rank).*

Condition 5.9-6. *There are only finitely many boundary ν-wnodes.*

Theorem 5.9-7. *Let x^ν be a bordering ν-wnode of a $(\nu - 1)$-wsection $\mathcal{S}_w^{\nu-1}$ with the weccentricity $e(x^\nu) = \omega^\nu \cdot k$, and let z be an internal wnode of $\mathcal{S}_w^{\nu-1}$ with the weccentricity $e(z) = \omega^\nu \cdot p$. Then, $|k - p| \leq 1$.*

Theorem 5.9-8. *The weccentricities of all the wnodes form a consecutive set of values in the finite set*

$$\{\omega^\nu \cdot p \colon 1 \leq p \leq 2m + 2\}$$

where p and m are natural numbers and m is the number of boundary ν-wnodes.

We define the *removal* of a pristine nonsingleton ν-wnode x^ν as the replacement of x^ν by singleton ν-wnodes, each containing exactly one of the $(\nu - 1)$-wtips of x^ν and with each such $(\nu - 1)$-wtip so assigned. The result of removing x^ν from $\mathcal{G}_w^\nu$ is denoted by $\mathcal{G}_w^\nu - x^\nu$. Then, a ν-*wblock* of $\mathcal{G}_w^\nu$ is a maximal ν-wconnected wsubgraph $\mathcal{H}$ of $\mathcal{G}_w^\nu$ such that, for each ν-wnode x^ν of $\mathcal{G}_w^\nu$, all the branches of $\mathcal{H}$ lie in the same component of $\mathcal{G}_w^\nu - x^\nu$. The ideas of "cut-wnodes," the "partition" of a ν-wgraph $\mathcal{G}_w^\nu$ by its ν-wblocks, the "separation" of a ν-wblock by its cut-wnodes, and the "end" and "non-end" $(\nu - 1)$-wsections are immediate analogs of the corresponding ideas for a ν-graph. All this leads to the following.

Theorem 5.9-9. *The wcenter of $\mathcal{G}_w^\nu$ lies in a ν-wblock in the sense that there is a ν-wblock containing all the wnodes of the wcenter.*

Theorem 5.9-10. *The ν-wblocks of $\mathcal{G}_w^\nu$ partition $\mathcal{G}_w^\nu$, and the cut-wnodes separate the ν-wblocks.*

A *wcycle* in $\mathcal{G}_w^\nu$ is by definition a closed ν-walk that passes through two or more $(\nu-1)$-wsections. If $\mathcal{G}_w^\nu$ has no wcycles, we call it *wcycle-free*. Also, $i(\mathcal{S}_w^{\nu-1})$ denotes the set of all internal maximal wnodes of the $(\nu-1)$-wsection $\mathcal{S}_w^{\nu-1}$.

Theorem 5.9-11. *The center of a wcycle-free ν-wgraph $\mathcal{G}_w^\nu$ having at least two $(\nu-1)$-wsections has one of the following forms:*

(a) A single boundary ν-wnode x^ν.
(b) The interior $i(\mathcal{S}_w^{\nu-1})$ of a single non-end $(\nu-1)$-wsection.
(c) The set $i(\mathcal{S}_w^{\nu-1}) \cup \{x^\nu\}$, where $i(\mathcal{S}_w^{\nu-1})$ is the interior of a single $(\nu-1)$-wsection and x^ν is one of the boundary ν-wnodes of that $(\nu-1)$-wsection $\mathcal{S}_w^{\nu-1}$.

Note that in the third case $\mathcal{S}_w^{\nu-1}$ may be an end $(\nu-1)$-wsection, as in Theorem 4.9-2.

5.10 Walk-based Transfinite Electrical Networks

Our objective in the rest of this chapter is to extend electrical network theory to networks obtained from wgraphs by assigning electrical parameters to the branches. We refer to such networks as *wnetworks*, just as we did for other entities relating to wgraphs. A theory for the electrical behavior of wnetworks based upon current flows in closed walks can be established in much the same way as was a previously presented theory for transfinite networks [51, Chapter 5] based upon current flows in loops, but there are significant differences. We will not repeat that prior theory here since it is not needed for any results obtained herein, but we will occasionally refer to it to point out differences. A notable result given in Sec. 5.15 below is that, in contrast to path-based networks, wnetworks possess unique wnode voltages with respect to a given ground wnode whenever wnode voltages exist.

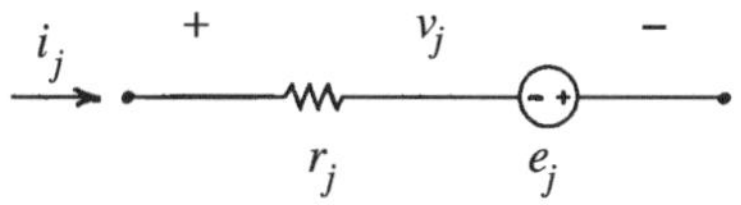

Fig. 5.8. Every branch b_j of $\mathbf{N}_w^\nu$ has this Thevenin form.

As always, $\mathbb{R}$ denotes the real line. We continue to assume that we have at hand a wconnected ν-wgraph $\mathcal{G}_w^\nu$, where $0 \le \nu \le \omega$. We assign to the branches of $\mathcal{G}_w^\nu$ some electrical parameters as follows. Let J be the set of indices j assigned to the branches b_j. J may be uncountable. All electrical entities relating to the branch b_j will also carry the same index j. We take it that every branch b_j of $\mathcal{G}_w^\nu$ is in the "Thevenin form" of a positive resistor $r_j \in \mathbb{R}$ in series with a pure voltage source $e_j \in \mathbb{R}$, as shown in Fig. 5.8. Thus, r_j is a positive real number, and e_j is any real number—possibly 0. Also, every branch b_j is assigned an orientation. The current $i_j \in \mathbb{R}$ and voltage $v_j \in \mathbb{R}$ on b_j are measured with respect to that branch's orientation. By Ohm's law, these quantities are related by

$$v_j = i_j r_j - e_j. \tag{5.29}$$

Thus, we are taking v_j to a voltage "drop" and e_j to be a voltage "rise" with respect to b_j's orientation, but these quantities can be either positive, negative, or zero. Also, $i = \{i_j\}_{j\in J}$, $v = \{v_j\}_{j\in J}$, and $e = \{e_j\}_{j\in J}$ will denote the *branch-current vector*, the *branch-voltage vector*, and the *branch-voltage-source vector*, respectively. On the other hand, the mapping $R: i_j \mapsto r_j i_j$ will denote *the branch resistance operator*. R maps i into a voltage vector Ri. Thus, Ohm's law can be written in this notation as

$$v = Ri - e. \tag{5.30}$$

We refer to $\mathcal{G}_w^\nu$ with these assigned branch parameters as a *ν-wnetwork* and denote it by $\mathbf{N}_w^\nu$. This, too, is assumed given and fixed in the following.

5.11 Tours and Tour Currents

Let γ be an ordinal such that $0 \leq \gamma \leq \omega$; thus, $\gamma \neq \vec{\omega}$. Remember that a closed γ-walk is a two-ended γ-walk such that its first wnode embraces or is embraced by its last wnode. (Thus, a γ-loop is a special case of a closed γ-walk.) Also, when $\gamma = 1$, a closed γ-walk will pass through any branch only finitely many times because each of the finitely many 0-walks between consecutive 1-nodes in it does so. (See the discussion in Sec. 5.2 of how a 1-walk passes through a 0-section.) However, for $\gamma > 1$, a closed γ-walk might pass through a branch infinitely often; this can result in infinite power dissipation for current regimes based on closed walks. We wish to disallow this because our fundamental Theorem 5.13-3, given below, is based upon finite-power regimes. For this reason, we restrict the allowable closed walks still further: A *γ-tour* is defined to be a closed γ-walk that passes through each branch at most finitely many times. As with any walk, every γ-tour is assigned an orientation (i.e., a direction for tracing it). A *tour* is a γ-tour of some unspecified rank γ.

Next, a *tour current* is a constant flow f of current passing along the tour in the direction of the tour's orientation; f is any real number. That flow f produces branch currents as follows: If the tour does not pass through a branch b_j, then the branch current i_j equals 0. If a tour passes through b_j just once, then $i_j = \pm f$, with the $+$ sign (resp. $-$ sign) used if the b_j's orientation and the tour's orientation agree (resp. disagree). If, however, the tour passes through a branch more than once, the corresponding branch current i_j is a multiple of f obtained by adding and/or subtracting f for each passage of the tour through the branch, addition (resp. subtraction) being used when the orientations of the tour's passage and of the branch agree (resp. disagree); that branch current may be 0 because of cancellations.

5.12 The Solution Space $\mathcal{T}$

In the following, the summation symbol $\sum$ will denote a sum over all branch indices j unless something else is explicitly indicated. Thus, each index appears just once for that summation. $\mathcal{I}$ will denote the set of current vectors such that each $i \in \mathcal{I}$ dissipates only a finite amount of power in the resistors:

$$\mathcal{I} = \{i : \sum i_j^2 r_j < \infty\}. \tag{5.31}$$

It follows from Schwarz's inequality that a linear combination of finite-powered branch-current vectors is again finite powered. Thus, $\mathcal{I}$ is a linear space over the field $\mathbb{R}$. We assign to $\mathcal{I}$ the inner product $(i, s) = \sum i_j s_j r_j$, where $i, s \in \mathcal{I}$. In the standard way, $\mathcal{I}$ can be shown to be complete under the corresponding norm $\|i\| = (\sum i_j^2 r_j)^{1/2}$, and thus $\mathcal{I}$ is a Hilbert space. Because of the restriction (5.31), we say that each $i \in \mathcal{I}$ is *finite-powered*. In general, a branch-current vector corresponding to a tour current may or may not be finite-powered in this sense.

No requirement concerning Kirchhoff's current law is being imposed on the members of $\mathcal{I}$. Nonetheless, we do wish to satisfy that law whenever possible—certainly at 0-nodes of finite degree. To this end, we construct a solution space $\mathcal{T}$ that will be searched for a current vector i such that i and its corresponding voltage vector v (as determined by Ohm's law (5.30)) satisfy Kirchhoff's laws whenever possible. Since each tour passes through any branch only finitely many times, we can define linear combinations of finitely many tour currents by taking linear combinations of the currents in each branch. By Schwarz's inequality again, the span $\mathcal{T}^o$ of all finite-powered tour currents will be a subspace of $\mathcal{I}$.

Finally, we let $\mathcal{T}$ be the closure of $\mathcal{T}^o$ in $\mathcal{I}$. Consequently, $\mathcal{T}$ is a Hilbert subspace of $\mathcal{I}$ with the same inner product. $\mathcal{T}$ will be the solution space that we will search for a unique branch-current vector satisfying Tellegen's equation (5.33), given below.

5.13 The Existence of a Unique Current-Voltage Regime

Next, we restrict the branch-voltage-source vector e to be of *finite total isolated power* by requiring that

$$\sum e_j^2 g_j < \infty \tag{5.32}$$

where $g_j = 1/r_j$ is the conductance of the jth branch. It follows from Ohm's law (5.29) and Schwarz's inequality that, for $i, s \in \mathcal{I}$ and e restricted by (5.32), $\sum |v_j s_j|$ is finite. Indeed,

$$\begin{aligned} \sum |v_j s_j| &\le \sum r_j |i_j s_j| + \sum |e_j s_j| \\ &= \sum |i_j| \sqrt{r_j} |s_j| \sqrt{r_j} + \sum |e_j| \sqrt{g_j} |s_j| \sqrt{r_j} \\ &\le \left(\sum i_j^2 r_j \sum s_j^2 r_j\right)^{1/2} + \left(\sum e_j^2 g_j \sum s_j^2 r_j\right)^{1/2} < \infty. \end{aligned}$$

We let $[w, s] = \sum w_j s_j$ denote the coupling between any voltage vector w and any current vector s whenever the sum exists. Then, *Tellegen's equation* along with Ohm's law (5.30) for our wnetwork $\mathbf{N}_w^\nu$ asserts that

$$[v, s] = [Ri - e, s] = 0 \tag{5.33}$$

where in the following we have $i \in \mathcal{T}$, $s \in \mathcal{T}$, and e restricted by (5.32). The imposition of (5.33) as an additional restriction ensures the existence of a unique current-voltage regime for $\mathbf{N}_w^\nu$, as we shall now show.

Lemma 5.13-1. *If e satisfies (5.32), then $e : s \mapsto [e, s]$ is a continuous linear functional on $\mathcal{I}$ and therefore on $\mathcal{T}$, too.*

Proof. We first show that $\sum |e_j i_j| < \infty$ whenever $i \in \mathcal{I}$. By Schwarz's inequality,

$$\sum |e_j i_j| = \sum |e_j| \sqrt{g_j} |i_j| \sqrt{r_j} \le \left[\sum e_j^2 g_j \sum i_j^2 r_j\right]^{1/2} < \infty. \tag{5.34}$$

So truly, $\sum e_j i_j$ is absolutely summable.

Now, let $i, s \in \mathcal{I}$ and let $a, b \in \mathbb{R}$. Then,

$$[e, ai + bs] = \sum (a e_j i_j + b e_j s_j). \tag{5.35}$$

Since $\sum e_j i_j$ and $\sum e_j s_j$ are both absolutely summable, we can rearrange the right-hand side of (5.35) to get

$$a \sum e_j i_j + b \sum e_j s_j = a[e, i] + b[e, s].$$

Thus, $[e, \cdot]$ is a linear functional on $\mathcal{I}$.

Finally, we have from (5.34) that, for every $i \in \mathcal{I}$,

$$|[e, i]| \le \sum |e_j i_j| \le \left[\sum e_j^2 g_j\right]^{1/2} \|i\|.$$

Consequently, the linear functional $[e, \cdot]$ is also continuous on $\mathcal{I}$.

Since $\mathcal{T}$ is a closed subspace of $\mathcal{I}$ with the same inner product, our conclusion holds for $\mathcal{T}$ as well. ♣

Our fundamental theorem is based upon the following standard result.

Theorem 5.13-2. *The Riesz-Representation Theorem: If f is a continuous linear functional on a Hilbert space H with the inner product $(\cdot, \cdot)$, then there exists a unique $y \in H$ such that $f(z) = (z, y)$ for every $z \in H$.*

We are now ready to state the fundamental theorem for a unique current-voltage regime in $\mathbf{N}_w^\nu$.

Theorem 5.13-3. *If e satisfies (5.32), then there exists a unique branch-current vector $i \in \mathcal{T}$ such that the corresponding unique branch-voltage vector $v = Ri - e$ satisfies Tellegen's equation (5.33) for every $s \in \mathcal{T}$. This equation insures the uniqueness of i and v even when s is restricted to $\mathcal{T}^o$.*

Proof. By Lemma 5.13-1 $[e, \cdot]$ is a continuous linear functional on $\mathcal{T}$, and by Theorem 5.13-2 there is a unique $i \in \mathcal{T}$ such that $[e, s] = (s, i)$ for every $s \in \mathcal{T}$. But, $(s, i) = \sum r_j s_j i_j = [Ri, s]$. Thus, (5.33) insures the existence and uniqueness of i and therefore of $v = Ri - e$. Finally, since $\mathcal{T}^o$ is dense in $\mathcal{T}$, we have from Lemma 5.13-1 again that a knowledge of $[e, \cdot]$ on $\mathcal{T}^o$ determines $[e, \cdot]$ on all of $\mathcal{T}$, whence the second conclusion. ♣

(Finally, let us note in passing that the solution space $\mathcal{T}$ for wnetworks can in general be expanded into a larger solution space $\mathcal{S}$ by allowing a thinning out of some current vectors as they spread out toward infinite extremities of the wnetwork. See the space $\mathcal{S}$ in [53, Sec. V-D]. This mimics how the span of "loop currents" is replaced by the span of "basic currents" for path-based transfinite networks [51, Sec. 5.2]. As a result, even stranger current distributions in wnetworks can arise, as is illustrated by [53, Fig. 10].)

Corollary 5.13-4. *Under the conditions of Theorem 5.13-3, the total power* $\sum i_j^2 r_j$ *dissipated in the resistors equals the total power* $\sum e_j i_j$ *generated by the sources and is no larger than the total isolated power* $\sum e_j^2 g_j$ *of the sources.*

Proof. The first conclusion is obtained by setting $s = i$ in (5.33). The second conclusion is then obtained by manipulating as in (5.34) and cancelling the factor $(\sum i_j^2 r_j)^{1/2}$. ♣

5.14 Kirchhoff's Laws

Let us now examine the possible satisfaction of Kirchhoff's laws under the current-voltage regime dictated by the fundamental Theorem 5.13-3.

Consider first the current law. A maximal 0-node x^0 is called *restraining* if the sum $\sum_{(x^0)} g_j$ is finite, where $\sum_{(x^0)}$ denotes a summation over all the branches b_j incident to x^0 and $g_j = 1/r_j$ denotes the positive conductance of that jth branch b_j. That x^0 is restraining implies that the degree of x^0 is at most countable because the sum of uncountably many positive conductances must be infinite.

Kirchhoff's current law applied to a maximal 0-node x^0 asserts that

$$\sum_{(x^0)} \pm i_j = 0 \qquad (5.36)$$

where the plus sign (resp. minus sign) is used if the jth branch incident to x^0 is oriented away from (resp. toward) x^0. Now, each tour current satisfies Kirchhoff's current law at every maximal 0-node because it passes through each branch only finitely many times and because its passage through a branch incident toward a maximal 0-node is accompanied by a passage through another branch incident away from that 0-node. As a result, every member of the span $\mathcal{T}^o$ also satisfies Kirchhoff's current law at every maximal 0-node.

Theorem 5.14-1. *If x^0 is a restraining maximal 0-node in $\mathbf{N}_w^\nu$, then, under the regime dictated by Theorem 5.13-3, Kirchhoff's current law (5.36) is satisfied. Moreover, the left-hand side of (5.36) is absolutely summable whenever the degree of x^0 is infinite (perforce, denumerable).*

Proof. Since x^0 is restraining and the current vector i for the regime is in $\mathcal{T}$, we have

$$\sum_{(x^0)} |i_j| \;=\; \sum_{(x^0)} g_j^{1/2} |i_j| r^{1/2} \;\le\; \left(\sum_{(x^0)} g_j \sum_{(x^0)} i_j^2 r_j \right)^{1/2} \;<\; \infty.$$

This demonstrates the asserted absolute summability. Consequently, we may rearrange the left-hand side of (5.36) in any fashion.

Since $\mathcal{T}^o$ is dense in $\mathcal{T}$, we can choose a sequence $\{i_m\}_{m=0}^{\infty}$ in $\mathcal{T}^o$ that converges in $\mathcal{T}$ to i. Let i_{mj} denote the jth component of i_m. Since every member of $\mathcal{T}^o$ satisfies Kirchhoff's current law at every maximal 0-node, $\sum_{(x^0)} \pm i_{mj} = 0$. Consequently,

$$\left| \sum_{(x^0)} \pm i_j \right| \;=\; \left| \sum_{(x^0)} \pm i_j - \sum_{(x^0)} \pm i_{mj} \right| \;\le\; \sum_{(x^0)} |i_j - i_{mj}|$$

$$\le\; \left(\sum_{(x^0)} g_j \sum_{(x^0)} (i_j - i_{mj})^2 r_j \right)^{1/2} \;\le\; \left(\sum_{(x^0)} g_j \right)^{1/2} \|i - i_m\| \;\to\; 0$$

as $m \to \infty$. ♣

The hypothesis of Theorem 5.3-1 is merely sufficient but not in general necessary. Kirchhoff's current law may still be satisfied at a nonrestraining node as a result of the configuration and conductance values of the branches not incident to the node.

We now turn to a consideration of Kirchhoff's voltage law:

$$\sum_{(S)} \pm v_j \;=\; 0 \tag{5.37}$$

where S denotes a tour with an assigned orientation, v_j is the voltage (5.29) on the jth branch in S, $\sum_{(S)}$ is a summation over the branches in S with each branch voltage appearing in the sum as many times as that branch is traversed by S, and the plus (resp. minus) sign is chosen as follows. If the jth branch is not traversed by S, its contribution to the sum (5.37) is 0. If the branch is traversed just once, its contribution to that sum is $\pm v_j$, where the plus (resp. minus) sign is used if the jth branch's orientation and the tour's direction of transversal through that branch agree (resp. disagree). If, however, the tour passes through the jth branch more than once, the term v_j appears as many times in the sum (5.37) as it is traversed by the tour S, and the same convention for the plus or minus sign is used for each traversal. By our definition of a tour, each v_j appears no more than finitely many times in (5.37).

Next, a tour S is said to be *permissive* if there exists a natural number n_S (depending on S) such that S passes through each branch no more than n_S times and if $\sum_{j\in J_S} r_j < \infty$, where J_S is the index set of those branches through which S passes. Thus, each r_j appears exactly once in that latter sum.

Theorem 5.14-2. *If S is a permissive tour in $\mathbf{N}_w^\nu$, then, under the current-voltage regime dictated by Theorem 5.13-3, Kirchhoff's voltage law (5.37) is satisfied around S, with the sum being absolutely summable.*

Proof. Since $v_j = r_j i_j - e_j$ for every branch b_j, we can check the absolute summability of the said sum by considering $\sum_{(S)} r_j |i_j|$ and $\sum_{(S)} |e_j|$ separately. By the Schwarz inequality,

$$\sum_{(S)} r_j |i_j| \leq \left(\sum_{(S)} r_j \sum_{(S)} i_j^2 r_j \right)^{1/2} < \infty$$

because $\sum_{(S)} r_j \leq n_S \sum_{j\in J_S} r_j < \infty$ and because $i \in \mathcal{I}$ so that

$$\sum_{(S)} i_j^2 r_j \leq n_S \sum_{j\in J_s} i_j^2 r_j \leq \infty.$$

Similarly,

$$\sum_{(S)} |e_j| = \sum_{(S)} \sqrt{r_j} |e_j| \sqrt{g_j} \leq \left(\sum_{(S)} r_j \sum_{(S)} e_j^2 g_j \right)^{1/2} < \infty$$

because e is of finite total isolated power so that

$$\sum_{(S)} e_j^2 g_j \leq n_S \sum_{j\in J_s} e_j^2 g_j \leq \infty.$$

Whence the asserted absolute summability.

Finally, according to Theorem 5.13-3, (5.33) holds. Let s be a tour current of value 1 A flowing around S. Upon substituting this s into (5.33), we obtain (5.37). ♣

Here too, our hypothesis is only a sufficient condition for the satisfaction of Kirchhoff's voltage law. That law may be satisfied around a nonpermissive tour if the rest of the network is structured properly.

5.15 The Uniqueness of Wnode Voltages

A shortcoming of electrical network theory for path-based ν-networks is that node voltages need not be unique even though they exist. That is, the node voltage determined by summing branch voltages along a path from the node to a given ground node may depend upon the path chosen. As a result, a special condition must be imposed to assure the uniqueness of node voltages; see Condition 5.4-1 and Theorem 5.5-4 of [51].

This difficulty does not exist for walk-based ν-wnetworks when wnode voltages are defined along "permissive" walks. Indeed, having chosen arbitrarily any wnode as the ground wnode x_g, we assign a wnode voltage to another wnode x if a "permissive" walk exists between those two wnodes. A walk is defined as being *permissive* in exactly the same way as a tour is defined as being permissive. (Just replace "tour" by "walk" in the definition preceding Theorem 5.14-2.) Then, the *wnode voltage* u at x is the sum $u = \sum_W \pm v_j$, where $\sum_W$ denotes the sum along the branches of a permissive walk W from x to x_g. Each time a branch b_j is traversed by the walk W, a term $\pm v_j$ is added with the plus sign (resp. minus sign) used if the branch orientation agrees (resp. disagrees) with the orientation of the walk through that branch. Then, two different permissive walks from x to x_g will yield the same wnode voltage at x. Indeed, we have the following.

Theorem 5.15-1. *Let $\mathbf{N}_w^\nu$ be a ν-wnetwork with a chosen ground wnode x_g, and let x be another wnode having at least two permissive walks W_1 and W_2 from x to x_g. Let u_1 (resp. u_2) be the wnode voltage assigned to x along W_1 (resp. W_2). Then, $u_1 = u_2$.*

Proof. Let $-W_1$ denote the walk W_1 but with the reverse orientation (i.e., $-W_1$ is oriented from x_g to x). Let $S = (-W_1) \cup W_2$ denote the permissive tour consisting of $-W_1$ followed by W_2. By the fundamental Theorem 5.13-3, $\sum v_j s_j = 0$, where s is a unit flow around S. As in the proof of Theorem 5.14-2, we can show that $\sum_W \pm v_j$ converges absolutely for $W = W_1$ and for $W = W_2$. Hence, we can write $u_1 = \sum_{W_1} \pm v_j = \sum_{W_2} \pm v_j = u_2$. ♣

Thus, we have a unique set of wnode voltages throughout $\mathbf{N}_w^\nu$ whenever there exists a set of permissive walks from any chosen ground wnode to all the wnodes in $\mathbf{N}_w^\nu$.

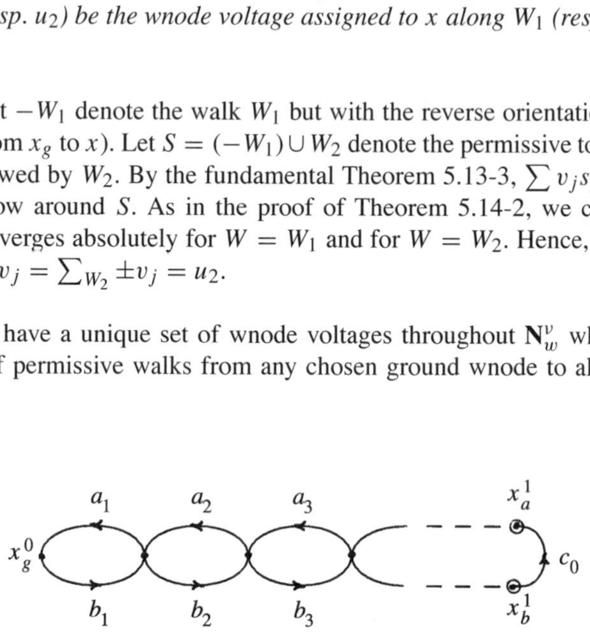

Fig. 5.9. The 1-wnetwork of Example 5.15-2.

Example 5.15-2. Let us create a 1-wnetwork $\mathbf{N}_w^1$ by first appending another branch c_0 to the 1-wgraph of Fig. 5.4, as is shown in Fig. 5.9, and then assigning electrical parameters. Specifically, the elementary tips of c_0 are shorted to the 0-tips

t_a^0 and t_b^0, thereby converting the 1-nodes x_a^1 and x_b^1 into nonsingleton 1-nodes. Every branch is oriented according to the shown arrows. Every branch to the left of the 1-nodes has a resistor of value 2^{-k} Ω, where $k = 1, 2, 3, \ldots$ counts pairs of parallel branches from the left. Moreover, branch c_0 is in the Thevenin form with a 1 V voltage source $e_0 = 1$ and a 1 Ω resistor $r_0 = 1$ in accordance with Fig. 5.8. A possible tour current s around a permissive tour S is a 1 A flow through all the branches in the direction of the arrows. Because of our choice of branch resistors, that tour current is of finite power. There are of course other possible tour currents, some of which are simply loop currents, each confined to a pair of parallel branches.

The current vector i for this 1-wnetwork dictated by Theorem 5.13-3 has 0 A in each branch to the left of the 1-nodes and 1 A flowing upward in c_0. Indeed, choose the tour S as a two-branch loop around any pair of parallel branches; then, the two voltage drops around that loop sum to 0, and, since there are no sources therein, each voltage drop therein must be 0, whence our assertion. On the other hand, upon choosing the tour current s as 1 A flowing around S, we must now conclude that the voltage drop v_0 in branch c_0 equals 0, that is, the rise in voltage across the source e_0 is cancelled by the drop in voltage across the resistor r_0. Then, with the ground node x_g^0 chosen as shown, the node voltage at x_a^1 determined by the path W_a along the a_k branches is equal to the node voltage at x_a^1 determined by the path W_b along the b_k branches followed by a transversal of c_0. That node voltage is 0. Note, however, that we now have an apparent violation of Kirchhoff's current law because 1 A flows toward x_a^1 while there are no other branch currents.

All this contrasts with the path-based theory with the current-voltage regime dictated by [51, Theorem 5.2-8] whereby the current in branch c_0 is 0. Thus, the node voltage at x_a^1 determined along W_a is 0, whereas the node voltage at x_a^1 determined along W_b is 1. We have that the node voltage at x_a^1 is not unique even though it exists for the two possible paths from x_g^0 to x_a^1. In other words, Kirchhoff's voltage law fails around the 1-loop consisting of those two paths in the path-based theory.

It appears that the restoration of Kirchhoff's voltage law obtained from this walk-based theory in place of the prior path-based theory is paid for by the collapse of Kirchhoff's current law—at least for this example. ♣

6

Hyperreal Currents and Voltages in Transfinite Networks

Kirchhoff's laws can fail to hold in various, conventionally infinite or transfinite, resistive networks. This was noted in Example 5.15-2. See also [50, Sec. 1.6] or [51, Examples 5.1-6 and 5.1-7] for several other examples of this anomaly. A basic problem is that standard calculus does not always allow the order of applying limiting processes to be reversed. Moreover, some severe restrictions have to be placed on the infinite networks in order to get convergent expressions for the voltages and currents. Such, for instance, is the requirement of finite total power generated or dissipated, but other restrictions are also used for the same purpose. See, for example, [11], [12], [13], [15], [16], [18], [19], [23], [28], [39], [41], [46], [50], [51], [54].

However, by borrowing a technique of nonstandard analysis, we can overcome all these difficulties and the restrictions they inspire. Specifically, given a "transfinite resistive network" $\mathbf{N}^\nu$ of rank ν (that is, a ν-graph whose branches are assigned resistances and sources), we stipulate in addition a sequence of finite networks that fill out $\mathbf{N}^\nu$. In order to maintain the connectivity between transfinitely distant branches, those finite networks cannot be simply node-induced or branch-induced subnetworks of $\mathbf{N}^\nu$. They have to be constructed by shorting some branches and opening others in a fashion specified by Procedure 6.2-4 and Theorem 6.2-7, given below. Otherwise, that sequence of finite networks may not reproduce the given transfinite network. Another difficulty is that some transfinite networks cannot be so reproduced from any sequence of finite networks whatsoever; see Theorem 6.2-6 in this regard. Those that can be reproduced from at least one such sequence will be called "restorable." It is for those networks that all the difficulties mentioned above can be overcome. How? Each branch of the transfinite network obtains a sequence of branch voltages, one voltage for each finite network, and that sequence determines a hyperreal voltage. Similarly, hyperreal branch currents are also generated in this way, and these are related to the hyperreal voltages through Ohm's law. Altogether, this yields for the transfinite network a *hyperreal operating point* (that is, a regime of hyperreal currents and voltages that satisfy Kirchhoff's laws and Ohm's law everywhere). That operating point will depend in general upon which restoring sequence of finite networks is chosen. Thus, under this approach, we view the transfinite graph as the end result of a specified restoration sequence of finite networks.

However, the possibly infinite sums of hyperreals that occur in Kirchhoff's laws are defined in a special way: The representative sequences of the hyperreals are restricted to those determined by the chosen restoration sequence. In effect, the restoration sequence determines not only the hyperreal branch currents and voltages but also the hyperreals representing sums of voltages along transfinite paths and sums of currents in infinite branch cuts.

This use of hyperreal voltages and currents also alleviates a variety of problems and restrictions regarding the extension of random walks to transfinite networks; this is accomplished in Sec. 6.6. Still other advantages accrue, as is shown in Chapter 7. Specifically, all our discussions of transfinite electrical network theory as in [51] and [54] were restricted to networks having only sources and linear or nonlinear resistors. Now, however, using hyperreal voltages and currents, we can establish a theory for transfinite RLC networks (i.e., networks having linear inductors and linear capacitors as well as sources and linear resistors).

The only feature of nonstandard analysis that we use for all of these results is the ultrapower construction of hyperreal currents and voltages. But, now a question arises. How about using more of nonstandard analysis, such as the transfer principle? Can nonstandard networks be obtained as enlargements of 0-graphs or more generally of transfinite graphs? The answer is "yes." There are several possible ways of doing so. They are briefly surveyed in Sec. 6.8, and then one of them is explored in Chapter 8.

We assume throughout this chapter that we have at hand a connected, countable (i.e., the branch set is countable), transfinite ν-network, possibly having ν-tips but no defined $(\nu+1)$-nodes for them. Also, we use boldface lower-case symbols to denote hyperreal-valued variables. Appendix A lists the ideas regarding nonstandard analysis that we use in this and the next two chapters.

6.1 Two Examples

We first examine two infinite networks, one demonstrating the possible failure of Kirchhoff's current law and the other the possible failure of Kirchhoff's voltage law in transfinite networks. We shall also suggest a means of reestablishing those laws by using the hyperreal numbers of nonstandard analysis in place of the real numbers used in standard analysis.

Example 6.1-1. Consider the infinite 0-network $\mathbf{N}^0$ of Fig. 6.1(a). It consists of countably many 1-Ω resistors r_j $(j = 1, 2, 3, \ldots)$, all connected in parallel with a source branch b_0 having a 1-V voltage source e_0 in series with a 1-Ω resistor r_0. We can use Theorem 5.13-3 to conclude that the current i_0 equals 1 A in b_0 whereas the current i_j equals 0 in all the other branches. Since a series $\sum_{j=1}^{\infty} i_j$ of 0's sums to 0, we have a violation of Kirchhoff's current law at both nodes x_1^0 and x_2^0.

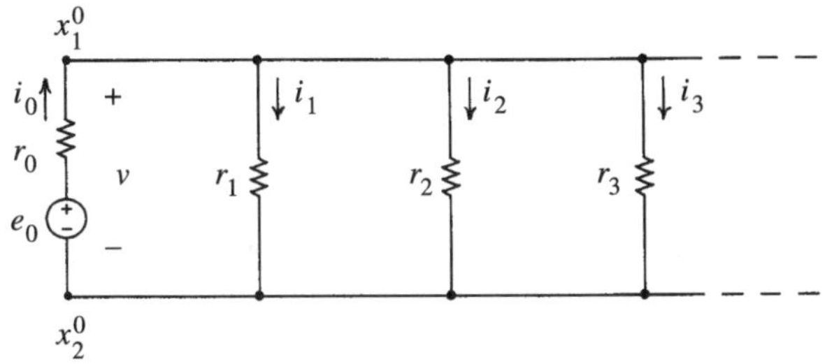

(a)

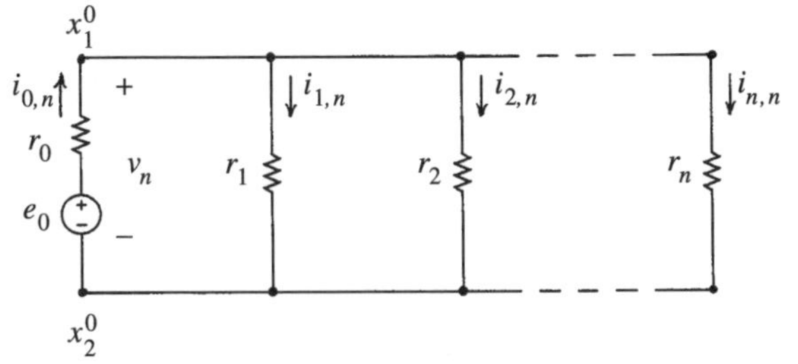

(b)

Fig. 6.1. (a) The 0-network $\mathbf{N}^0$, wherein a source branch in Thevenin form feeds power into countably many resistors all in parallel. Here, $e_0 = 1$ and $r_j = 1$ $(j = 0, 1, 2, \ldots)$. (b) A finite truncation $\mathbf{N}_n^0$ of $\mathbf{N}^0$ having n resistors in parallel with the source branch.

A physical argument also yields the same current-voltage regime: Since an infinite parallel circuit of 1-Ω resistors is effectively a short circuit, v should be 0. Hence, $i_0 = 1$ and $i_j = 0$ for $j > 0$.[1]

However, if the currents i_j $(j > 0)$ are taken to be infinitesimals, it becomes possible to have $i_0 = \sum_{j=1}^{\infty} i_j$, a nonstandard version of Kirchhoff's current law. One way of getting hyperreals is to use an ultrapower construction, which yields hyperreals as equivalence classes of sequences of reals. This in turn can be achieved if we view the network $\mathbf{N}_0$ of Fig. 6.1(a) as the "finished" result of a building process whereby the resistors r_j are "appended" one by one in parallel with the source branch b_0. Thus, at the nth step $(n \geq 1)$[2] of that process, we would have the finite network $\mathbf{N}_n^0$ shown in Fig. 6.1(b), where now we have $v_n = 1/(n+1)$, $i_{0,n} = n/(n+1)$, and $i_{j,n} = 1/(n+1)$ for $1 \leq j \leq n$ and $i_{j,n} = 0$ for $n < j$. This yields a sequence of real numbers for each current and voltage in $\mathbf{N}^0$. These sequences can be identified as

[1]Even if we were to assume that v is a nonzero real number, we would still get a violation of Kirchhoff's current law at x_1^0 and at x_2^0. Indeed, in this case, $i_0 = 1 - v$ and $i_j = v$ for $j > 0$. Hence, $\sum_{j=1}^{\infty} i_j = \infty \neq i_0$.

[2]There is no essential alteration imposed by restricting n to $n \geq 1$ instead of $n \geq 0$.

representatives of hyperreals, the corresponding equivalence classes being specified in terms of a chosen nonprincipal ultrafilter $\mathcal{F}$ (see Appendix A). In this way, we have the hyperreal voltages and currents for $\mathbf{N}^0$ as being $\mathbf{v} = [v_n] = [(n+1)^{-1}]$, $\mathbf{i}_0 = [i_{0,n}] = [n(n+1)^{-1}]$, and $\mathbf{i}_j = [i_{j,n}]$, where $i_{j,n}$ is as stated above. Kirchhoff's laws are now satisfied in a nonstandard way when we sum the representatives of the hyperreals componentwise using the indicated representatives. More specifically, Kirchhoff's current law is satisfied in a particular way, as follows. In the source branch we have

$$\mathbf{i}_0 = [\,\frac{1}{2},\ \frac{2}{3},\ \frac{3}{4},\ \ldots\,].$$

On the other hand, for the sum of hyperreal currents in the other branches, we have

$$\begin{aligned}\sum_{j=1}^{\infty} \mathbf{i}_j &= [\, i_{1,1},\ i_{1,2},\ i_{1,3},\ \ldots\,] \\ &+ [\, i_{2,1},\ i_{2,2},\ i_{2,3},\ \ldots\,] \\ &+ [\, i_{3,1},\ i_{3,2},\ i_{3,3},\ \ldots\,] \\ &+ \ \ldots\end{aligned}$$

and thus

$$\begin{aligned}\sum_{j=1}^{\infty} \mathbf{i}_j &= [\,\frac{1}{2},\ \frac{1}{3},\ \frac{1}{4},\ \ldots\,] \\ &+ [\,0,\ \frac{1}{3},\ \frac{1}{4},\ \ldots\,] \\ &+ [\,0,\ 0,\ \frac{1}{4},\ \ldots\,] \\ &+ \ \ldots .\end{aligned}$$

Summing columnwise, we obtain

$$\sum_{j=1}^{\infty} \mathbf{i}_j = [\,\frac{1}{2},\ \frac{2}{3},\ \frac{3}{4},\ \ldots\,],$$

satisfying thereby a nonstandard version of Kirchhoff's current law.

Let us emphasize that there is an implicit restriction in this summing of hyperreal branch currents. The representative of each branch current is restricted to the one dictated by the sequence of finite networks that build toward the infinite parallel circuit. Any other choices of representatives will in general result in a violation of Kirchhoff's current law. In short, Kirchhoff's current law is satisfied only in a specialized way. This issue is discussed in Sec. 6.4 and also in Appendix A11.

In any case, Ohm's law is satisfied: $[v_n] = [r_j i_{j,n}]$ for $j > 0$ and similarly $[v_0] = [e_0 - r_0 i_{0,n}]$ when $j = 0$. So, too is Kirchhoff's voltage law. Thus, we may say that $\mathbf{N}^0$ has a "hyperreal operating point."

However, this is not the only way of reestablishing Kirchhoff's laws. We could have appended resistors two at a time. In this case, the hyperreal voltage and currents for $\mathbf{N}^0$ turn out to be $\mathbf{v}' = [(2n+1)^{-1}]$, $\mathbf{i}'_0 = [2n(2n+1)^{-1}]$, and $\mathbf{i}'_j = [i_{j,n}]$ where $i_{j,n} = (2n+1)^{-1}$ for $1 \le j \le 2n$ and $i_{j,n} = 0$ for $j > 2n$. $\mathbf{v}'$ and $\mathbf{i}'_j$ for $1 \le j \le n$ are smaller infinitesimals than $\mathbf{v}$ and $\mathbf{i}$, but Kirchhoff's laws are still satisfied in a nonstandard way.

There are, of course, many different ways of appending resistors finitely many at a time, yielding many different hyperreal operating points for $\mathbf{N}^0$. We shall refer to each of them as a "finitely built version" of $\mathbf{N}^0$, there being many such versions. ♣

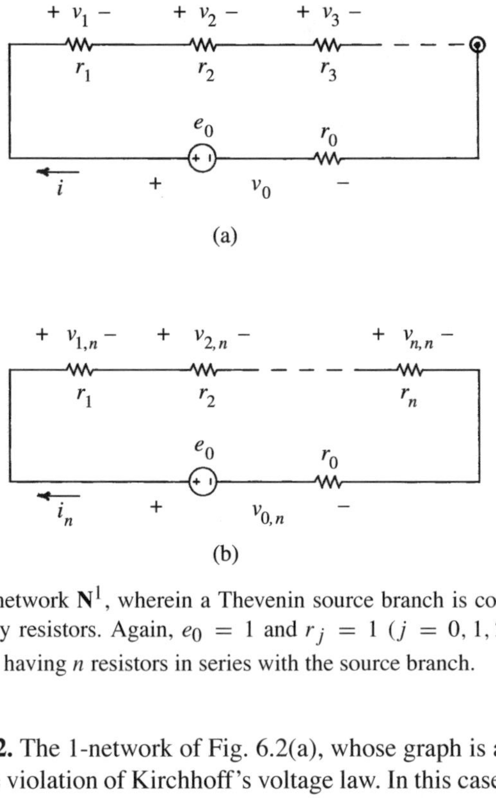

Fig. 6.2. (a) The 1-network $\mathbf{N}^1$, wherein a Thevenin source branch is connected in a 1-loop with countably many resistors. Again, $e_0 = 1$ and $r_j = 1$ $(j = 0, 1, 2, \ldots)$. (b) A finite truncation $\mathbf{N}_n^0$ of $\mathbf{N}^1$ having n resistors in series with the source branch.

Example 6.1-2. The 1-network of Fig. 6.2(a), whose graph is a 1-loop, provides an example for the violation of Kirchhoff's voltage law. In this case, Theorem 5.13-3 dictates that $i = 0$, $v_0 = 1$, and $v_j = 0$ $(j > 0)$. Again a physical argument yields the same results: Since the total series resistance of the 1-loop is infinite whereas $e_0 = 1$, we have $i = 0$, from which the voltage values follow. Kirchhoff's voltage law is violated because $v_0 = 1$ while $\sum_{j=1}^{\infty} v_j = 0$.

Here too, we can obtain the satisfaction of Kirchhoff's voltage law by replacing the real-valued currents and voltages by hyperreal ones obtained by viewing the 1-loop as the result of an increasing sequence of finite 0-loops, but now we have to "insert" branches, instead of appending them. However, a complication arises here. In order to get a 1-loop (instead of a loop of higher rank), we must be careful of how the insertions are made. If each additional resistor r_n is inserted between r_0 and the last inserted one r_{n-1}, then at the nth step ($n \geq 1$) we will have the finite loop shown in Fig. 6.2(b). In this case, the hyperreal voltages and current are $\mathbf{i} = [i_n] = [(n+1)^{-1}]$, $\mathbf{v}_0 = [v_{0,n}] = [n(n+1)^{-1}]$, and $\mathbf{v}_j = [v_{j,n}]$, where $v_{j,n} = [(n+1)^{-1}]$ for $1 \leq j \leq n$ and $v_{j,n} = 0$ for $n < j$. Once this process of inserting resistors is finished, we will have the 1-network $\mathbf{N}^1$ of Fig. 6.2(a). Note that at each step of the process we only have a finite 0-network, but, when the process is finished, the 1-network $\mathbf{N}^1$ with its 1-node leaps into view. Note also that we have to insert resistors (instead of appending them) in order to maintain the connections for a finite loop and thereby the connection through the 1-node in the finished network $\mathbf{N}^1$. Nevertheless, Kirchhoff's voltage law is now satisfied in a special nonstandard way that is quite analogous to the satisfaction of Kirchhoff's current law in the preceding example; that is, the representatives of the hyperreal branch voltages must be those dictated by the increasing sequence of finite networks and then columnwise summation is used. On the other hand, Kirchhoff's current law and Ohm's law are satisfied everywhere in a nonstandard way without any additional restrictions for this simple example. Here again, we may say that the network has a "hyperreal operating point."

Here too, there are many ways of building $\mathbf{N}^1$ by inserting resistors finitely many at a time, thereby obtaining many different hyperreal operating points. Moreover, in contrast to the appending of resistors, which can only yield infinite 0-networks, the insertion of resistors can yield transfinite networks of higher ranks.

$m =$	1	2	3	4	5	$\cdots$
$k = 1$	11	21	31	41	51	$\cdots$
2		22	32	42	52	$\cdots$
3			33	43	53	$\cdots$
4				44	54	$\cdots$
5					55	$\cdots$
$\vdots$						

Fig. 6.3. The array of indices discussed in Example 6.1-2.

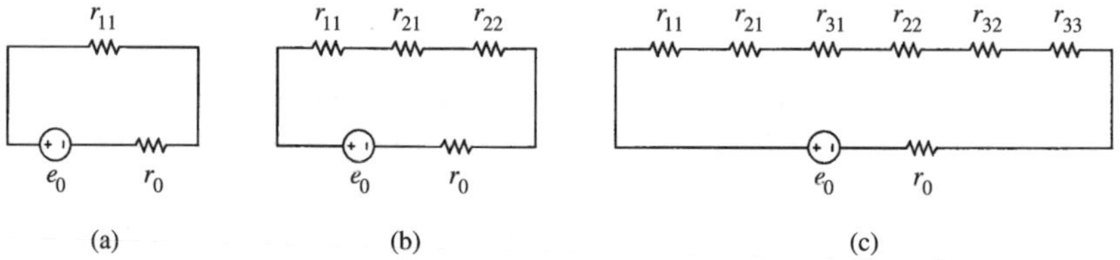

Fig. 6.4. The first three 0-loops of a sequence of finite 0-loops through which the transfinite 1-loop of Fig. 6.5 is built.

For instance, here is one way of building a 2-network: Consider a countable collection of resistors r_{mk} with double subscripts m, k where $m = 1, 2, 3, \ldots$ and

$k = 1, 2, \ldots, m$. Thus, these indices form an array as shown in Fig. 6.3. Now, start with the network of Fig. 6.4(a) with r_{11} already inserted to form a loop with the source branch b_0. Then, insert r_{21} and r_{22} between r_{11} and b_0 as shown in Fig. 6.4(b). Next, proceeding down the third column ($m = 3$) of Fig. 6.3, insert r_{31}, r_{32}, r_{33} as shown in Fig. 6.4(c). Next, insert the resistors of the fourth column ($m = 4$) to get altogether a series circuit of resistors in the sequential order

$$r_{11}, r_{21}, r_{31}, r_{41}, r_{22}, r_{32}, r_{42}, r_{33}, r_{43}, r_{44}.$$

In general, at the mth step of this procedure, the order of the series circuit of resistors is obtained by proceeding along the first row of Fig. 6.3 up to m, then along the second row up to m, and so on up to the mth row. After all that, this is repeated with m replaced by $m + 1$, and so on for increasing m. The result is the 2-network of Fig. 6.5 obtained when this insertion process is finished.

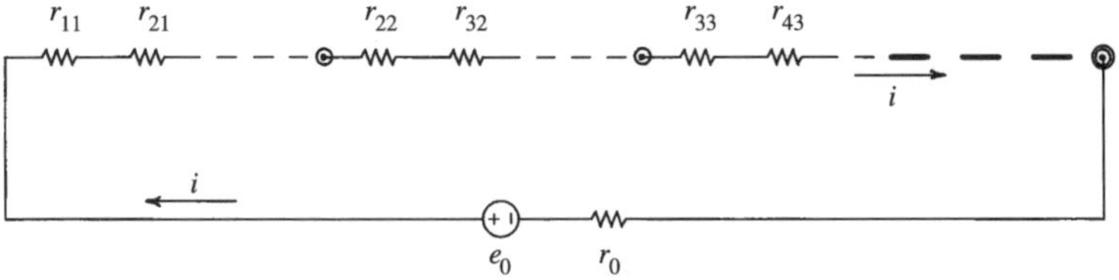

Fig. 6.5. The 2-network resulting from the insertion process corresponding to Fig. 6.4 discussed in Example 6.1-2.

In this simple case, the insertion of resistors one at a time in the sequence indicated yields the hyperreal loop current $\mathbf{i} = [(n + 1)^{-1}]$. Thus, electrically this 2-network has the same hyperreal loop current as that of Fig. 6.2(a) for resistors inserted as stated before, but its graph is radically different.

More complicated insertion procedures will yield loops of even higher transfinite ranks. ♣

As we shall note in the next section, a variety of transfinite graphs other than loops can be built by alternating in some fashion the appending and inserting branches. Such a graph along with a specified sequence of appending and inserting branches might be called a *finitely built* graph since only finite graphs will be at hand before the process is finished—and similarly we might call the final result a *finitely built* network.

6.2 Restorable Networks

A *transfinite network* (more specifically, a *transfinite electrical network*) is a transfinite graph whose branches are assigned electrical parameters. In this chapter, we restrict those parameters to resistors and independent sources. In the next chapter, we will also allow inductors and capacitors.

There are many different ways of building a transfinite graph by appending and inserting branches sequentially in different ways. If we do this arbitrarily, we will in general have no idea what transfinite graph might arise. However, we may obtain some guidance as to what may ensue if we start with a transfinite network $\mathbf{N}^\nu$, "short" and "open" all but finitely many branches, and then "restore" them finitely many at a time. The electrical operation of "opening" (resp. "shorting") a branch corresponds to—but is not the same as— the graphical operation of "removing" (resp. "contracting") a branch. To *remove* a branch, delete its two elementary tips from its incident 0-node or 0-nodes and discard the branch. If that 0-node is incident to only that branch, the 0-node disappears. To *contract* a branch, first remove it and then, if it is not a self-loop, replace its two incident maximal nodes by a single node that embraces all the tips embraced by those two nodes other than the deleted ones.[3] The difference between these two pairs of operations is that, in the electrical case of opening and shorting a branch, the branch is not removed or contracted but instead its electrical parameters are assigned certain extreme values. The restoration process will correspond to a building process, and under certain circumstances, we can be assured of obtaining the transfinite network $\mathbf{N}^\nu$ with which we started.

So, let us now specify what we mean by "opening" or "shorting" a branch when the branch has a positive resistance and possibly a source. Any branch can be represented either by its Norton circuit, shown in Fig. 6.6(a), or equivalently by its Thevenin circuit, shown in Fig. 6.6(b).

To *open* the branch will mean that, with respect to its Norton representation, the branch conductance g, current source h, and current i are all set equal to 0: $g = h = i = 0$. The branch voltage v cannot be determined from Ohm's law, $i + h = gv$, and its value will not be needed until the branch is *restored* (i.e., its original parameter values g and h are reestablished).[4]

To *short* the branch will mean that with respect to its Thevenin representation, the branch resistance r, voltage source e, and voltage v are all set equal to 0: $r = e = v = 0$. The branch current i cannot be determined from Ohm's law, $v + e = ri$, and its value will not be needed until the branch is *restored* (i.e., its original parameter values r and e are reestablished).[5]

Example 6.2-1. To establish some ideas with respect to the sought-for restoration procedure, let us now consider the 1-network of Fig. 6.7(a) consisting of a one-way infinite ladder network connected at its infinite extremity to a resistor r through two 1-nodes x_1^1 and x_2^1. Here, x_1^1 (resp. x_2^1) embraces the 0-tip of the upper (resp. lower) path of horizontal branches and one of the elementary tips of r. After numbering all

[3] When restricting the allowable graphs to "simple" ones—that is, to graphs having no self-loops or parallel branches—there is a final step for a contraction, namely, remove all the other self-loops and combine parallel branches into single branches when these are created by the contraction.

[4] In fact, when specifying a sequence for the hyperreal voltage $\mathbf{v}$ for that branch, we may and will arbitrarily set $v = 0$ until the branch is restored.

[5] Here, too, when specifying a sequence for the hyperreal current $\mathbf{i}$, we may and will set $i = 0$ before the branch is restored.

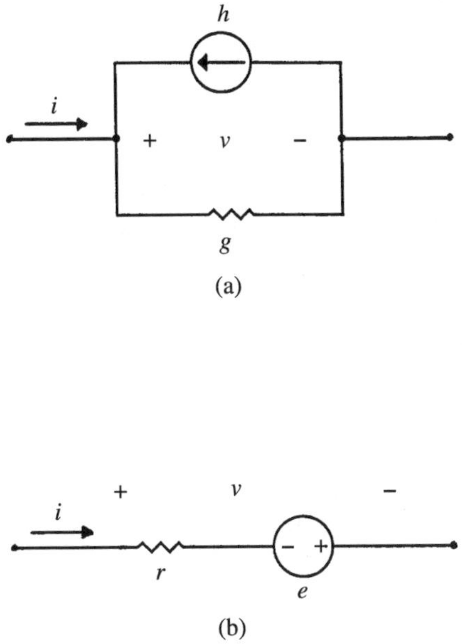

Fig. 6.6. (a) The Norton circuit representing a branch with a positive conductance g in parallel with a current source h. The branch current i and the branch voltage v are related by Ohm's law: $i + h = gv$. (b) The Thevenin circuit representing a branch with a positive resistance r in series with a voltage source e. Now, Ohm's law has the form $v + e = ri$. When $r = 1/g$ and $e = -hr$, these two circuits are equivalent.

the branches, we short some of them and open the others and then restore them in sequence according to the numbering. Were we to open all the branches and short none of them, r would be disconnected from the rest of the network at each stage of the restoration, and thus would remain so in the final restored network, as shown in Fig. 6.7(b). In order to recover the connections to r, we may choose two representative 0-paths for the ladder's 0-tips in x_1^1 and x_2^1 and short all their branches. The other branches are opened. Then, after all branches are restored, we will have recovered the original network of Fig. 6.7(a). ♣

For the general case of any countably infinite network, we will always short the branches of a representative path for each nonopen nonelementary tip in order to maintain connections through transfinite nodes. This is not done for the open tips. All but finitely many of the other branches are opened before the restoration begins.

Example 6.2-2. Let us now consider the 2-network obtained by replacing every branch of the ladder in Fig. 6.7(a) by an endless path. Every 0-node becomes a 1-node and the two 1-nodes become two 2-nodes x_1^2 and x_2^2 that embrace the 0-nodes of r. This is shown in Fig. 6.8. The resulting network is still countable. Number its branches in any fashion using the natural numbers. To obtain an expanding sequence of (electrically finite) networks that fill out this 2-network, we choose two represen-

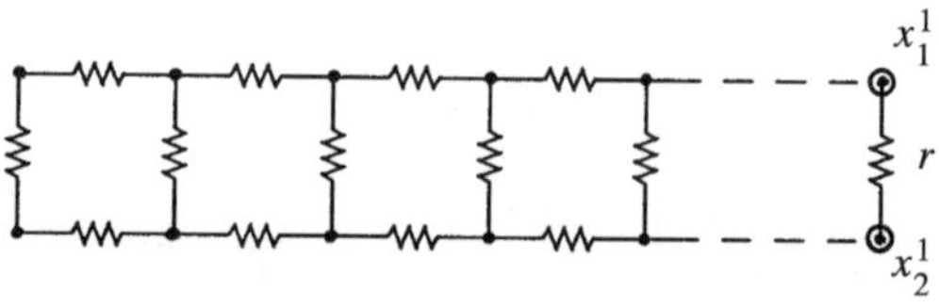

Fig. 6.7a. A one-way infinite ladder connected at its infinite extremity to a resistor r through two 1-nodes x_1^1 and x_2^1 (shown by the small circles). The elementary tips of r are embraced by the 1-nodes. Any of the branches may have sources; these are not shown in the figure.

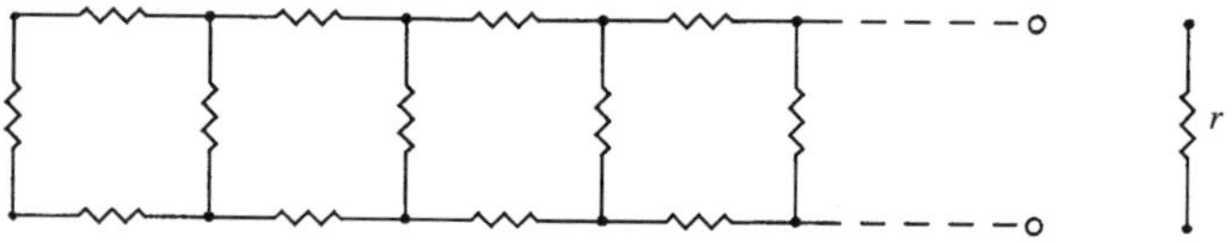

Fig. 6.7b. The ladder of part (a) except that the resistor r is now disconnected from the ladder. The 1-nodes no longer embrace the elementary tips of r.

tative one-ended 1-paths[6] for the two 1-tips in x_1^2 and x_2^2 and short their branches. We then choose representative one-ended 0-paths for all the 0-tips of all the 1-nodes and short their branches too. Some of those branches will already be shorted in the first step. At this point, we can and do choose the representative paths of the two 0-tips of each vertical endless path to be disjoint. Next, we open all the other branches. Finally, we restore the branches one-by-one in accordance with the branch numbering. At each stage, we have in effect a finite electrical network because the opened and shorted branches can be removed and contracted, respectively, without altering the electrical behavior. Moreover, after all the branches are restored, we will have recovered the original 2-network because connections through the 1-nodes and 2-nodes will have been maintained.

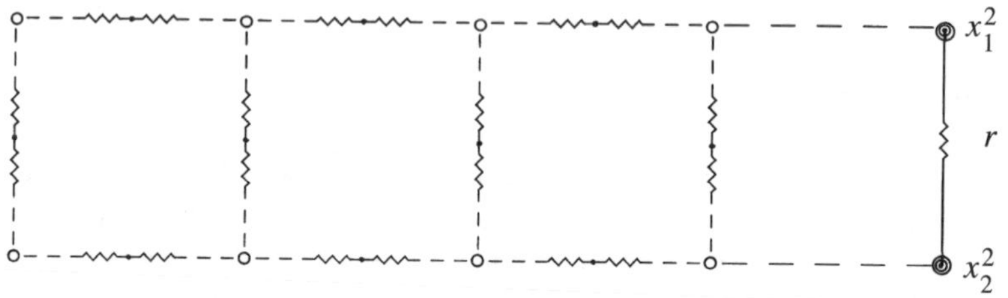

Fig. 6.8. A one-way infinite ladder consisting of endless 0-paths. The ladder's connections are now made through 1-nodes (the small circles). The ladder is connected at its infinite extremity to a resistor r through two 2-nodes (the double circles). Those 2-nodes embrace the elementary tips of r.

[6]These can be any one-ended 1-subpaths of the upper and lower horizontal 1-paths.

The reason for choosing disjoint representative paths in each of the vertical endless paths is to insure that x_1^2 and x_2^2 do not coalesce into a single 2-node. Indeed, if an infinity of those vertical endless paths had been entirely shorted, then at every stage of the restoration the two 2-nodes would have been coalesced into a single 2-node. As a result, they would be replaced by a single 2-node in the final restored network. Thus, it appears that some care must be taken in the way the transfinite network is restored from a sequence of finite networks. ♣

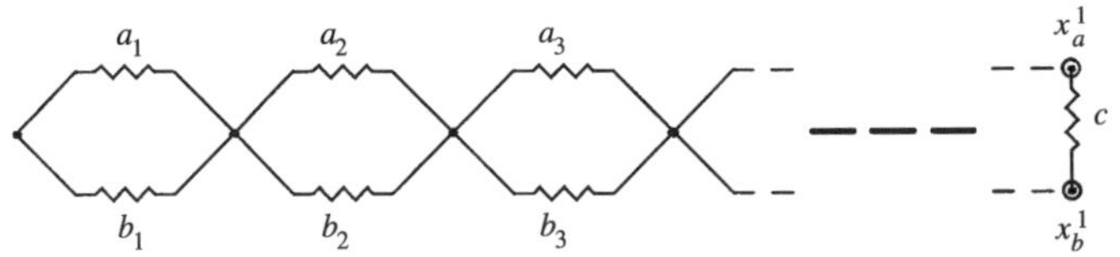

Fig. 6.9. A 1-network. The pairs of parallel branches extend infinitely to the right. The 1-node x_a^1 (resp. x_b^1) embraces the 0-tip of the one-ended 0-path along the upper branches a_k (resp. lower branches b_k) and an elementary tip (resp. the other elementary tip) of the branch c. The two 0-tips are nondisconnectable. Every branch has a positive resistor and possibly a source.

Example 6.2-3. In fact, there is still another problem: Not all transfinite networks can be restored through a sequence of finite networks. For instance, consider the 1-network of Fig. 6.9 having two nondisconnectable 0-tips in two different 1-nodes x_a^1 and x_b^1. Such a network can also occur as part of a larger transfinite network. We wish to maintain the connection provided by the 1-node x_a^1 between the 0-tip of the 0-path of the a_k branches and an elementary tip of branch c. In order to ensure this when building up the 1-network from a sequence of finite networks, we have to short (the branches of) a representative path for that 0-tip. For a similar reason, we have to short a representative path for the 0-tip in x_b^1. We open all the other branches. Those two shorted representative paths meet infinitely often, and thus the network obtained through any sequence of branch restorations will yield the different 1-network of Fig. 6.10(a), in which x_a^1 and x_b^1 are coalesced into a single 1-node x_c^1. That is, at each step of the restoration sequence, x_a^1 and x_b^1 coalesce into a single 1-node x_c^1 through the shorting of some representative paths for their 0-tips. As a result, x_a^1 and x_b^1 have the same hyperreal voltage after the restoration process is completed. Electrically, x_a^1 and x_b^1 have been shorted, and we draw the restored network accordingly with x_a^1 and x_b^1 replaced by the single 1-node x_c^1.

On the other hand, were we to open all branches in Fig. 6.9, we would lose the connections through x_a^1 and x_b^1 and would end up with the disconnected 1-network of Fig. 6.10(b). We can conclude that the network of Fig. 6.9 cannot be restored through a sequence of finite networks. ♣

Now, consider the general case for some given ν-network $\mathbf{N}^\nu$. We wish to generate an expanding sequence of finite graphs that fills out $\mathbf{N}^\nu$ and at the same time maintain the connections that the transfinite nodes provide in $\mathbf{N}^\nu$. To this end, we proceed as follows. Remember that the branch set $\mathcal{B}$ of $\mathbf{N}^\nu$ is countably infinite.

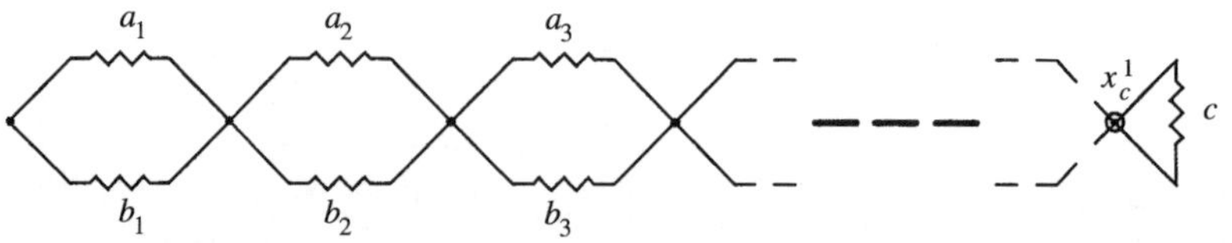

Fig. 6.10a. The network resulting from any sequence of restorations of the branches in Fig. 6.9, wherein the shorting between the upper 0-tip and an elementary tip of branch c and the shorting between the lower 0-tip and the other elementary tip of branch c are maintained by initially shorting representative paths of those 0-tips.

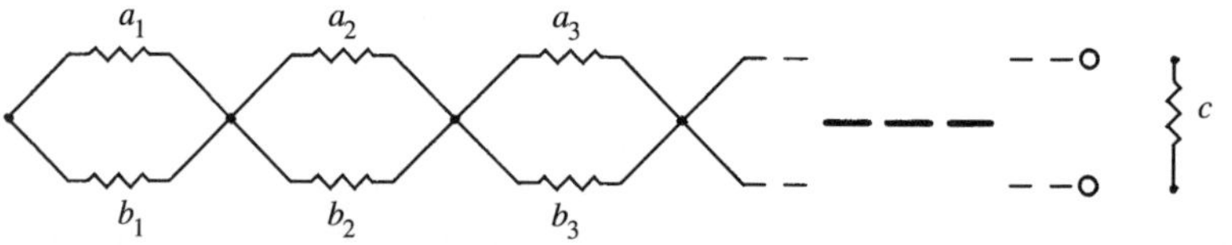

Fig. 6.10b. The network obtained after all the branches are opened and then restored sequentially.

Procedure 6.2-4.

1. Number all the branches using the natural numbers $j = 0, 1, 2, \ldots$.
2. For each nonopen nonelementary tip, choose one of its representative paths and short every branch in that path. Open all other branches.
3. Restore branches finitely many at a time, starting with the lowest numbered branches and restoring others in accordance with increasing branch numbers. Number these steps of restoring finitely many branches at a time using the natural numbers $n = 0, 1, 2, \ldots$.

Note that there are many ways of following Procedure 6.2-4 because of the many ways of numbering the branches, choosing shorted representative paths, and restoring branches finitely many at a time.

Just after the nth step ($n = 0, 1, 2, \ldots$) of the restoration process (item 3 in Procedure 6.2-4), let $\mathbf{M}_n^\nu$ be the infinite network having the same graph as $\mathbf{N}^\nu$ but with the electrical parameters existing at that point of restoration; thus, finitely many of the branches of $\mathbf{M}_n^\nu$ will have their original electrical-parameter values and the other branches of $\mathbf{M}_n^\nu$ will be either opened or shorted. For a given $\mathbf{M}_n^\nu$, a maximal set of maximal nodes in $\mathbf{M}_n^\nu$ that are pairwise connected through shorted paths will be called a *proximity*. Also, a maximal node that is not connected to any other maximal node through a shorted path will comprise a *proximity* in $\mathbf{M}^\nu$, albeit a singleton one. Every maximal node x^α in $\mathbf{M}_n^\nu$ will belong to some proximity in $\mathbf{M}_n^\nu$, namely, the nth *proximity of* x^α, and the proximities in $\mathbf{M}_n^\nu$ partition the set of maximal nodes. We also say that the embraced tips of a maximal node lie in that node's proximity.

As n increases, the proximities of a given maximal node shrink, in general, and never increase because shorted branches are being restored. That is, upon denoting the nth proximity of a given maximal node x^α by $\mathcal{P}_n(x^\alpha)$, we have $\mathcal{P}_n(x^\alpha) \supseteq \mathcal{P}_{n+1}(x^\alpha)$ for every n.

Lemma 6.2-5. *For a fixed choice of branch shortings (part 2 of Procedure 6.2-4), let* $\langle \mathcal{P}_n(x^\alpha) \rangle$ *and* $\langle \mathcal{Q}_n(x^\alpha) \rangle$ *be two sequences of proximities of a maximal node* x^α *resulting from two different choices of branch numberings (part 1 of Procedure 6.2-4). For each* $n \in \mathbb{N}$*, there will be a sufficiently large* $m \in \mathbb{N}$ *such that* $\mathcal{Q}_m(x^\alpha) \subseteq \mathcal{P}_n(x^\alpha)$.

Proof. The number of restored branches in $\mathbf{M}_n(x^\alpha)$ for the first choice of branch numberings is finite. Therefore, we need only choose m large enough to insure the restoration of these branches under the second choice of branch numberings. ♣

As $n \to \infty$, every maximal node in $\mathcal{P}_n(x^\alpha)$ other than x^α may eventually disappear from the $\mathcal{P}_n(x^\alpha)$. (See Examples 6.2-1 and 6.2-2.) If this happens for every maximal node x^α in $\mathbf{N}^\nu$, we say that $\mathbf{N}^\nu$ is *restorable*, and we also say that $\{\mathbf{M}_n^\nu\}_{n=0}^\infty$ is a sequence that *restores* $\mathbf{N}^\nu$. However, this may not happen; there may be one or more other maximal nodes that remain in $\mathcal{P}_n(x^\alpha)$ for all n. (See Examples 6.2-2 and 6.2-3.) As a result, there will be tips that are not shorted together (i.e., are not embraced by the same node) in $\mathbf{N}^\nu$ but remain shorted together in $\mathbf{M}_n^\nu$ through paths of shorted branches for every n. If the latter case always occurs whatever be the choices of representative paths for the nonopen nonelementary tips of ranks less than ν, we say that $\mathbf{N}^\nu$ is *not restorable*. Lemma 6.2-5 implies that the restorability or nonrestorability of $\mathbf{N}^\nu$ is independent of the branch numbering. In the following, we assume that the branch numbering is arbitrarily chosen but fixed.

Theorem 6.2-6. *A necessary condition for* $\mathbf{N}^\nu$ *to be restorable is that Condition 3.1-2 holds for* $\mathbf{N}^\nu$*; namely, if any two tips in* $\mathbf{N}^\nu$ *of ranks less than* ν *are nondisconnectable, then either they are shorted together (i.e., are embraced by the same node) or at least one of them is open (i.e., is the sole member of a maximal singleton node).*

Note. We can take tips of rank ν to be open because they do not contribute to connections within $\mathbf{N}^\nu$.

Proof. Indeed, if this condition is not satisfied by some pair of tips, then those tips will not be open and will be embraced by different maximal nodes. Moreover, the two shorted representative paths chosen for them in part 2 of Procedure 6.2-4 will meet for every n because only finitely many shorted branches are restored at every step of the restoration. Hence, two maximal nodes containing those two tips will be in the same proximity for every n. Thus, $\mathbf{N}^\nu$ is not restorable. ♣

Given two maximal nodes x^α and y^β in $\mathbf{N}^\nu$, we will say that their sequences of proximities, namely, $\langle \mathcal{P}_n(x^\alpha) : n \in \mathbb{N} \rangle$ and $\langle \mathcal{P}_n(y^\beta) : n \in \mathbb{N} \rangle$, are *eventually disjoint* if there exists a natural number m such that $\mathcal{P}_n(x^\alpha) \cap \mathcal{P}_n(y^\beta) = \emptyset$ for every $n \geq m$.

Theorem 6.2-7. $\mathbf{N}^\nu$ *is restorable if and only if it is possible to choose a shorted representative path for every nonopen nonelementary tip in* $\mathbf{N}^\nu$ *such that, for every pair of maximal nodes, the corresponding sequences of proximities in* $\mathbf{M}^\nu$ *in which those two nodes reside are eventually disjoint.*

Proof. *Only if:* Assume it is impossible to choose shorted representative paths as stated. Then, for some pair of distinct maximal nodes in $\mathbf{N}^\nu$, their embraced tips will all be shorted together in $\mathbf{M}_n^\nu$ for every n. Thus, $\mathbf{N}^\nu$ is not restorable.

If: Assume shorted representatives can be chosen as stated. If two tips belong to the same maximal node in $\mathbf{N}^\nu$, they will remain in the same proximity for each $\mathbf{M}_n^\nu$ and therefore will be shorted together for all n. If two tips belong to different maximal nodes in $\mathbf{N}^\nu$, they will eventually belong to disjoint proximities. Thus, they will eventually not be shorted together. So, no additional shortings of tips persist throughout the restoration process. $\mathbf{N}^\nu$ is restorable. ♣

Let us note that every countable, conventionally infinite network can be viewed as a restorable 1-network in the following two ways. One way is to assume that all the 0-tips (if there are any) are open. In this case, we assign to each 0-tip a singleton 1-node. Therefore, no representative path needs to be shorted. All branches are first opened and then restored sequentially. We might say that such a 1-network is "open at infinity."

As the other way, we assume the other extreme of a countable, conventionally infinite network that is "shorted at infinity." That is, all the 0-tips of such a network are all shorted together in a single 1-node. Hence, the proximities of that 1-node are eventually disjoint from every 0-node, and thus we have a restorable 1-network again.

6.3 Hyperreal Currents and Voltages; A Hyperreal Operating Point

Let us now review some of our ideas. We have assumed that the ν-network $\mathbf{N}^\nu$ is connected and that no $(\nu+1)$-nodes have been defined for it. Let us also assume that it is restorable and that a choice of Procedure 6.2-4 has been made which restores $\mathbf{N}^\nu$. Let $\langle \mathbf{M}_n^\nu : n \in \mathbb{N}\rangle$ be that restoration sequence, and let $\langle \mathbf{N}_n^0 : n \in \mathbb{N}\rangle$ be the corresponding sequence of finite networks $\mathbf{N}_n^0$ resulting from the removals and contractions of the opened and shorted branches in the $\mathbf{M}_n^\nu$. Then, each $\mathbf{N}_n^0$ has a unique current-voltage regime, and this determines eventually a hyperreal current and voltage in every branch of $\mathbf{N}^\nu$. In particular, for the jth branch b_j in $\mathbf{N}^\nu$, there will be a natural number n_j such that b_j is restored for all $n > n_j$ and has the branch current $i_{j,n}$ and branch voltage $v_{j,n}$ for all $n > n_j$. This determines a *hyperreal current* $\mathbf{i}_j = [i_{j,n}]$ and a *hyperreal voltage* $\mathbf{v}_j = [v_{j,n}]$ for b_j, where we are free to set $i_{j,n} = v_{j,n} = 0$ for $0 \le n \le n_j$. Moreover, Ohm's law $v_{j,n} = e_j + r_j i_{j,n}$ (using the Thevenin form of the branch—see Fig. 6.6(b)) will be satisfied for $n > n_j$, and this yields the nonstandard expression, $\mathbf{v}_j = [e_j] + [r_j]\mathbf{i}_j$, for Ohm's law.[7] We shall show in the next section that all the hyperreal branch voltages $\{\mathbf{v}_j : j \in \mathbb{N}\}$ together satisfy Kirchhoff's voltage law around every (finite or transfinite) loop in $\mathbf{N}^\nu$ in a

[7] $[e_j]$ (resp. $[r_j]$) is the hyperreal representation of the real number e_j (resp. r_j); see Appendix A5.

certain restricted way and that all the hyperreal branch currents $\{\mathbf{i}_j : j \in \mathbb{N}\}$ together satisfy Kirchhoff's current law at every 0-node (of finite or infinite degree) and at certain cuts that isolate the proximities of transfinite nodes again in a certain restricted way. When Ohm's law and Kirchhoff's laws are so satisfied, we shall refer to the set $\{\{\mathbf{i}_j, \mathbf{v}_j\} : j \in \mathbb{N}\}$ of all such hyperreal current-voltage pairs as a *hyperreal operating point* for $\mathbf{N}^\nu$ and will also write it as $\{\mathbf{i}, \mathbf{v}\}$, where $\mathbf{i} = \{\mathbf{i}_j : j \in \mathbb{N}\}$ $\mathbf{v} = \{\mathbf{v}_j : j \in \mathbb{N}\}$.

In general, $\mathbf{i}_j$ and $\mathbf{v}_j$ will depend upon the choices made in the Procedure 6.2-4, although two different sets of choices may yield the same hyperreal operating point. The fundamental idea here is that $\mathbf{N}^\nu$ should not be viewed simply as a given network whose graph is defined as in Chapter 2. Instead, it should be taken to be a given network $\mathbf{N}^\nu$ coupled with a specified expanding sequence $\langle \mathbf{N}_n^0 : n \in \mathbb{N} \rangle$ of finite networks $\mathbf{N}_n^0$ that produces $\mathbf{N}^\nu$ as the end result of a building process. One way of getting such a building process is to choose a restoration sequence according to Procedure 6.2-4 when $\mathbf{N}^\nu$ is restorable. To denote this coupling of $\mathbf{N}^\nu$ with a chosen restoration sequence, we use the symbols $\mathbf{N}_r^\nu = \{\mathbf{N}^\nu, \langle \mathbf{N}_n^0 : n \in \mathbb{N} \rangle\}$ or more simply $\mathbf{N}_r^\nu = \{\mathbf{N}^\nu, \langle \mathbf{N}_n^0 \rangle\}$. We call $\mathbf{N}_r^\nu$ a *restoration* of $\mathbf{N}^\nu$, and we refer to both $\langle \mathbf{M}_n^\nu \rangle$ and $\langle \mathbf{N}_n^0 \rangle$ as *restoration sequences* for $\mathbf{N}^\nu$.

6.4 Eventual Connectedness, Eventual Separability, and Kirchhoff's Laws

We now argue that the hyperreal voltages and currents in a restorable network $\mathbf{N}^\nu$ obtained from a properly chosen restoration procedure satisfy in a certain way Kirchhoff's voltage law around every (finite or transfinite) loop and Kirchhoff's current law at every branch cut determined by proximities.

Let P^α be any two-ended path in $\mathbf{N}^\nu$; thus, its rank α is no larger than ν. Also, let $\langle \mathbf{M}_n^\nu \rangle$ be a restoration sequence for $\mathbf{N}^\nu$. In $\mathbf{M}_n^\nu$, P^α becomes a path of shorted, opened, and restored branches, but only finitely many of those branches will be restored. Now, P^α has an image P_n^0 in $\mathbf{N}_n^0$ consisting of restored branches, but that image need not be a path because it may have gaps produced by opened branches in $\mathbf{M}_n^\nu$. We now argue that eventually that image becomes connected and thus a path.

Lemma 6.4-1. *Given any two-ended path P^α in $\mathbf{N}^\nu$, there is a natural number q such that, for each $n \geq q$, the image P_n^0 of P^α in $\mathbf{N}_n^0$ is a path.*

Proof. Assume that $\alpha > 0$. (The argument for $\alpha = 0$ is much simpler.) Since P^α is two-ended, it contains only finitely many α-nodes and therefore traverses only finitely many $(\alpha - 1)$-tips. Now, in $\mathbf{M}_n^\nu$ a representative one-ended path for each of the $(\alpha - 1)$-tips has been shorted (i.e., every branch in each representative path is shorted). What remains unshorted are finitely many, maximally long, unshorted, two-ended subpaths of P^α, each of rank less than α. Let P^β $(\beta < \alpha)$ be one of them. It will traverse only finitely many $(\beta - 1)$-tips, and in $\mathbf{M}_n^\nu$ a representative path for each these will be shorted. This yields finitely many maximally long, unshorted, two-ended subpaths of P^β, each of rank less than β. Continuing in this way for each subpath so generated, we finally arrive at finitely many maximally long, unshorted, two-ended 0-subpaths of P^α. Since those 0-subpaths are two-ended, they will each have only finitely many branches. Those branches are either restored or opened. Thus, the opened branches in all those 0-subpaths will be finitely many and will all be restored for some $\mathbf{N}_q^\nu$, q sufficiently large. Thus, for each $n \geq q$, all the branches of P^α as a path in $\mathbf{M}_n^\nu$ will be either shorted or restored. Upon contracting the shorted branches in P^α, we see that the image of P^α in $\mathbf{N}_n^0$ is a path whenever $n \geq q$. ♣

Lemma 6.4-2. *Given any loop L of any positive rank in $\mathbf{N}^\nu$, there exists a natural number q such that, for each $n \geq q$, the image L_n^0 of L in $\mathbf{N}_n^0$ is a loop.*

Note. If L is a loop of rank 0 in $\mathbf{N}^\nu$, it has finitely many branches, and thus our conclusion holds as soon as its opened branches are restored.

Proof. This is established exactly as is Lemma 6.4-1 since the number of nodes of maximum rank in L is finite. ♣

We can now solve for the currents and voltages in each finite network $\mathbf{N}_n^0$. Clearly, Kirchhoff's voltage law is satisfied around any finite loop in $\mathbf{N}^\nu$ for all n sufficiently large. So, let L be any loop of positive rank in $\mathbf{N}^\nu$. L will have denumerable many branches. Let L_n^0 be its image in $\mathbf{N}_n^0$, as before. For all n sufficiently large, say, for $n \geq q$, Kirchhoff's voltage law will be satisfied around L_n^0 and also around L as a loop in $\mathbf{M}_n^\nu$ because the voltage of any shorted branch is 0. Let us now orient each branch in L in the same direction around L, and let us measure branch voltage accordingly. Also, let us number the branches in L in any fashion using the natural numbers $k \in \mathbb{N}$. Within the network $\mathbf{M}_n^\nu$, this yields a voltage $v_{k,n}$ for the kth branch in L for all $n \geq q$. (q depends upon the choice of L.) This in turn determines a hyperreal branch voltage

$$\mathbf{v}_k = [v_{k,0}, v_{k,1}, v_{k,2}, \ldots]$$

for the kth branch in L, where we are free to set the initial entries $v_{k,n}$ equal to 0 for $n < q$; this will not change the hyperreal $\mathbf{v}_k$.

Now, with the $\mathbf{v}_k$ in hand, we can define a *total loop voltage* for L in $\mathbf{M}_n^\nu$ as the sum $\sum_{k \in \mathbb{N}} v_{k,n}$ for each $n \geq q$. We can then define a *hyperreal total loop voltage* for L in $\mathbf{N}^\nu$ through the representative sequence $\langle \sum_{k \in \mathbb{N}} v_{k,n} : n \in \mathbb{N} \rangle$. Each term of that representative sequence will equal 0 according to Kirchhoff's voltage law for $\mathbf{M}_n^\nu$ with $n \geq q$ and is 0 by choice for $n < q$. So, that hyperreal total loop voltage is equal to $\mathbf{0} = [0]$. It is in this restricted sense that we can assert the following.

Theorem 6.4-3. *The hyperreal total loop voltage equals $\mathbf{0}$ for every loop in $\mathbf{N}_r^\nu = \{\mathbf{N}^\nu, \langle \mathbf{N}_n^0 \rangle\}$. In this particular way, Kirchhoff's voltage law is satisfied in $\mathbf{N}_r^\nu$.*

The reason we define the total loop voltage for L as a loop in $\mathbf{M}_n^\nu$ before defining the hyperreal total loop voltage for L is that the sum $\sum_{k\in\mathbb{N}} \mathbf{v}_k$ of hyperreal branch voltages on L in $\mathbf{N}_r^\nu$ is not well-defined, as we shall see in a moment.[8] First note however that our way of summing the hyperreal branch voltages for L amounts to the following. We may write

$$\begin{aligned}\sum_{k\in\mathbb{N}} \mathbf{v}_k &= [v_{0,0}, v_{0,1}, v_{0,2}, \ldots] \\ &+ [v_{1,0}, v_{1,1}, v_{1,2}, \ldots] \\ &+ [v_{2,0}, v_{2,1}, v_{2,2}, \ldots] \\ &+ \cdots\end{aligned} \tag{6.1}$$

We have set all the entries to the left of the qth column (i.e., for $n < q$) equal to 0; this does not change any of the $\mathbf{v}_k$. Note also that, for any $n \geq q$, $v_{k,n} = 0$ if the kth branch b_k is a short. As we have shown above, all but finitely many of the branches of L in $\mathbf{M}_n^\nu$ are shorts. This means that, for the nth column with $n \geq q$ in this array, there are only finitely many nonzero entries. Therefore, we can sum columnwise to get the sum of voltages around the image of the loop L in $\mathbf{M}_n^\nu$. By Kirchhoff's voltage law, each such column sum equals zero. Thus, $\sum_{k\in\mathbb{N}} \mathbf{v}_k = 0$.

In general, however, the sum (6.1) *by itself* is not well-defined and in fact can be made equal to any hyperreal, say, $[a_n]$ without altering the $\mathbf{v}_k$. Indeed, just change the diagonal entries to $v_{n,n} = a_n - \sum_{k\neq n} v_{k,n}$. This leaves the $\mathbf{v}_k$ unchanged, but $\sum_{k\in\mathbb{N}} \mathbf{v}_k$ becomes $[a_n]$. Our specification of how the hyperreal total loop voltage is to be determined avoids this problem. Indeed, the chosen restoration sequence not only determines which representative is to be used for each hyperreal branch voltage and which representative for the total loop voltage but moreover indicates how the array (6.1) is to be summed (i.e., columnwise) in order to get the representative of the total loop voltage from the representatives of the branch voltages. Let us emphasize that the summation (6.1) along with the specification of the representative of each branch voltage in the nonstandard loop as dictated by the networks $\mathbf{N}_n^0$ in the way stated is a uniquely determined hyperreal.[9]

We now turn our attention to Kirchhoff's current law, which will be applied to a sequence of branch cuts that separates a maximal node of $\mathbf{N}^\nu$ from all the other maximal nodes of $\mathbf{N}^\nu$. We again assume that $\mathbf{N}^\nu$ is restorable, that a restoration sequence has been chosen for it so that we can view $\mathbf{N}^\nu$ as being represented by $\mathbf{N}_r^\nu = \{\mathbf{N}^\nu, \langle \mathbf{N}_n^0 \rangle\}$, and that $\mathbf{M}_n^\nu$ and $\mathbf{N}_n^0$ denote the networks defined before. Given any maximal node x of any rank in $\mathbf{N}^\nu$, let $\mathcal{P}_n(x)$ be its proximity in $\mathbf{M}_n^\nu$. Also, let $\mathcal{C}_n(x)$ be the set of branches in $\mathbf{M}_n^\nu$, each having exactly one tip in $\mathcal{P}_n(x)$ and one tip

[8] This issue is also discussed in Appendix A11.

[9] Another way of avoiding the ambiguity in the summation (6.1) is to abandon the hyperreals and simply use instead the sequence of voltages for each branch as determined by the chosen restoration sequence. Indeed, a theory for restorable networks can be constructed based strictly on such sequences for branch voltages and currents, with Kirchhoff's laws being satisfied through sums of sequences defined as above. But, this would ignore the fact those sequences represent hyperreal voltages and currents in the restored transfinite network.

not in $\mathcal{P}_n(x)$. The branches of $\mathcal{C}_n(x)$ are either restored or opened, and all but finitely many of them are opened. Moreover, for all n sufficiently large $\mathcal{C}_n(x)$ will have at least one restored branch. The restored branches will be incident to the image x_n^0 of x in $\mathbf{N}_n^0$, and Kirchhoff's current law will be satisfied at x_n^0 and thereby on $\mathcal{C}_n(x)$ because the current in any opened branch is 0. Furthermore, given any maximal node y different from x, $\mathcal{C}_n(x)$ separates x from y eventually in the sense that any path connecting x and y passes through $\mathcal{C}_n(x)$ for all n sufficiently large.

By a *cut around* x we will mean a set $\mathcal{C}_n(x)$ for some $n \in \mathbb{N}$. Fix $q \in \mathbb{N}$ such that $\mathcal{C}_q(x)$ has at least one restored branch. Assume $\mathcal{C}_q(x)$ has infinitely many branches (the finite case is simpler). $\mathcal{C}_q(x)$ consists only of restored and opened branches, and all but finitely many of them are opened. Number the branches of $\mathcal{C}_q(x)$ by the natural numbers $k \in \mathbb{N}$ in any fashion, and orient those branches away from x (i.e., away from the proximity $\mathcal{P}_q(x)$). The solution of $\mathbf{N}_n^0$ for each $n \geq q$ yields a current in every restored branch in $\mathcal{C}_q(x)$, and the currents in the opened branches of $\mathcal{C}_q(x)$ are all 0. We denote all these currents by $i_{k,n}$. Again, for each $n < q$, we are free to set $i_{k,n} = 0$ without altering the hyperreal current $\mathbf{i}_k$ in the kth branch. All this yields an infinite series of hyperreal branch currents, which can be displayed as in the above array for hyperreal voltages but with $\mathbf{v}$ and v replaced by $\mathbf{i}$ and i. Since for each n all but finitely many branches in $\mathcal{C}_q(x)$ are opened and have zero currents, each column in that array has only finitely many nonzero entries and sums to 0 because of Kirchhoff's current law applied to $\mathbf{N}_n^0$. More specifically, for each n, we can define a *total cut current* for $\mathcal{C}_q(x)$ in $\mathbf{M}_n^\nu$ as being $\sum_{k \in \mathbb{N}} i_{k,n}$; by Kirchhoff's current law, this will be 0 for $n \geq k$ and will be 0 by choice for $n < k$. Thus, we have a representative sequence of 0's for the *hyperreal total cut current*; that is, $\sum_{k \in \mathbb{N}} \mathbf{i}_k = [\sum_{k \in \mathbb{N}} i_{k,n}] = [0]$ when that sum is determined as stated. Note once again that the restoration sequence dictates which representative is to be chosen for each hyperreal branch current when summing them. In summary, we may assert the following.

Theorem 6.4-4. *For each maximal node x of $\mathbf{N}_r^\nu = \{\mathbf{N}^\nu, \langle \mathbf{N}_n^0 \rangle\}$, the hyperreal total cut current for any cut $\mathcal{C}_q(x)$ around x equals $\mathbf{0} = [0]$. In this particular way, Kirchhoff's current law is satisfied in $\mathbf{N}_r^\nu$.*

In view of Theorems 6.4-3 and 6.4-4 and the fact that Ohm's law is satisfied for each branch after its restoration, we can summarize our discussions in this and the preceding section as follows.

Theorem 6.4-5. *Given any $\mathbf{N}_r^\nu = \{\mathbf{N}^\nu, \langle \mathbf{N}_n^0 \rangle\}$ corresponding to a restorable ν-network $\mathbf{N}^\nu$ and a sequence $\langle \mathbf{N}_n^0 \rangle$ of finite networks that restores $\mathbf{N}^\nu$ (chosen perhaps according to Procedure 6.2-4), $\mathbf{N}_r^\nu$ has a hyperreal operating point $\{\mathbf{i}, \mathbf{v}\}$ that satisfies Ohm's law and also Kirchhoff's laws in the way stated in Theorems 6.4-3 and 6.4-4.*

The phrase "In this particular way" appearing in Theorems 6.4-3 and 6.4-4 expresses an important idea. For the sake of emphasis, let us state it again. The chosen restoration sequence dictates not only a representative sequence for each hyperreal branch voltage and each hyperreal branch current, it also dictates a representative sequence for the sum of the hyperreal voltages around any loop and a representative sequence for the sum of the hyperreal currents in any cut. The latter two representative sequences are obtained by columnwise summing of arrays, such as (6.1), whose entries are determined by the finite networks in which the image of a loop or a cut appears and are taken to be 0 before that image appears.

An immediate corollary of Lemma 6.4-1 and Theorem 6.4-5 is that the choices of a ground node x_g and of a restoration sequence $\langle \mathbf{N}_n^0 \rangle$ uniquely determines a hyperreal node voltage $\mathbf{u}_x$ at every node x in $\mathbf{N}^\nu$. Indeed, we may choose any path P from x to x_g and sum the hyperreal branch voltages along P to get the node voltage $\mathbf{u}_x$ at x. Here, too, that sum can be obtained through a columnwise summation of an array such as that of (6.1), but now the column sums will in general be different from 0. To be more specific, let q be the natural number mentioned in Lemma 6.4-1. Then, all the entries of the array to the left of the qth column are set equal to 0. Also, for $n \geq q$, the entries of the nth column are determined by $\mathbf{N}_n^0$; such an entry is 0 if the corresponding branch is a short and is the voltage determined by $\mathbf{N}_n^0$ if that branch is restored. Furthermore, the node voltage $\mathbf{u}_x$ will be independent of the choice of P because of the satisfaction of Kirchhoff's voltage law in each $\mathbf{N}_n^0$. We can restate these facts as follows.

Corollary 6.4-6. *Assume the hypothesis of Theorem 6.4-5, and choose any maximal node x_g of $\mathbf{N}^\nu$ as the ground node. Then, with respect to x_g, every node x of $\mathbf{N}^\nu$ has a hyperreal node voltage $\mathbf{u}_x$ determined by the operating point of $\mathbf{N}_r^\nu = \{\mathbf{N}^\nu, \langle \mathbf{N}_n^\nu \rangle\}$; $\mathbf{u}_x$ is uniquely determined once $\langle \mathbf{N}_n^\nu \rangle$ is chosen.*

One final comment may be worth mentioning: Theorem 6.4-5 has been argued for the case of a linear network $\mathbf{N}^\nu$, that is, when every branch resistance is a positive constant so that Ohm's law holds in the form: $v_j = e_j + r_j i_j$ for each restored branch b_j. However, our analysis immediately extends to any nonlinear restorable ν-network whose every branch has a nonlinear resistance characteristic $v_j = R_j(i_j)$ where R_j is a continuous, strictly monotonic bijection of $\mathbb{R}$ onto $\mathbb{R}$ with $R_j(0)$ not necessarily equal to 0. Thus, $R_j(i_j) \to +\infty$ as $i_j \to +\infty$, and $R_j(i_j) \to -\infty$ as $i_j \to -\infty$. If $R_j(0) \neq 0$, the branch b_j has an implicit voltage source $R_j(0)$ in series with a passive resistor $R_j(i_j) - R_j(0)$. With every branch resistance characteristic so specified, Theorem 6.4-5 again holds exactly as stated with the understanding that Ohm's law is replaced by $\mathbf{v}_j = {}^*R_j(\mathbf{i}_j)$, in particular, by $v_{j,n} = R_j(i_{j,n})$ for $n > n_j$, where n_j is the index at which the branch b_j is restored. Here, ${}^*R_j$ is the standard member of ${}^*V(\mathbb{R})$ corresponding to the function R_j; see Appendix A20. Duffin [20] proved that such a standard nonlinear finite network has an operating point. (See also [54, Sec. 6.4] for an exposition of Duffin's result.)

6.5 Three Examples Involving Ladder Networks

By way of examples, we now determine some hyperreal operating points for certain 1-networks with specified restoration sequences. Other examples can be found in [57]. Also, examples for infinite 0-networks are given in [55].

In the following, we will be using the Fibonacci numbers $F(k)$; see Appendix B for a short summary of what we use.

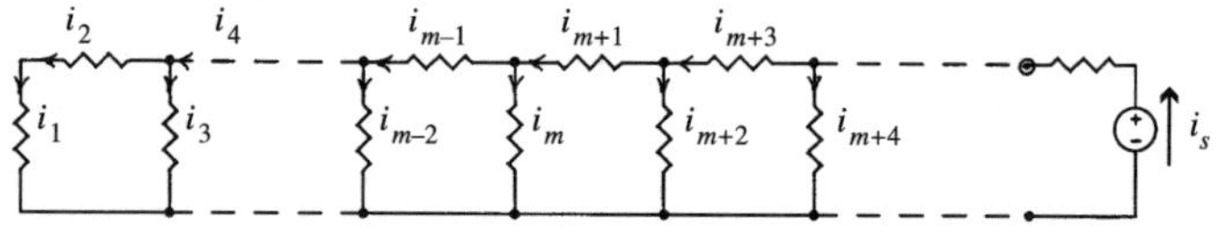

Fig. 6.11a. A one-way infinite, purely resistive ladder excited at infinity by a 1 V voltage source in series with a resistor. All resistors are 1 Ω. Here, m is an odd positive integer.

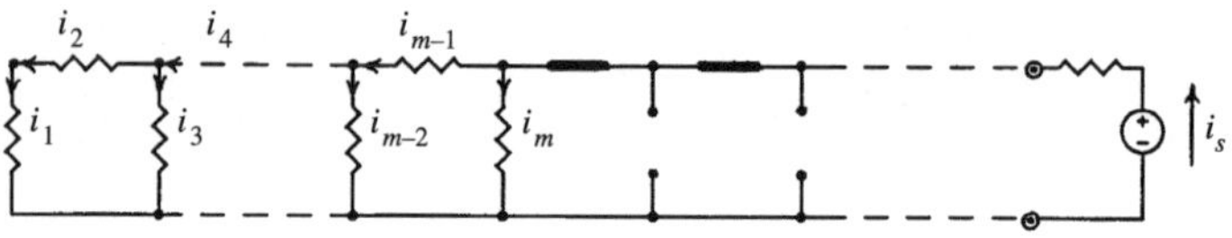

Fig. 6.11b. The truncated ladder. All series (resp. shunt) resistors beyond the mth resistor are shorts (resp. opens).

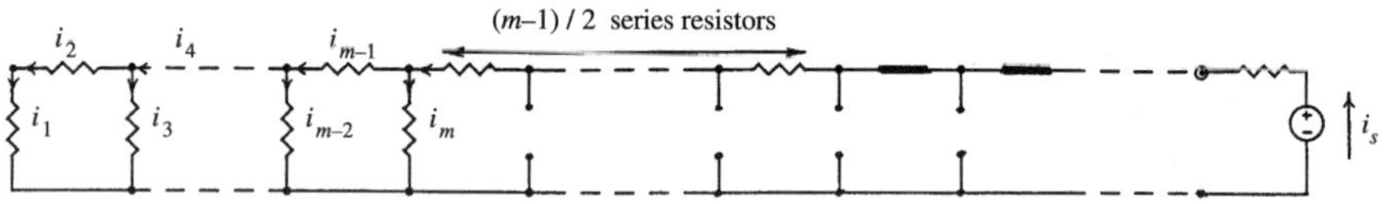

Fig. 6.11c. The network obtained just after restoring a shunt resistor during the second way of building up the transfinite network of Fig. 6.11(a). Here again, m is an odd positive integer.

Example 6.5-1. *A one-way infinite ladder with a source at infinity:*

Let us now consider a transfinite network for which a standard analysis provides only a trivial current-voltage regime. The 1-network $\mathbf{N}^1$ we examine is the one-way infinite ladder of 1 Ω resistors excited at its infinite extremity by a Thevenin branch with a 1 V voltage source and a 1 Ω resistor. See Fig. 6.11(a). This network satisfies the conditions that allow a standard analysis to be applied in accordance with Theorem 5.13-3. Every 1-loop or 1-tour passing through the one and only voltage source is transfinite with an infinite sum for its resistors. Hence, the solution space $\mathcal{T}$ has no nonzero 1-loop or 1-tour current. Since there are no other sources, the only solution the standard analysis gives is the one where every branch current is 0, and in this case Kirchhoff's voltage law is violated around every transfinite 1-loop and 1-tour.

(Note that this does not violate Theorem 5.15-1 because wnode voltages are defined only along permissive walks and no walk passing through the 1-node is permissive.)

Far more interesting are the results provided by nonstandard analyses. They restore Kirchhoff's voltage law, albeit with hyperreal values. Now, however, there are many different solutions depending upon how the network is restored from finite ones. One way is to truncate the ladder after the mth resistor, m odd, by shorting (resp. opening) all subsequent series (resp. shunt) resistors. This is indicated in Fig. 6.11(b). The Thevenin branch at infinity remains unchanged. A straightforward recursive analysis[10] shows that the ladder's currents in this finite network, as shown in Fig. 6.11(b), have the values

$$i_{k,m} = \frac{F(k-1)}{F(m+1)}, \quad k = 1, \ldots, m, \tag{6.2}$$

where k denotes the indices of the branches up to the mth branch and $F(k)$ is the kth Fibonacci number. (In that figure, with m fixed we have set $i_k = i_{k,m}$.) Also, $i_s = F(m)/F(m+1)$. If instead of truncating after a shunt resistor we were to truncate after the series resistor of index $m+1$, the current values would be

$$i_{k,m+1} = \frac{F(k-1)}{F(m+2)}, \quad k = 1, \ldots, m+1, \tag{6.3}$$

and $i_s = F(m)/F(m+2)$.

Now, we can obtain a restoration sequence $\langle \mathbf{N}_n^0 \rangle$ by alternately restoring the shunt and series resistors one at a time. Thus, the current in the kth resistor is the hyperreal $\mathbf{i}_k = [i_{k,n}]$, where k is fixed, n is the index for the representative sequence, and $i_{k,n}$ is given by (6.2) for $n = m$ odd and by (6.3) for $n = m+1$ even. $\mathbf{i}_k$ is an infinitesimal. On the other hand, the hyperreal source current is $\mathbf{i}_s = [i_n]$, where $i_n = F(n)/F(n+1)$ for n odd and $i_n = F(n-1)/F(n+1)$ for n even. $\mathbf{i}_s$ is limited but not infinitesimal, that is, it is appreciable. In terms of these hyperreals summed as stated in Sec. 6.4, Kirchhoff's laws are satisfied everywhere in $\mathbf{N}_r^\nu = \{\mathbf{N}^1, \langle \mathbf{N}_n^0 \rangle\}$ (including the voltage law around transfinite loops) since they are satisfied in each of the finite networks $\mathbf{N}_n^0$.

Let us consider another way of restoring the transfinite network of Fig. 6.11(a). After opening all the shunt resistors and shorting all the series resistors, we restore them starting at the left by restoring the first shunt resistor, then the first and second series resistors, then the second shunt resistor, then the third and fourth series resistors, and so forth alternately restoring one shunt and then two series resistors. In general, just after restoring a shunt resistor, we will have the network of Fig. 6.11(c). Also, just after restoring two series resistors, we will have the same network except for $(m+3)/2$ series resistors in place of the indicated $(m-1)/2$ series resistors. A recursive analysis once again yields the following results. Just after restoring a shunt resistor, we have

[10] Compute the sequence of driving-point resistances of the ladder to the left of each resistor, working from the left to the right. Then, compute the branch currents working from right to left.

$$i_{k,m} = \frac{2F(k-1)}{2F(m-1)+(m+1)F(m)}, \quad k = 1,\ldots,m, \tag{6.4}$$

$$i_s = \frac{2F(m)}{2F(m-1)+(m+1)F(m)}.$$

Just after restoring two series resistors, we have

$$i_{k,m} = \frac{2F(k-1)}{2F(m-1)+(m+5)F(m)}, \quad k = 1,\ldots,m, \tag{6.5}$$

$$i_s = \frac{2F(m)}{2F(m-1)+(m+5)F(m)}.$$

Now, $\mathbf{N}_r^\nu$ corresponding to this sequence $\langle \mathbf{N}_n^0 \rangle$ of restorations has the following hyperreal currents. For the kth resistor and with n being the index for the representative sequence as before, $\mathbf{i}_k = [i_{k,n}]$, where $i_{k,n}$ is given by (6.4) for $n = m$ odd and by (6.5) for $n = m+1$ even (i.e., replace m by $n-1$ in (6.5)). The hyperreal current in the series circuit of $(m-1)/2$ resistors or $(m+3)/2$ resistors is the same as the source current $\mathbf{i}_s$, which is given by $\mathbf{i}_s = [i_n]$, where

$$i_n = \frac{2F(n)}{2F(n-1)+(n+1)F(n)}, \quad n \text{ odd},$$

$$i_n = \frac{2F(n-1)}{2F(n-2)+(n+4)F(n-1)}, \quad n \text{ even}.$$

In the present case, $\mathbf{i}_k$ is a smaller infinitesimal than it was for the first method of restoring resistors because of the larger denominators. Also, $\mathbf{i}_s$ is now an infinitesimal, in contrast to the appreciable source current obtained previously. Of course, Kirchhoff's laws are satisfied here as well in a nonstandard way. ♣

Example 6.5-2. *A one-way infinite ladder with a resistor at infinity:* Consider now the ladder of Fig. 6.12(a). It is excited at its input by a pure current source of real value h A. All resistors are 1 Ω including the resistor r_ω connected to the ladder at its infinite extremity. Under a standard analysis, the real-valued branch currents converge to 0 as infinity is approached. As a result, we have to conclude that the real current i_ω in r_ω is 0. Under a nonstandard analysis, we can determine a nonzero hyperreal current $\mathbf{i}_\omega$ in r_ω due to a nonzero hyperreal input current $\mathbf{h}$, and can do so whether $\mathbf{h}$ is infinitesimal, appreciable, or unlimited.

We first have to specify how the ladder is restored from finite ones. Let us assume that it is restored by inserting "el-sections," each consisting of a shunt resistor followed by a series resistor. That is, each of the finite networks have the form shown in Fig. 6.12(b), which starts with a shunt resistor and ends with a series resistor before the opened and shorted branches begin. Here, k and m are odd positive integers

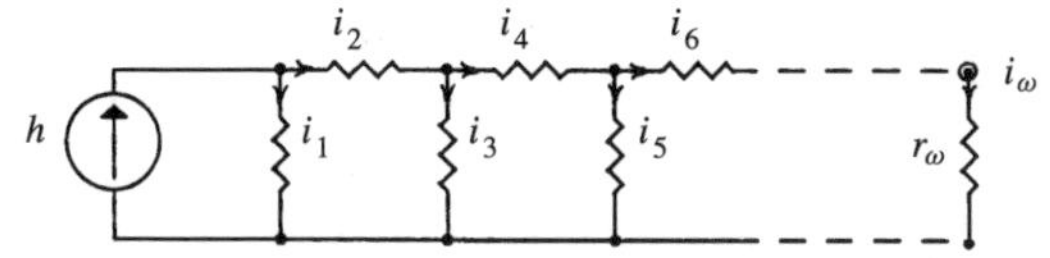

Fig. 6.12a. A one-way infinite ladder connected at infinity to a resistor r_ω and excited by a pure current source h at its input. All resistors, including r_ω, are 1 Ω.

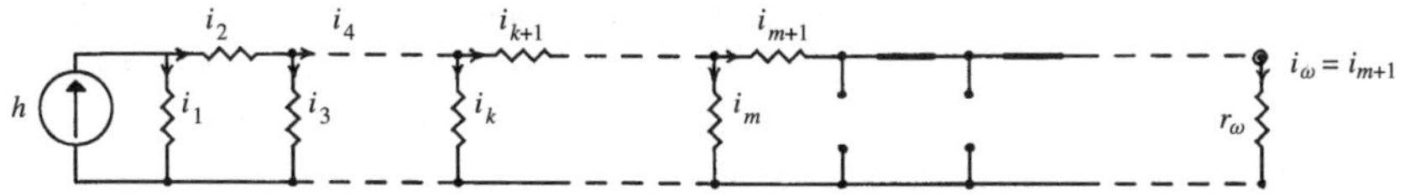

Fig. 6.12b. A finite truncation of the ladder. The ladder is built up with el-sections restored one at a time.

with $k = 1, 3, \ldots, m$. The sequence of finite truncations is obtained by increasing m according to $m = 1, 3, 5, \ldots$. In this case, we have

$$i_k = h\frac{F(m+2-k)}{F(m+2)}, \quad i_{k+1} = h\frac{F(m+1-k)}{F(m+2)}, \quad i_\omega = i_{m+1} = h\frac{1}{F(m+2)}.$$

In order to have n as the index for each step of the expanding sequence of finite ladders ($n = 1, 2, 3, \ldots$), we set $m = 2n-1$. Thus, for the corresponding restoration of the ladder we have the following hyperreal currents, where $\mathbf{h}$ is the hyperreal input source current.

$$\mathbf{i}_k = \mathbf{h}\left[\frac{F(2n+1-k)}{F(2n+1)}\right], \quad \mathbf{i}_{k+1} = \mathbf{h}\left[\frac{F(2n-k)}{F(2n+1)}\right], \quad \mathbf{i}_\omega = \mathbf{h}\left[\frac{1}{F(2n+1)}\right].$$

From Appendix B, we have that, as $k \to \infty$, $F(k) \sim \lambda_1^{k+1}/\sqrt{5}$, and thus

$$\frac{F(2n+1-k)}{F(2n+1)} \sim \frac{1}{\lambda_1^k}, \quad \frac{F(2n-k)}{F(2n+1)} \sim \frac{1}{\lambda_1^{k+1}}, \quad \frac{1}{F(2n+1)} \sim \frac{\sqrt{5}}{\lambda_1^{2n+2}} = \frac{A}{\lambda_1^{2n}}$$

where $A = \sqrt{5}/\lambda_1^2$. Thus, $\mathbf{i}_k$ and $\mathbf{i}_{k+1}$ are infinitesimal (resp. appreciable, resp. unlimited) whenever $\mathbf{h}$ is infinitesimal (resp. appreciable, resp. unlimited). On the other hand, if $\mathbf{h}$ has a representative that is $o(\lambda_1^{2n})$ as $n \to \infty$, then $\mathbf{i}_\omega$ is infinitesimal. If $\mathbf{h}$ has a representative that is asymptotic to $B\lambda_1^{2n}$, where B is a nonzero constant, then $\mathbf{i}_\omega$ is appreciable. Finally, if $\mathbf{h}$ has a representative $\langle h_n \rangle$ such that $h_n\lambda_1^{-2n} \to \infty$, then $\mathbf{i}_\omega$ is unlimited. ♣

Example 6.5-3. *Two one-way infinite ladders in cascade:* As our third example, let us consider a network that is more substantially transfinite than the networks we have considered so far. In particular, let that network be a cascade connection of two ladders identical to that of Fig. 6.12(a) except that the second ladder replaces r_ω. That is, the infinite extremity of the first ladder is connected through a 1-node to the input of the second ladder, which in turn has at its infinite extremity a resistor $r_{\omega \cdot 2}$ connected through another 1-node. We shall now denote the currents with double subscripts, the first subscript being 1 for the first ladder and 2 for the second ladder. We shall also truncate both ladders in the same way with the same number of el-sections. Thus, the last el-section in each ladder has the branches with second-subscript indices m and $m+1$, where m is an odd positive integer. Also, we use the odd positive integers $k = 1, 3, \ldots, m$ and $p = 1, 3, \ldots, m$ to index shunt branches in the first and second ladders respectively, and $k+1$ and $p+1$ for the series branches. The resulting finite network can be analyzed exactly as before except that now we have twice as many el-sections for each m. Next, we restore el-sections simultaneously; that is, for a transition from m to $m+2$, we restore two el-sections, one at the end of the first finite ladder and the other at the end of the second finite ladder. Finally, to get the current expressions as sequences indexed by $n = 1, 2, 3, \ldots$, we set $m = 2n-1$. Altogether then, the following expressions are obtained:

$$\mathbf{i}_{1,k} = \mathbf{h}\left[\frac{F(4n+1-k)}{F(4n+1)}\right], \quad \mathbf{i}_{1,k+1} = \mathbf{h}\left[\frac{F(4n-k)}{F(4n+1)}\right], \quad \mathbf{i}_{\omega} = \mathbf{h}\left[\frac{F(2n+1)}{F(4n+1)}\right],$$

$$\mathbf{i}_{2,p} = \mathbf{h}\left[\frac{F(2n+1-p)}{F(4n+1)}\right], \quad \mathbf{i}_{2,p+1} = \mathbf{h}\left[\frac{F(2n-p)}{F(4n+1)}\right], \quad \mathbf{i}_{\omega \cdot 2} = \mathbf{h}\left[\frac{F(1)}{F(4n+1)}\right].$$

As before, we can use the asymptotic behavior of the Fibonacci numbers, $F(n) \sim \lambda_1^{n+1}/\sqrt{5}$ as $n \to \infty$, to determine the character of each current (whether it is infinitesimal, appreciable, or unlimited) given the character of the hyperreal input-source current $\mathbf{h}$. ♣

6.6 Random Walks on Restorable Transfinite Networks

As another application of our theory of restorable networks, we now show how it can be used to define and analyze random walks on transfinite networks using hyperreal transition probabilities. We continue to assume that $\mathbf{N}^\nu$ is a connected transfinite network with countably many branches and that $\mathbf{N}^\nu$ is restorable through an expanding sequence $\langle \mathbf{N}_n^0 \rangle$ of finite 0-networks $\mathbf{N}_n^0$ chosen as in Procedure 6.2-4. We use the symbol $\{\mathbf{N}^\nu, \langle \mathbf{N}_n^0 \rangle\}$ to represent all this, as before. Furthermore, we now assume that each branch b has a positive conductance g_b $(0 < g_b < \infty)$ but no source. Because each $\mathbf{N}_n^0$ is is a finite network, we can readily lift many standard theorems concerning random walks on finite networks into a nonstandard setting. We shall list some of them in this section and illustrate them with the transfinite network of Fig. 6.13(a).

Our prior development of random walks on transfinite networks [51, Chap. 7], [54, Chap. 8] used standard analysis and required the imposition of many restrictions on the networks. Moreover, the standard probability of a random walk leaving a transfinite node is 0. To obtain nontrivial results, only the exceptional cases where the random walker does leave transfinite nodes was admitted; this is the "roving" assumption [51, page 197], [54, page 156], which conditions probabilities on those

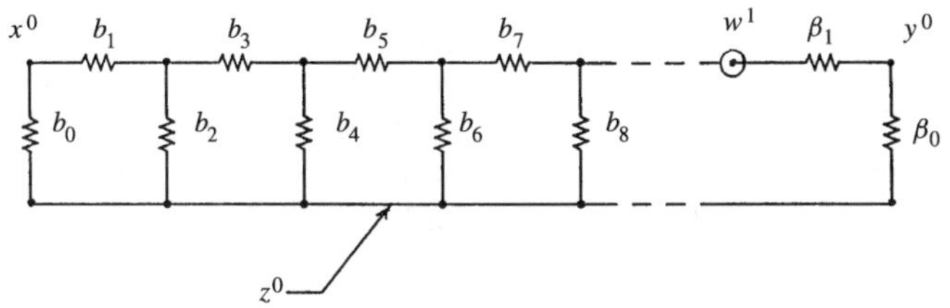

Fig. 6.13a. The 1-network on which the random walk occurs.

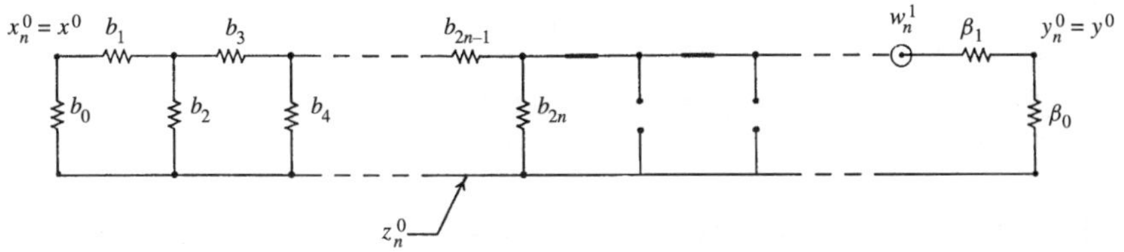

Fig. 6.13b. One stage of the restoration of that 1-network.

exceptional cases. No such restrictions on $\mathbf{N}^\nu$ are now needed other than the restorability of $\mathbf{N}^\nu$, but now probabilities are hyperreals, possibly infinitesimals.

In the following, we assume that a random walker wanders on a finite network with conductances g_b in accordance with the *nearest-neighbor rule*: The probabilities of proceeding from a node x to the nodes adjacent to x are proportional to the conductances between x and the adjacent nodes. A slight complication arises for us. When constructing the finite networks $\mathbf{N}_n^0$ in accordance with Procedure 6.2-4, some $\mathbf{N}_n^0$ may be found to have parallel branches and/or self-loops. The nearest-neighbor rule as stated does not encompass these. To overcome this problem, we combine parallel branches into a single branch by summing conductances. Also, we delete self-loops; in effect, we will be ignoring the steps a random walker may take around self-loops and are only analyzing the random walks that are reduced in this way. This conforms with the fact that, when an electrical source is appended to two nodes, the currents in self-loops are 0, and thus node voltages are unaffected by the deletion of self-loops.

The first theorem we lift is the *Nash-Williams rule* [31]. Let x^0, y^0 and z^0 be three nodes in a finite network $\mathbf{N}^0$. Let prob(sx^0,ry^0,bz^0) denote the probability that a random walker, after starting from the node x^0 and following the nearest-neighbor rule, will reach y^0 before reaching z^0.[11] The Nash-Williams rule asserts that prob(sx^0,ry^0,bz^0) is equal to the voltage at x^0 when y^0 is held at 1 volt and z^0 is held at 0 volt. That voltage at x^0 is determined by Kirchhoff's laws and Ohm's law applied to the electrical network $\mathbf{N}^0$ with an additional branch for the voltage source appended to y^0 and z^0 and whose electrical branch parameters are the branch conductances g_b except for the voltage source.

[11]More generally, we can replace y^0 and z^0 by two disjoint sets of nodes and then apply the following analysis by shorting together the nodes of each set.

Now, let x^α, y^β, and z^γ be three maximal nodes in $\mathbf{N}^\nu$.[12] As a consequence of restorability, the three nodes x^α, y^β, and z^γ will eventually appear as different 0-nodes x_n^0, y_n^0, and z_n^0 in $\mathbf{N}_n^0$. Moreover, by Lemma 6.4-1, x_n^0, y_n^0, and z_n^0 will eventually be connected in $\mathbf{N}_n^0$. Consequently, the Nash-Williams rule can eventually be applied to $\mathbf{N}_n^0$ to get the following nonstandard result.[13]

Theorem 6.6-1. *The hyperreal probability that a random walker on* $\{\mathbf{N}^\nu, \langle \mathbf{N}_n^0 \rangle\}$, *after starting from* x^α, *will reach* y^β *before reaching* z^γ *is*

$$\mathbf{prob}(\mathrm{s}x^\alpha, \mathrm{r}y^\beta, \mathrm{b}z^\gamma) \;=\; [\mathrm{prob}(\mathrm{s}x_n^0, \mathrm{r}y_n^0, \mathrm{b}z_n^0)] \tag{6.6}$$

where, for each n, the term within the brackets on the right-hand side is given by the Nash-Williams rule applied to $\mathbf{N}_n^0$ *for all n sufficiently large.*

Example 6.6-2 Consider the 1-network $\mathbf{N}^1$ of Fig. 6.13(a). Every branch is assigned a conductance of 1 siemens. A random walker on $\mathbf{N}^1$ starts at x^0 and wanders, possibly through the 1-node w^1, to reach y^0. What is the hyperreal probability $\mathbf{prob}(\mathrm{s}x^0,\mathrm{r}y^0,\mathrm{b}z^0)$ that the random walker reaches y^0 before reaching z^0? To answer this, we have to specify the restoration sequence $\langle \mathbf{N}_n^0 \rangle$ we are using.

Short and open branches as in the first part of Example 6.5-1; this is illustrated in Fig. 6.13(b). Specifically, let $\mathbf{N}_0^0$ be the loop consisting of branches b_0, β_0, and β_1. Then, restore branches two at a time according to b_{2n-1} and b_{2n} $(n = 1, 2, 3, \ldots)$. Thus, $\mathbf{N}_n^0$ is the ladder network of Fig. 6.13(b) restored up to the branch b_{2n}. Then, according to the Nash-Williams rule for the finite network $\mathbf{N}_n^0$, the probability of the random walker, starting at x^0 and reaching the node y^0 before reaching the node z_n^0 is the voltage $u_n(x^0)$ at x^0 when y^0 is held at 1 volt and z_n^0 is held at 0 volt. A straightforward computation gives $u_n(x^0) = 1/F(2n+2)$. Thus,

$$\mathbf{prob}(\mathrm{s}x^0, \mathrm{r}y^0, \mathrm{b}z^0) \;=\; \left[\frac{1}{F(2n+2)}\right]. \tag{6.7}$$

If instead we use the nonstandard network obtained by restoring b_1, b_2, b_3, then b_4, b_5, b_7, then b_6, b_9, b_{11}, and so one (that is, we restore one vertical and two horizontal branches at each step $n = 1, 2, 3, \ldots$) we obtain

$$\mathbf{prob}(\mathrm{s}x^0, \mathrm{r}y^0, \mathrm{b}z^0) \;=\; \left[\frac{1}{F(2n+2) + nF(2n+1)}\right]. \tag{6.8}$$

Both of these results are infinitesimals, but (6.8) is smaller than (6.7), as is to be expected since the random walker must pass through more resistance in $\mathbf{N}_n^0$ in order to reach y^0 for the second case as compared to the first case. ♣

[12] Here too, we can get a more general result by replacing y^β and z^γ by disjoint sets of nodes.

[13] As was mentioned before, we will use boldface lower-case symbols to denote hyperreals.

Another standard result concerns the *escape probability* $p_{\rm esc}(x^0 \rightarrow y^0)$ in a finite network $\mathbf{N}^0$. This is the probability of a random walker, after starting from x^0, reaches y^0 before returning to x^0. The result [7, page 304] asserts that $p_{\rm esc}(x^0 \rightarrow y^0) = c_{\rm eff}(x^0, y^0)/c_{x^0}$, where $c_{\rm eff}(x^0, y^0)$ is the input conductance between x^0 and y^0 and c_{x^0} is the total conductance incident to x^0. We have $c_{\rm eff}(x^0, y^0) = 1/u(x^0)$, where $u(x^0)$ is the node voltage at x^0 with respect to y^0 chosen as ground (i.e., choose $u(y^0) = 0$) when a current source injects 1 ampere at x^0 and extracts 1 ampere at y^0.

This, too, can be immediately lifted into the following result.

Theorem 6.6-3. *The hyperreal escape probability that a random walker on* $\{\mathbf{N}^\nu, \langle \mathbf{N}_n^0\rangle\}$, *after starting from a node* x^α, *reaches another node* y^β *before returning to* x^α *is*

$$\mathbf{p_{esc}}(x^\alpha \rightarrow y^\beta) = [p_{\rm esc}(x_n^0 \rightarrow y_n^0)] = \left[\frac{c_{\rm eff}(x_n^0, y_n^0)}{c_{x_n^0}}\right]. \tag{6.9}$$

If the right-hand side is an infinitesimal, $\{\mathbf{N}^\nu, \langle \mathbf{N}_n^0\rangle\}$ may be called *recurrent from* x^α *to* y^β, and, if it is appreciable, $\{\mathbf{N}^\nu, \langle \mathbf{N}_n^0\rangle\}$ may be called *transient from* x^α *to* y^β.

The corresponding standard concept relates to a conventionally infinite network whose 0-tips are all shorted together into a single 1-node y^1 (i.e., the network is "shorted at infinity"). Then, that network is called recurrent if $p_{\rm esc}(x^0 \rightarrow y^1) = 0$ and is called transient if $p_{\rm esc}(x^0 \rightarrow y^1) > 0$. In the nonstandard case for that conventionally infinite network shorted at infinity, the hyperreal $\mathbf{p_{esc}}(x^0 \rightarrow y^1)$ depends not only on the choice of x^0 but also on the choice of the restoration sequence $\langle \mathbf{N}_n^0\rangle$; however, the different $\mathbf{p_{esc}}(x^0 \rightarrow y^1)$ for fixed x^0 but different $\langle \mathbf{N}_n^0\rangle$ will lie within the same halo. Moreover, in the recurrent case, the standard escape probability $p_{esc}(x^0 \rightarrow y^1) = 0$ will not depend upon the choice of x^0. However, in the nonstandard case, the hyperreal $\mathbf{p_{esc}}(x^0 \rightarrow y^1)$ will, in general, take on different infinitesimal values depending on the choice of x^0 even when the restoration sequence $\langle \mathbf{N}_n^0\rangle$ is fixed, and thus we can compare the sizes of these infinitesimal escape probabilities for different choices of x^0 but fixed $\langle \mathbf{N}_n^0\rangle$.

Example 6.6-4. For the standard 1-network $\mathbf{N}^1$ of Fig. 6.13(a) wherein all branch conductances are 1 siemens, short and open branches and then restore them two at a time as in the first part of Example 6.6-2. We determine the escape probability from x^0 to y^0. In order to get $c_{\rm eff}(x_n^0, y_n^0)$, we need to determine the node voltage $u(x_n^0)$ when $u(y_n^0) = 0$ and a current source injects 1 ampere into x_n^0 and extracts 1 ampere from y_n^0. This is easily computed by using superposition; first apply the current source from z_n^0 to x_n^0, then apply the current source from y_n^0 to z_n^0; finally, add the results. We get

$$c_{\rm eff}(x_n^0, y_n^0) = \frac{1}{u(x_n^0)} = \frac{F(2n+3)}{2(F(2n+2)-1)}.$$

Since $c_{x_n^0} = 2$, we obtain

$$\mathbf{p}_{\mathrm{esc}}(x^0 \to y^0) \;=\; \left[\frac{F(2n+3)}{4(F(2n+2)-1)}\right].$$

The right-hand side is an appreciable hyperreal less than **1**. Its shadow is 0.4045…, as can be seen by using the asymptotic expression for the Fibonacci numbers (see Appendix B). It is not infinitesimal because the random walker on $\mathbf{N}_n^0$ can go from x^0 to y^0 in two steps by passing through z^0. Thus, $\{\mathbf{N}^\nu, \langle \mathbf{N}_n^0 \rangle\}$ is transient from x^0 to y^0.

Similarly, to get the escape probability from w^1 to y^0 with respect to the same restoration sequence, we determine $c_{\mathrm{eff}}(w_n^0, y_n^0) = F(2n+1)/F(2n)$. Now, $c_{w_n^0} = 3$. Thus,

$$\mathbf{p}_{\mathrm{esc}}(w^1 \to y^0) \;=\; \left[\frac{F(2n+3)}{3F(2n+2)}\right].$$

This has the shadow 0.5393…. This is a larger escape probability because w^1 is closer to y^0 than is x^0. ♣

Some other standard results concern "times" and "transversals" [7, Secs. IX.2 and IX.3]. Let $s_{z^0}(x^0 \to y^0)$ be the *sojourn time* (that is, the expected number of occurrences) that the random walker is at node z^0 before it reaches node y^0, given that it starts at x^0; x^0 and z^0 may be the same node, counting the start as one occurrence. Then, $s_{z^0}(x^0 \to y^0) = c_{z^0} u(z^0)$ where c_{z^0} is the total conductance incident to z^0 and $u(z^0)$ is the node voltage at z^0 when a current of 1 ampere is injected into x^0 and extracted at y^0 and with y^0 taken as the ground node (i.e., $u(y^0) = 0$). In particular, $s_{x^0}(x^0 \to y^0) = c_{x^0}/c_{\mathrm{eff}}(x^0, y^0)$. Furthermore, let $\vec{b}$ be a branch with an orientation, and let $e_{\vec{b}}(x^0 \to y^0)$ be the expected difference between the number of occurrences that the random walker traverses b in the direction of b's orientation minus the number of occurrences it traverses b in the reverse direction, given that it starts at x^0 and stops when it first reaches y^0. Then, $e_{\vec{b}}(x^0 \to y^0)$ is equal to the current in b measured in the direction of b's orientation, when again 1 ampere is injected at x^0 and extracted at y^0.

These results, too, can be immediately lifted for $\{\mathbf{N}^\nu, \langle \mathbf{N}_n^0 \rangle\}$. By virtue of Lemma 6.4-1, given any two nodes x^α and y^β in $\mathbf{N}^\nu$ and their corresponding nodes x_n^0 and y_n^0 in $\mathbf{N}_n^0$, we can inject 1 ampere into x_n^0 and extract 1 ampere at y_n^0 for all sufficiently large n. Similarly, any branch b of $\mathbf{N}^\nu$ will be restored for all sufficiently large n. Also, let z^γ be another node of $\mathbf{N}^\nu$ (possibly, $z^\gamma = x^\alpha$), and let z_n^0 be its corresponding node in $\mathbf{N}_n^0$. Then, a node voltage $u(z_n^0)$ and a branch current $i_{\vec{b},n}$ will exist for all but finitely many values of n. So, we can state the following nonstandard versions for sojourn times and branch transversals.

Theorem 6.6-5. *Let x^α, y^β, and z^γ be three nodes in $\mathbf{N}^\nu$, let x_n^0, y_n^0, and z_n^0 be their images in $\mathbf{N}_n^0$ (possibly, $x^\alpha = z^\gamma$), and let $\vec{b}$ be an oriented branch in $\{\mathbf{N}^\nu, \langle \mathbf{N}_n^0 \rangle\}$. Inject 1 ampere into x^α and extract 1 ampere from y^β. Then, for the random walker starting at x^α and stopping when it first reaches y^β, the hyperreal sojourn time at z^γ is*

$$\mathbf{s}_{z^\gamma}(x^\alpha \to y^\beta) \;=\; [c_{z_n^0} u(z_n^0)] \tag{6.10}$$

and the hyperreal expected difference in b's transversal numbers with respect to b's orientation is

$$\mathbf{e}_{\vec{b}}(x^\alpha \to y^\beta) \;=\; [i_{\vec{b},n}] \tag{6.11}$$

where, with respect to $\mathbf{N}_n^0$ *and all n sufficiently large,* $u(z_n^0)$ *is the node voltage at* z_n^0 *with respect to* y_n^0 *taken as ground,* $c_{z_n^0}$ *is the total conductance incident at* z_n^0*, and* $i_{\vec{b},n}$ *is the current in* $\vec{b}$ *measured with respect to b's orientation. If* $z^\gamma = x^\alpha$*, then*

$$\mathbf{s}_{x^\alpha}(x^\alpha \to y^\beta) \;=\; \left[\frac{c_{x_n^0}}{c_{\rm eff}(x_n^0, y_n^0)}\right]. \tag{6.12}$$

Example 6.6-6. Consider again the 1-network of Fig. 6.13(a) with all branch conductances being 1 siemens, and also let $\langle \mathbf{N}_n^0 \rangle$ be the restoration sequence specified in Example 6.6-4 (see the first part of Example 6.6-2). Orient b_1 and β_1 from left to right. Then, straightforward calculations yield

$$\mathbf{s}_{w^1}(x^0 \to z^0) \;=\; \left[\frac{2}{F(2n+3)}\right], \tag{6.13}$$

$$\mathbf{s}_{x^0}(x^0 \to z^0) \;=\; \left[\frac{F(2n+2)}{F(2n+3)}\right], \tag{6.14}$$

$$\mathbf{e}_{\vec{b}_1}(x^0 \to z^0) \;=\; \left[\frac{F(2n+1)}{F(2n+3)}\right], \tag{6.15}$$

$$\mathbf{e}_{\vec{\beta}_1}(x^0 \to z^0) \;=\; \left[\frac{1}{F(2n+3)}\right]. \tag{6.16}$$

As is to be expected from the locations of the nodes and branches, (6.13) and (6.16) are infinitesimal and (6.14)and (6.15) are appreciable. The shadows of (6.14) and (6.15) are $1/\lambda_1 = 0.618\ldots$ and $1/\lambda_1^2 = 0.381\ldots$, respectively. ♣

As for the *mean return time* $h(x^0, x^0)$ and the *commute time* $c(x^0, y^0)$, we have the following for a finite network, all of whose branch conductances are 1 siemens. $h(x^0, x^0)$ is the expected number of branch transversals that a random walker makes after starting from x^0 and then returning to x^0 for the first time. For this we have $h(x^0, x^0) = 2|\mathcal{B}|/d(x^0)$, where $|\mathcal{B}|$ is the number of branches in the network and $d(x^0)$ is the degree of x^0. Also, $c(x^0, y^0)$ is the expected number of branch transversals the random walker makes in going from x^0 to y^0 and returning to x^0 for the first time. In this case, $c(x^0, y^0) = 2|\mathcal{B}|/c_{\rm eff}(x^0, y^0)$. Thus, we have the following for $\{\mathbf{N}^\nu, \langle \mathbf{N}_n^0 \rangle\}$, where $|\mathcal{B}_n|$ denotes the number of branches in $\mathbf{N}_n$, and x_n^0 and y_n^0 are the images of x^α and y^β in $\mathbf{N}_n^0$.

Theorem 6.6-7 *Assume that every branch of* $\mathbf{N}^\nu$ *has a conductance of 1 siemens. Then the hyperreal mean return time* $\mathbf{h}(x^\alpha, x^\alpha)$ *and the mean commute time* $\mathbf{c}(x^\alpha, y^\beta)$ *are given by*

$$\mathbf{h}(x^\alpha, x^\alpha) = \left[\frac{2|\mathcal{B}_n|}{d(x_n^0)}\right] \tag{6.17}$$

and

$$\mathbf{c}(x^\alpha, y^\beta) = \left[\frac{2|\mathcal{B}_n|}{c_{\text{eff}}(x_n^0, y_n^0)}\right] \tag{6.18}$$

where $|\mathcal{B}_n|$ *is the number of branches in* $\mathbf{N}_n^0$.

Example 6.6-8. Continuing Example 6.6-6, we have the hyperreals

$$\mathbf{h}(x^0, x^0) = \mathbf{h}(y^0, y^0) = [2n+3],$$

$$\mathbf{h}(w^1, w^1) = \left[\frac{4n+6}{3}\right],$$

$$\mathbf{h}(z^0, z^0) = \left[\frac{4n+6}{n+2}\right],$$

$$\mathbf{c}(x^0, y^0) = \left[4(2n+3)\frac{F(2n+2)-1}{F(2n+3)}\right].$$

♣

6.7 Appending and Inserting Branches; Buildable Graphs

Our Procedure 6.2-4 starts with a given transfinite network $\mathbf{N}^\nu$, then reduces it to a finite network by shorting and opening branches, and finally rebuilding it by restoring its electrical parameters to finitely many branches at a time. In other words, it uses the given $\mathbf{N}^\nu$ as a guide for developing a sequence of finite networks that fill out $\mathbf{N}^\nu$ and moreover uses its electrical parameters in an essential way.

All this suggests that there may be a strictly graph-theoretic approach that builds transfinite graphs from finite ones simply by "appending" and "inserting" branches (as was suggested in Examples 6.1-1 and 6.1-2) without making any use of electrical parameters. This is truly so. To explain this, we should define terms more explicitly. We shall define "appending' and "inserting" branches to any transfinite graph $\mathcal{G}^\nu$ (not just to finite ones). So, let $\mathcal{G}^\nu$ be given.

The first operation, "appending a branch," is the inverse operation of removing a branch. The *appending* is done by first creating b as a new branch; then, each tip of b is either joined to a node of $\mathcal{G}^\nu$ as an additional embraced elementary tip or is made an elementary tip of a newly created maximal 0-node. If b is a self-loop, both of its tips are appended to one maximal node of $\mathcal{G}^\nu$, or the newly created maximal 0-node has only those two tips. If b is not a self-loop, then either a tip of b is appended to a maximal node of $\mathcal{G}^\nu$ or a new singleton 0-node is created that contains that tip, and similarly for the other tip of b. Thus, we may have b's tips appended to existing nodes of $\mathcal{G}^\nu$, or one tip might be appended to an existing node and the other contained in a newly created singleton 0-node, or both might be in two newly created singleton 0-nodes—one tip to each. Thus, appending a branch is the inverse operation of removing a branch (see Sec. 6.2 for the definition of the latter).

The second operation will be called *inserting* a branch b to $\mathcal{G}^\nu$. To insert a newly created branch b into $\mathcal{G}^\nu$, choose a nonsingleton maximal node x^γ, partition its embraced tips into two nonempty sets to obtain two nodes x^α and x^β, and then join one elementary tip of b to x^α and the other to x^β. Thus, inserting a branch is the inverse operation of contracting a branch (see Sec. 6.2 again for the definition of the latter).

A transfinite graph can be built by starting from a finite 0-graph $\mathcal{H}_0^0$ and appending or inserting branches, one branch at a time, to get an "increasing" sequence $\langle \mathcal{H}_j^0 : j \in \mathbb{N} \rangle$ of finite 0-graphs. Here, j also numbers the branches as they are introduced, and in this sense $\langle \mathcal{H}_j^0 : j \in \mathbb{N} \rangle$ *increases*. This construction need not lead to a transfinite graph because no nonsingleton node of rank greater than 0 need arise. On the other hand, it might produce a transfinite graph, as Example 6.1-2 suggests. Let us call a transfinite graph *buildable* if it can be obtained from a finite graph by appending and/or inserting branches infinitely often but only finitely many at a time. A result that seems apparent is that "the graph of any restorable network is buildable." This is indeed so, but there are some matters that should be proven. Those details are presented in [56, Sec. 10].

Building a graph without any guidance provided by the restoration of a given transfinite graph leaves the result indeterminate. We are again faced with the ancient Aristotelian dichotomy of a potential infinity and an actual infinity. No matter how finitely many steps we take in building a graph, we still have on hand only a finite graph. The infinite process must be finished[14] before a transfinite graph is achieved, as will certainly be the case when restoring a restorable one. What may arise from an arbitrary but finished infinite building process? Can we get something more than what is presently available from our theory of transfinite graphs? Let us consider this more closely.

There does not seem to be any restriction on the countable ranks that can be achieved through an appropriate building process—or is there? Can we obtain any countable rank, such as a rank that is a nameless and unidentified countable ordinal larger than any specific countable ordinal that has heretofore been examined [36]? We may be able to reach higher ranks by appending and inserting uncountably many branches to transfinite graphs (rather than finite ones) at each step of the building process. After all, the appending and inserting of branches were defined above on any transfinite graph. For an example of the appending of uncountably many branches

[14] We say "finished" instead of "completed" because "to complete" has another technical meaning for metric spaces, which was used to create transfinite networks in a different way [13], [54, Chapter 4].

all at once, consider the infinite binary tree; it has uncountably many 0-tips, each of which can be made the sole member of a singleton 1-node; then, branches can be appended to those 1-nodes simultaneously, one branch to each 1-node. Furthermore, a transfinite path, rather than a single branch, can be inserted all at once into a transfinite node by first partitioning the node. So, can we generate a sequence of buildable graphs whose countable ranks increase unboundedly? Can we in this way obtain transfinite graphs of arbitrarily large countable ranks? More puzzlingly, what about an $\aleph_1$-graph? Still further, can a transfinite graph of any uncountable rank be constructed or defined somehow? We are groping through uncharted territory with these speculations.

6.8 Other Ideas: Nonstandard Graphs and Networks

So far, we have used nonstandard analysis only to determine a hyperreal operating point in a given restorable transfinite network $\mathbf{N}^\nu$. This is done by choosing an expanding sequence $\langle \mathbf{N}_n^0 \rangle$ of finite 0-networks $\mathbf{N}_n^0$ that fills out $\mathbf{N}^\nu$, and the hyperreal operating point depends upon the choice of $\langle \mathbf{N}_n^0 \rangle$, a fact suggested by the symbols $\{\mathbf{N}^\nu, \langle \mathbf{N}_n^0 \rangle\}$ used in place of $\mathbf{N}^\nu$.

However, we might consider a substantially different use of nonstandard analysis that generates "nonstandard graphs and networks," as follows. Given any conventionally infinite 0-graph $\mathcal{G}^0$, we let its 0-nodes along with the real numbers be the individuals. From them we construct nonstandard 0-nodes as equivalence classes of sequences of 0-nodes in $\mathcal{G}^0$ (modulo a given nonprincipal ultrafilter). The adjacencies between 0-nodes in $\mathcal{G}^0$ determined by the branches determine in turn nonstandard branches. These results then yield a "nonstandard graph" ${}^*\mathcal{G}^0$ as an enlargement of $\mathcal{G}^0$. Thus, each 0-graph has its own enlargement ${}^*\mathcal{G}^0$. (All this is roughly analogous to the enlargement of the set of natural numbers into the set of hypernatural numbers; see Appendix A15.) We do all this in Chapter 8 using ultrapower constructions. Moreover, using the transfer principle, we obtain nonstandard versions of several properties of standard graphs.

More generally, some of these ideas can be extended to "nonstandard transfinite graphs" by constructing nonstandard versions of transfinite tips and thereby nonstandard transfinite nodes. Still more generally, in place of a single ν-graph, we can start with a sequence $\langle \mathcal{G}_n^{\nu_n} \rangle$ of ν_n-graphs $\mathcal{G}_n^{\nu_n}$, and then consider sequences of nodes $\langle x_n^{\rho_n} \rangle$, where $x_n^{\rho_n}$ $(\rho_n \leq \nu_n)$ is a ρ_n-node in $\mathcal{G}_n^{\nu_n}$, to generate thereby nonstandard nodes $\mathbf{x}^\rho$ of rank ρ (i.e., $\rho = [\rho_n]$). Such is done in Sec. 8.8, but only at the first rank of transfiniteness.

Furthermore, by replacing the transfinite graphs $\mathcal{G}_n^{\nu_n}$ by transfinite electrical networks $\mathbf{N}_n^{\nu_n}$, the nonstandard branches resulting from the ultrapower construction acquire hyperreal branch parameters, and the corresponding nonstandard network acquires a "nonstandard operating point" under certain circumstances. This done in Sec. 8.9, but again for the first rank of transfiniteness only.

We should also mention some earlier works on nonstandard graphs and networks. In the book [24] (see Sec. 19.1) R. Goldblatt obtained a nonstandard graph $\{{}^*V, {}^*E\}$

as an enlargement of a conventional graph $\{V, E\}$ defined as a set V of vertices along with a set E of edges determined by an irreflexive symmetric relation of V. The transfer principle was then used to establish in a nonstandard way the standard theorem that, if every finite subgraph of a conventionally infinite graph G has a coloring with k many colors (k a fixed natural number), then G itself can be colored with k colors.

A very different kind of nonstandard graph has been constructed by Thayer [40]. His work is based on certain metric spaces satisfying a polynomial growth condition on the volume of balls. Nonstandard analysis is used to show that such spaces can be viewed as parts of hyperfinite graphs whose branches have infinitesimal lengths, that is, whose adjacent nodes are infinitesimally close.

Another early work [52] examined electrical resistive networks consisting of finitely many one-ports, each being internally a conventionally infinite or transfinite network and having a hyperreal Thevenin or Norton representation. The standard formula for a nodal analysis of a finite network was lifted into the hyperreal realm and made applicable to any so-structured network.

Apart from graphs, we can use the transfer principle to enlarge distributed transmission lines and cables and derive their hyperreal-valued transients as internal functions on ${}^*\mathbb{R}_+ \times {}^*\mathbb{R}_+$. This is done in Sec. 7.6.

Let us emphasize that these possible nonstandard generalizations of graphs and networks are different from what we have been doing in this chapter.

7

Hyperreal Transients in Transfinite RLC Networks

All prior analyses of transfinite electrical networks, as in Chapters 5 and 6 and in [50], [51], [54], have been restricted to purely resistive networks. That is, the branch parameters have only been (linear or nonlinear) resistors along with voltage and current sources. However, the technique employed in Chapter 6, whereby the transfinite network $\mathbf{N}^{\nu}$ is viewed as the end result of a sequence of finite networks that fill out $\mathbf{N}^{\nu}$, now makes it possible to analyze transfinite electrical networks whose branch parameters are also allowed to be inductors and capacitors. This is certainly so if the resistors, inductors, and capacitors are linear and positive. By an RLC *network* we will mean an electrical network whose parameters are so restricted.

As before, when the transfinite RLC network is restorable and viewed as the result of a restoration sequence $\langle \mathbf{N}_n^0 \rangle$ of finite RLC networks $\mathbf{N}_n^0$, we again denote it by $\mathbf{N}_r^{\nu} = \{\mathbf{N}^{\nu}, \langle \mathbf{N}_n^0 \rangle\}$. The transient responses of the $\mathbf{N}_n^0$ on the real time line $\mathbb{R}_+$ together determine a hyperreal-valued transient response on the hyperreal time line $^*\mathbb{R}+$, as we shall see in Sec. 7.2. As special cases, we determine the hyperreal transients on transfinite ladder networks whose series branches contain resistors and inductors and whose shunt elements are capacitors and possibly conductors. This is the so-called "artificial line," which is the discrete version of a transmission line occurring when the distributed parameters of the latter are approximated by the lumped parameters of the former. These results are presented in Secs. 7.4 and 7.5. We show therein that an artificial wave on the artificial line with no shunt conductors—or alternatively an artificial diffusion on an artificial cable—can "pass through infinity" and penetrate transfinite extensions of the line with appreciable values during unlimited hyperreal times.

In the latter half of this chapter, we turn our attention to distributed lines, the better model of a transmission line, and again examine the transfinite versions of it. We take it that the transmission line extends "beyond infinity" in a manner specified in Secs. 7.7 and 7.9. We determine how a hyperreal traveling wave in the case of a transmission line or a hyperreal diffusion in the case of a cable can pass "through infinity" to produce hyperreal voltages within the transfinite extensions of the line or cable. The difficulty that must be overcome in this case is that the techniques that worked for lumped networks must be modified in a substantial way in order to make them applicable to distributed lines and cables. We exploit the fact that the voltages along a conventional, one-way infinite line or cable are known. So, we represent the transfinite line or cable as the end result of an expanding sequence of conventionally infinite lines or cables that "fill out" the transfinite line or cable. This is accomplished by assigning sample points along the transfinite line or cable and then truncating it by removing infinite parts between some of the sample points to obtain a conventionally infinite line or cable. Then, upon sequentially restoring the removed parts in steps, we can obtain a sequence of time-varying voltages at each sample point, which in turn can be identified as a hyperreal voltage variation at that sample point. We thus obtain hyperreal voltage transients at all the sample points of the transfinite line or cable.

In the same way, we can determine hyperreal current transients at the sample points, but we skip doing this since the technique is exactly the same.

7.1 Hyperreal Transients on the Hyperreal Time Line

As always, $\mathbb{R}_+ = \{t \in \mathbb{R} : t \geq 0\}$ denotes the nonnegative part of $\mathbb{R}$, and ${}^*\mathbb{R}_+$ denotes its enlargement (see Appendix A15). We take $\mathbb{R}_+$ to be the *conventional time line* (starting at $t = 0$), and ${}^*\mathbb{R}_+$ to be the *hyperreal time line*; $\mathbf{t} = [t_n] \in {}^*\mathbb{R}_+$ will be *hyperreal time*. Thus, the hyperreal time line starts at 0, passes through the infinitesimals, then through the appreciable hyperreals (whose shadows comprise the positive conventional time line), and finally through the unlimited hyperreals. Both ${}^*\mathbb{R}$ and ${}^*\mathbb{R}_+$ are internal sets (see Appendix A12).

Next, let $\langle f_n : n \in \mathbb{N} \rangle$ be a sequence of standard (i.e., conventional, real-valued) functions mapping $\mathbb{R}_+$ into $\mathbb{R}$. This is a representative of the internal function $\mathbf{f} = [f_n]$ defined on ${}^*\mathbb{R}_+$ by

$$\mathbf{f}(\mathbf{t}) = [f_n(t_n)], \quad \mathbf{t} = [t_n] \in {}^*\mathbb{R}_+.$$

(See Appendix A14.) We will view such an $\mathbf{f}$ as a *hyperreal transient* defined on the hyperreal time line.

As an example, let

$$f_0(t) = 1, \quad f_n(t) = 1 - e^{-t/n}, \quad t \in \mathbb{R}_+, \quad n = 1, 2, 3, \ldots$$

For $n > 0$, each f_n is a strictly increasing function with $f_n(0) = 0$ and $\lim_{t\to\infty} f_n(t) = 1$. Then, for $\mathbf{f} = [f_n]$, we have $\mathbf{f}(\mathbf{t}) = [1 - e^{-t_n/n}]$. The hyperreal function $\mathbf{f}$ is strictly increasing; indeed, if $\mathbf{t}_1 = [t_{1,n}] < \mathbf{t}_2 = [t_{2,n}]$, we have $t_{1,n} < t_{2,n}$ a.e. and $1 - e^{-t_{1,n}/n} < 1 - e^{-t_{2,n}/n}$ a.e., and thus $\mathbf{f}(\mathbf{t}_1) < \mathbf{f}(\mathbf{t}_2)$. Furthermore, $\mathbf{f}(\mathbf{t})$ is a positive infinitesimal when $\mathbf{t} = [t_n]$ is positive and limited; indeed, $0 < t_n < T$ a.e. for some fixed real $T \in \mathbb{R}_+$, and therefore

$$0 < f_n(t_n) < 1 - e^{-T/n} \to 0, \quad n \to \infty.$$

Thus, on the limited hyperreal time line, $\mathbf{f}(\mathbf{t})$ increases but remains infinitesimal. However, on the unlimited hyperreal time line, $\mathbf{f}(\mathbf{t})$ continues to increase and eventually increases through limited hyperreals but remains less than 1; indeed, we can choose t_n such that t_n/n approaches any positive real τ so that $1-e^{-t_n/n}$ approaches $1-e^{\tau}$. Furthermore, if $t_n/n \to \infty$, then $1-e^{-t_n/n} \to 1$ so that $\mathbf{f}(\mathbf{t})$ is in the halo of $\mathbf{1}$ but remains less than $\mathbf{1}$.

7.2 Hyperreal Transients in Restorable RLC Networks

We start with a restorable transfinite RLC ν-network $\mathbf{N}^\nu$ with a chosen restoration sequence $\langle \mathbf{N}_n^0 \rangle$ of finite networks $\mathbf{N}_n^0$. The branch parameters of $\mathbf{N}^\nu$ are resistors, inductors, capacitors (all linear and positive) along with independent voltage or current sources. We allow neither dependent sources nor mutual coupling between branches.

Then, with the sources being functions of time $t \in \mathbb{R}_+$ and with initial conditions assigned to the inductor currents and capacitor voltages, we obtain a transient regime of real-valued currents and voltages in each $\mathbf{N}_n^0$. Thus, for each branch b and each $t \in \mathbb{R}_+$, we have a sequence $\langle i_{b,n}(t) \colon n \in \mathbb{N} \rangle$ of branch currents, where $i_{b,n}(t)$ is the current in b as a branch in $\mathbf{N}_n^0$. For any real $t \geq 0$, $[i_{b,n}(t)]$ is a hyperreal current in b with $\langle i_{b,n}(t) \colon n \in \mathbb{N} \rangle$ being a representative sequence for it. (If b has not yet been restored in $\mathbf{N}_n^0$, we can assign any value to $i_{b,n}(t)$ without affecting the hyperreal current for b.)

More generally, we can take time to be hyperreal as well: $\mathbf{t} = [t_n] \in {}^*\mathbb{R}_+$. Then, $\mathbf{i}_b(\mathbf{t}) = [i_{b,n}(t_n)] \in {}^*\mathbb{R}_+$ is a hyperreal current for the chosen $\mathbf{t} \in {}^*\mathbb{R}_+$; it is defined even for unlimited $\mathbf{t}$. Also, $\mathbf{i}_b = [i_{b,n}]$ is an internal function mapping $\mathbf{t} = [t_n] \in {}^*\mathbb{R}_+$ into $\mathbf{i}(\mathbf{t}) = [i_{b,n}(t_n)] \in {}^*\mathbb{R}$. Hence, $\mathbf{i}_b(\mathbf{t})$, where $\mathbf{t} \in {}^*\mathbb{R}_+$, is well-defined. In fact, $\mathbf{i}_b$ is an internal function defined on any internal subset of ${}^*\mathbb{R}_+$, such as the set $A = \{\mathbf{t} = [t_n] \colon 0 \leq t_n \leq \epsilon_n\}$ where ϵ_n is any positive real number for each n. In short, $\mathbf{i}_b \colon \mathbf{t} \mapsto \mathbf{i}_b(\mathbf{t})$ is a *hyperreal-valued transient* mapping the hyperreal time line ${}^*\mathbb{R}_+$ into the set ${}^*\mathbb{R}$ of hyperreals.

In the same way, we have a hyperreal-valued transient for each branch voltage and for each node voltage with respect to a chosen ground node. The set of all these hyperreal transients is a *hyperreal transient regime* for $\mathbf{N}_r^\nu = \{\mathbf{N}^\nu, \langle \mathbf{N}_n^0 \rangle\}$.

In the rest of this paper, we shall illustrate these ideas by examining an artificial (i.e., lumped parameter) RC cable and an artificial RLC transmission line that extend transfinitely ("spatially beyond infinity," so to speak). We will find that hyperreal-valued artificial diffusions and waves can "penetrate beyond infinity" and can pass on to transfinite extensions of artificial cables and lines during unlimited hyperreal times with appreciable hyperreal values.

7.3 A Transfinite RLC Ladder

Let us now consider a transfinite ladder in the form of Fig. 7.1(a), where now every branch is a one-port consisting internally of finitely many resistors, inductors, and/or capacitors, the series branches all being the same and the shunt branches all being the same but in general different from the series branches. The only source is a voltage source $e(t)$, $t \in \mathbb{R}_+$, at the input of the ladder. The initial currents and voltages on the

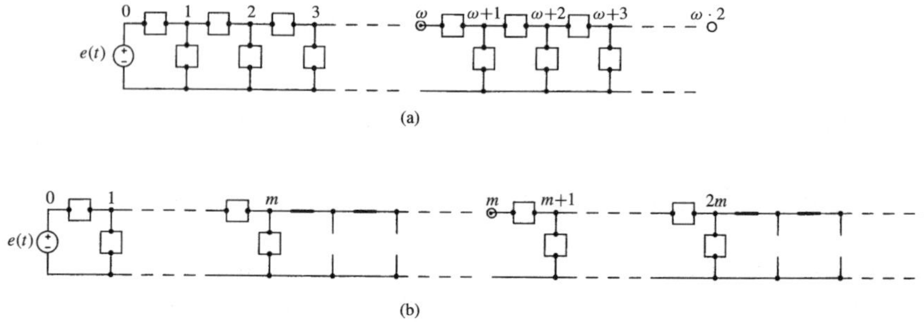

Fig. 7.1. (a) A transfinite network consisting of two, one-way infinite, uniform ladders in cascade driven at the input by a voltage source $e(t)$. The boxes denote 1-ports, each containing finitely many inductors, capacitors and resistors. The series boxes are all the same, and the shunt boxes are all the same. The small circles denote 1-nodes. The 1-node in the middle connects the infinite extremity of the first ladder to the input of the second ladder. The upper nodes are indexed first by the natural numbers 0, 1, 2, ... and then by the transfinite ordinals $\omega, \omega+1, \omega+2, \ldots, \omega \cdot 2$. (b) A finite network obtained by shorting and opening branches in the same way in both ladders.

inductors and capacitors respectively are assumed to be 0. Upon applying the Laplace transformation, we obtain the transformed circuit of Fig. 7.2(a), where Z denotes an impedance $s \mapsto Z(s)$, Y denotes an admittance $s \mapsto Y(s)$, and $E : s \mapsto E(s)$ is the transformed input voltage; here, s is a complex variable. Later on, we set $x(s) = Z(s)Y(s)$ but drop the argument "(s).". By an *el-section* we will mean a series Z followed by a shunt Y.

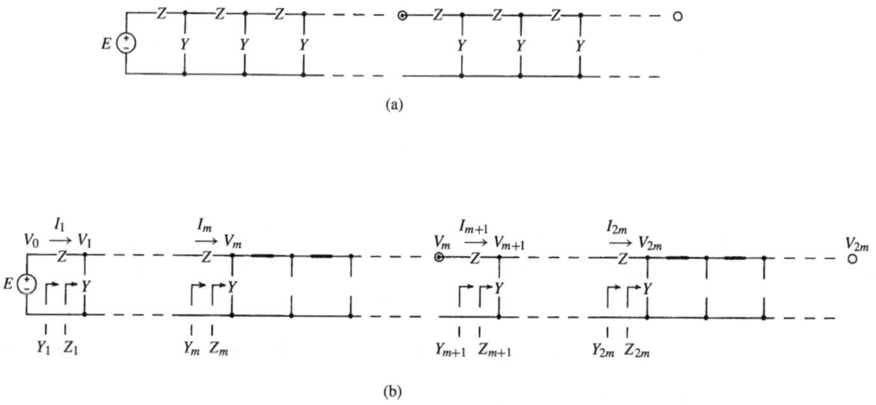

Fig. 7.2. (a) The Laplace-transform-domain representation of the RLC network of Fig. 1(a). Z and Y are the impedance and admittance of the series and shunt boxes, respectively. (b) The finite network having the same graph as that of Fig. 1(b). Each V_j and I_j is the Laplace transform of the time-dependent node voltage and branch current. The Y_j and Z_j are driving-point admittances and impedance, respectively, for the networks to the right of the places where the Y_j and Z_j appear.

To obtain a particular restoration sequence $\langle \mathbf{N}_n^0 \rangle$ for this network, we short and open branches as shown in Fig. 7.1(b) to obtain a finite ladder and then restore el-sections two at a time, one for each conventional ladder, proceeding from left to right in each ladder. One stage of this restoration process is shown in Fig. 7.2(b), where the natural numbers $j = 0, 1, \ldots, m, m+1, \ldots, 2m$ serve as indices for the nodes that have restored incident branches.

Given $V_0 = E$, we wish to determine the resulting node voltages V_j. Perhaps the easiest way of doing this is to set $V_{2m} = 1$, compute the V_j working from right to left using Kirchhoff's and Ohm's law to get finally V_0, and then multiply all the obtained V_j by E/V_0 to get the node voltages corresponding to the input voltage E. In the following, I_j is the current in the series impedance flowing from node $j-1$ to node j. So, upon setting $V_{2m} = 1$ and $x = ZY$, we get (with $Q_0 = 1$),

$$\begin{aligned}
I_{2m} &= YV_{2m} = Y = YQ_0,\\
V_{2m-1}(x) &= ZY + 1 = x + 1 = P_1(x),\\
I_{2m-1}(x) &= YV_{2m-1}(x) + I_{2m}(x) = (x+2)Y = YQ_1(x),\\
V_{2m-2}(x) &= ZI_{2m-1}(x) + V_{2m-1}(x) = x^2 + 3x + 1 = P_2(x),\\
I_{2m-2}(x) &= YV_{2m-2}(x) + I_{2m-1}(x) = (x^2 + 4x + 3)Y = YQ_2(x).
\end{aligned}$$

Continuing in this fashion, we obtain the following recursive equations, wherein $k = 1, 2, \ldots, 2m$:

$$\begin{aligned}
P_0(x) &= 1, \quad Q_0(x) = 1,\\
P_k(x) &= xQ_{k-1}(x) + P_{k-1}(x),\\
Q_k(x) &= P_k(x) + Q_{k-1}(x),\\
V_{2m} &= 1, \quad I_{2m} = Y,\\
V_{2m-k}(x) &= P_k(x),\\
I_{2m-k}(x) &= YQ_k(x).
\end{aligned}$$

Both P_k and Q_k are polynomials in x of degree k:

$$\begin{aligned}
P_k(x) &= p_{k,0} + p_{k,1}x + \cdots + p_{k,k-1}x^{k-1} + p_{k,k}x^k,\\
Q_k(x) &= q_{k,0} + q_{k,1}x + \cdots + q_{k,k-1}x^{k-1} + q_{k,k}x^k.
\end{aligned}$$

It follows from these recursive equations that the first and last coefficients are $p_{k,0} = 1$, $p_{k,k} = 1$, $q_{k,0} = k+1$, and $q_{k,k} = 1$. (Values for other coefficients up to $k = 10$ are listed in Tables 1 and 2 of [49, page 605].) Thus, with $j = 0, 1, \ldots, 2m$ being the indices for the nodes as in Fig. 7.2(b), we have $j = 2m - k$ (with $k = 0$ when $j = 2m$). Then, the voltage transfer function from node 0 to node j is

$$\begin{aligned}
\frac{V_j(x)}{E} &= \frac{V_{2m-k}(x)}{V_0(x)} = \frac{P_k(x)}{P_{2m}(x)}\\
&= \frac{1 + p_{k,1}x + \cdots + p_{k,k-1}x^{k-1} + x^k}{1 + p_{2m,1}x + \cdots + p_{2m,2m-1}x^{2m-1} + x^{2m}},
\end{aligned} \tag{7.1}$$

and the transfer admittance from the input voltage to the current in the jth series branch is

$$\frac{I_j(x)}{E} = \frac{I_{2m-k}(x)}{V_0(x)} = \frac{YQ_k(x)}{P_{2m}(x)} = \frac{Y(k+1+q_{k,1}x+\cdots+q_{k,k-1}x^{k-1}+x^k)}{1+p_{2m,1}x+\cdots+p_{2m,2m-1}x^{2m-1}+x^{2m}}.$$

7.4 A Transfinite Artificial Cable

As our first example of a hyperreal transient on a transfinite RLC network, we examine a transfinite artificial cable consisting of two conventional one-way-infinite artificial cables in cascade, one being a transfinite extension of the other, as shown in Fig. 7.2(a). Each series element is a resistor of value r, and each shunt element is a capacitor of value c. Thus, $x(s) = rcs$. Let us assume that the source at the input provides a unit-step of voltage ($e(t) = 1$ for $t \geq 0$ and $e(t) = 0$ for $t < 0$), so that $E(s) = 1/s$. Consider the standard voltage $v_j(t)$ at node j ($1 \leq j \leq 2m$) for the truncated finite ladder of Fig. 7.2(b). It follows from (7.1) and the initial-value theorem that the node voltages have, for $t \to 0+$, the asymptotic values

$$v_j(t) \sim \frac{1}{(rc)^j}\cdot\frac{t^j}{j!}, \qquad j = 0, \ldots, 2m. \tag{7.2}$$

In fact, these asymptotic expressions are also bounds on the $v_j(t)$ for all $t \in \mathbb{R}_+$ [60]:

$$|v_j(t)| \leq \frac{1}{(rc)^j}\cdot\frac{t^j}{j!}. \tag{7.3}$$

For each j, this bound holds for all truncations of the transfinite ladder wherein the last capacitors restored are at nodes j and $j+m$, where now $1 \leq j \leq m$. Furthermore, all the poles of $V_j(s)$ are real and negative [43, page 332] except for a simple pole at the origin due to the factor $E(s) = 1/s$. Consequently, we may use (7.1) and the final-value theorem to write $\lim_{t\to\infty} v_j(t) = 1$ (which indeed is physically obvious).

Next, we argue that $v_j(t)$ is strictly monotonically increasing for all $t > 0$. Indeed, the voltage transfer function $V_j(s)/E(s)$ in the Laplace-transform domain can be written as a product of the driving-point admittances and impedances, shown in Fig. 7.2(b), measured toward the right:

$$\frac{V_j}{E} = \frac{I_1}{E}\cdot\frac{V_1}{I_1}\cdot\frac{I_2}{V_1}\cdot\frac{V_2}{I_2}\cdots\frac{I_j}{V_{j-1}}\cdot\frac{V_j}{I_j} = Y_1Z_1Y_2Z_2\cdots Y_jZ_j. \tag{7.4}$$

For each index i ($1 \leq i \leq j$), all the poles and zeros of Z_i are simple and alternate along the nonpositive real axis, with a pole at the origin and a zero at infinity [43, page 412]. As for the product Y_iZ_i, we have

$$Y_iZ_i = \frac{1}{1+\frac{r}{Z_i}},$$

from which it follows that all the poles and zeros of $Y_i Z_i$ are simple and alternate along the negative real axis; closest to the origin is a pole, not a zero. Also, $Y_i(s)Z_i(s)$ tends to 0 as $|s| \to \infty$. Thus, the residues of the poles are positive, and the unit-impulse response of $Y_i Z_i$ is a finite sum of the form $\sum_{k=1}^{K} a_k\, e^{-p_k t}$, where $a_k > 0$ and $p_k > 0$ for all k. So, upon applying the convolution integral repeatedly according to the inverse Laplace transform of (7.4), we can conclude that the unit-impulse response is positive for all $t > 0$. It follows that the unit-step response for $v_j(t)$ (i.e., with $E(s) = 1/s$) is continuous and strictly monotonically increasing, as asserted.

This is also true for the voltage $v_{j,\infty}(t)$ at the jth node of a one-way-infinite artificial RC cable; namely, $v_{j,\infty}(t)$ is a continuous, strictly monotonically increasing function, starting at $t = 0$ and approaching 1 as $t \to \infty$ (see Appendix C). Moreover, [2, Formula 9.6.7] implies that $v_{j,\infty}(t)$ has the asymptotic form (7.2).

Let us now consider the hyperreal voltage transient at a fixed node for the restoration $\{\mathbf{N}^{\nu}, \langle \mathbf{N}_n^0 \rangle\}$ of the transfinite artificial RC cable resulting from the restoration of el-sections two at a time, one for each of the two conventional ladders. That is, the restoration proceeds through the finite ladders of Fig. 7.2(b) toward the transfinite ladder of Fig. 7.2(a). Let us number these restoration stages by $n = 0, 1, 2, \ldots$. Thus, $n = m$, and we have $2n = 2m$ el-sections in Fig. 7.2(b). When $n = 0$, we have only the source $e(t)$, which again we take to be a unit-step of voltage.

Going to the hyperreal case corresponding to the stated restoration process, let $\mathbf{v}_j(\mathbf{t})$ be the hyperreal transient voltage at the jth node in the first ladder. As $\mathbf{t}$ increases through ${}^*\mathbb{R}_+$, $\mathbf{v}_j(\mathbf{t})$ does too; indeed, for $0 < \mathbf{t}_1 = [t_{1,n}] < \mathbf{t}_2 = [t_{2,n}]$, we have $t_{1,n} < t_{2,n}$ a.e. and $v_{j,n}(t_{1,n}) < v_{j,n}(t_{2,n})$ a.e., so that

$$0 \;<\; \mathbf{v}_j(\mathbf{t}_1) \;=\; [v_{j,n}(t_{1,n})] \;<\; [v_{j,n}(t_{2,n})] \;=\; \mathbf{v}_j(\mathbf{t}_2). \tag{7.5}$$

When $\mathbf{t}$ increases through the positive infinitesimals, $\mathbf{v}_j(\mathbf{t})$ also remains infinitesimal. This follows directly from (7.3). Indeed, for any positive infinitesimal $\mathbf{t} = [t_n]$, we have

$$|\mathbf{v}_j(\mathbf{t})| \;=\; [|v_{j,n}(t_n)|] \;\le\; \frac{1}{(rc)^j\, j!}[t_n^j].$$

The right-hand side is infinitesimal.

On the other hand, as soon as $\mathbf{t}$ becomes appreciable, $\mathbf{v}_j(\mathbf{t})$ becomes appreciable, too. This follows from (7.2). Indeed, we have

$$v_j(t) \;\ge\; \frac{t^j}{2(rc)^j\, j!}$$

for all $t \in \mathbb{R}_+$ sufficiently small. Thus, for $\mathbf{v}_j(\mathbf{t}) = [v_{j,n}(t_n)]$ and $\mathbf{t} = [t_n]$, as $\mathbf{t}$ increases through all the positive infinitesimals, so too does $\mathbf{v}_j(\mathbf{t})$ because

$$\mathbf{v}_j(\mathbf{t}) \;\ge\; \frac{\mathbf{t}^j}{2(rc)^j\, j!}$$

when $\mathbf{t}$ is infinitesimal. By the overflow principle (see Appendix A16), $\mathbf{v}_j(\mathbf{t})$ achieves appreciable values whenever $\mathbf{t}$ does. Indeed, let $\mathbf{t}$ increase throughout the internal interval $[I_n]$, where $I_n = \{t_n : 0 \leq t_n \leq \epsilon\}$ for each n and where ϵ is any fixed positive real number. Then, $\mathbf{v}_j([I_n])$ is an internal interval containing all the infinitesimals, and by overflow, $\mathbf{v}_j([I_n])$ also contains appreciable values. Since this is so for every positive real ϵ, no matter how small, our assertion follows.

Furthermore, $\mathbf{v}_j(\mathbf{t})$ is infinitesimally close to 1 (more specifically, $1 - \mathbf{v}_j(\mathbf{t})$ is a positive infinitesimal), for all sufficiently large unlimited $\mathbf{t}$. Indeed, let us choose $\mathbf{w} = [w_n]$ as a positive infinitesimal. Then, because of the continuity and strictly increasing monotonicity of $v_{j,n}$, there is, for each n, a unique T_n such that $1 - v_{j,n}(T_n) = w_n$, and $0 < 1 - v_{j,n}(t_n) \leq 1 - v_{j,n}(T_n)$ for all $t_n \geq T_n$. So, for all $\mathbf{t} = [t_n] \geq \mathbf{T} = [T_n]$, we have $0 < 1 - \mathbf{v}_j(\mathbf{t}) \leq 1 - \mathbf{v}_j(\mathbf{T}) = \mathbf{w}$, as asserted.

Altogether, the hyperreal transient $\mathbf{v}_j(\mathbf{t})$ behaves much like the standard transient $v_{j,\infty}(t)$.

However, for a fixed node in the second ladder, we have a different situation. Let that be the pth node therein ($0 \leq p \leq \omega$). With respect to the node numbering shown in Fig. 7.1(a), that node's index is $\omega + p$. As $\mathbf{t}$ increases, so too does $\mathbf{v}_{\omega+p}(\mathbf{t})$ (replace j by $\omega + p$ in (7.5)). As $\mathbf{t}$ increases first through the infinitesimals and then through all the appreciable hyperreals, the hyperreal voltage $\mathbf{v}_{\omega+p}(\mathbf{t})$ remains infinitesimal, even for all appreciable $\mathbf{t}$. This, too, follows from (7.3) because now the number of el-sections preceding the node $\omega + p$ increases indefinitely during the restoration process. Thus, for any appreciable $\mathbf{t} = [t_n]$, there is a $T \in \mathbb{R}_+$ such that $0 < t_n < T$ a.e. Then,

$$0 \leq v_{\omega+p,n}(t_n) \leq \frac{1}{(rc)^{n+p}} \cdot \frac{T^{n+p}}{(n+p)!}$$

for almost all n. The right-hand side tends to 0 as $n \to \infty$, whence our assertion.

On the other hand, as $\mathbf{t} = [t_n]$ increases through the unlimited hyperreals, $\mathbf{v}_{\omega+p}(\mathbf{t})) = [v_{\omega+p,n}(t_n)]$ remains infinitesimal for a while but then increases through appreciable values, eventually getting infinitesimally close to 1 but remaining less than 1. For instance, for $\mathbf{t} = [t_n]$ and $n > 1$, we can choose

$$\frac{t_n}{rc} = ((n+p-1)!)^{1/(n+p)} \tag{7.6}$$

so that $\mathbf{t} = [t_n]$ is unlimited. Then, by (7.3) again and with ω replaced by n at the nth truncation, we have

$$|v_{\omega+p,n}(t_n)| \leq \frac{(n+p-1)!}{(n+p)!} = \frac{1}{n+p} \to 0, \quad n \to \infty.$$

So, $\mathbf{v}_{\omega+p}(\mathbf{t}) = [v_{\omega+p,n}(t_n)]$ is infinitesimal. On the other hand, this choice of the t_n corresponds to an unlimited $\mathbf{t} = [t_n]$, as can be seen from Stirling's asymptotic formula for the gamma function [2, page 257]. Indeed, the right-hand side of (7.6) is asymptotic to $(n+p)/e$ as $n \to \infty$.

However, for sufficiently large unlimited $\mathbf{t}$, $\mathbf{v}_{n+p}(\mathbf{t})$ becomes appreciable. To see this, let $u \in \mathbb{R}_+$ be such that $0 < u < 1$. Then, for each n, there is a unique t_n such that $v_{\omega+p,n}(t_n) = u$. By what we have already shown, $\mathbf{t} = [t_n]$ must be unlimited. Also, $\mathbf{v}_{n+p}([t_n]) = [u]$ is appreciable. Furthermore, if we replace u by $1 - w_n$ with $\mathbf{w} = [w_n]$ being a positive infinitesimal, we can argue as before to see that $\mathbf{v(t)} = [v_{\omega+p,n}(t_n)]$ is infinitesimally close to $\mathbf{1}$ but less than $\mathbf{1}$ for all sufficiently large unlimited $\mathbf{t}$.

To summarize all this heuristically, let us first note that voltage transmission along an artificial RC cable corresponds to a discrete version of diffusion. So, upon applying a unit-step of voltage at the input to the transfinite artificial cable, we have, for appreciable values of time, appreciable values of voltage artificially diffusing throughout the first ladder, no matter how small appreciable time may be. However, the second ladder (the transfinite extension) has only infinitesimal voltages, no matter how large appreciable time may be and even for some initial unlimited values of $\mathbf{t}$. It is only when hyperreal time becomes sufficiently unlimitedly large that appreciable voltage diffuses into the second ladder. Finally, at each node of both ladders, the voltages become infinitesimally close to 1 for all sufficiently large unlimited time.

By using more elaborate truncations, we can derive similar results for transfinite ladders that are cascades of many, conventionally infinite, artificial RC cables, even for higher ranks of transfiniteness. Then, for any node, we need merely choose $[t_n]$ as a sufficiently large unlimited hyperreal in order to get an appreciable voltage.

7.5 A Transfinite Artificial Transmission Line

The arguments needed in this section are quite similar to those of the preceding one, and therefore we will merely summarize most of them.

For this second example, we first convert the artificial cable just considered into an artificial transmission line by inserting an inductor l in series with each resistor r. Thus, we now have that $x(s) = (ls + r)cs$. With a unit-step of voltage applied at the input so that $E(s) = 1/s$, consider the voltage $v_j(t)$ at the jth node ($j \geq 1$) of the truncated finite ladder of Fig. 7.2(b). In general, $v_j(t)$ is not monotonic, although it may be if l is small enough. The initial-value theorem and (7.1) yield the asymptotic estimate

$$v_j(t) \sim \frac{1}{(lc)^j} \cdot \frac{t^{2j}}{(2j)!}, \qquad t \to 0+, \tag{7.7}$$

which also happens to be a bound on the transient for all $t \in \mathbb{R}_+$ [60]:

$$|v_j(t)| \leq \frac{1}{(lc)^j} \cdot \frac{t^{2j}}{(2j)!}. \tag{7.8}$$

Also, because of the presence of the series resistors, all the poles of the Laplace transform V_j of $v_j(t)$ are in the open left-half s-plane [43, pages 311 and 332] except for the simple pole at the origin due to $E(s)$. Thus, the final-value theorem can be applied to (7.1) to get $v_j(t) \to 1$ as $t \to \infty$.

Now let us restore el-sections as in the preceding section. At a fixed jth node in the first ladder ($0 < j < m$) and with $n = m$ as the index of the restoration procedure, we have the hyperreal voltage $\mathbf{v}_j(\mathbf{t}) = [v_{j,n}(t_n)]$ defined for hyperreal time $\mathbf{t} = [t_n] \in {}^*\mathbb{R}_+$. By arguments similar to those given in Sec. 7.4, we have the following results:

It follows from (7.8) that $\mathbf{v}_j(\mathbf{t})$ is infinitesimal when $\mathbf{t}$ is. The overflow principle coupled with (7.7) shows that, as $\mathbf{t}$ increases through the positive infinitesimals and then becomes appreciable, $\mathbf{v}_j(\mathbf{t})$ also increases through the positive infinitesimals and then becomes appreciable. But now, for some larger appreciable values, $\mathbf{v}_j(\mathbf{t})$ might vary through 0 and its halo because of the possible oscillatory nature of the $v_{j,n}(t)$.

Finally, for all sufficiently large unlimited $\mathbf{t}$, $\mathbf{v}_j(\mathbf{t})$ is infinitesimally close to 1. For this last assertion, we have to adjust some inequalities to account for the fact that v_j may be oscillatory. In particular, we choose $\mathbf{w} = [w_n]$ as a positive infinitesimal and choose T_n as the minimum time for which $|1 - v_{j,n}(t_n)| \leq w_n$ for all $t_n \geq T_n$. So, for all $\mathbf{t} = [t_n] \geq \mathbf{T} = [T_n]$, we have $|1 - \mathbf{v}_j(\mathbf{t})| \leq \mathbf{w}$.

Here, too, we can examine transients in the second ladder, something that could not be done heretofore when only standard analyses of transfinite networks was available. As before, let $\omega + p$ denote the index of a fixed node in the second ladder ($0 \leq p \leq \omega$). Also, as before, let us restore the transfinite ladder using two el-sections at each step n of restoration, one for each ladder. Thus, $n = m$, again as before. For a limited hyperreal $\mathbf{t} = [t_n]$, $\mathbf{v}_{\omega+p}(\mathbf{t}) = [v_{\omega+p,n}(t_n)]$ remains infinitesimal, as can be seen from (7.8) and an argument similar to that in the preceding section. However, as $\mathbf{t} = [t_n]$ increases through the unlimited hyperreals, $\mathbf{v}_{\omega+p}(\mathbf{t})$ remains infinitesimal at first but then increases through appreciable hyperreal values and finally gets infinitesimally close to 1 for sufficiently large hyperreal values of $\mathbf{t}$. This, too, is argued as in the preceding section with some adjustment of inequalities.

To summarize, the artificial wave propagates down the first ladder with appreciable values for all appreciable values of time. However, it propagates into the second ladder with appreciable values only for sufficiently large unlimited hyperreal times.

Finally, let us consider the most general case where the transmission line also has a conductance g in parallel with each capacitor c. Now, $x(s) = (ls + r)(cs + g)$. The asymptotic expression (7.7) and the bound (7.8) still hold for the truncated ladder of Fig. 7.2(b) [60]. However, the final value theorem now yields a different result. Again, all the poles of $V_j(x)/E$ are in the open left-half s-plane; but, as $s \to 0$, $x(s)$ approaches rg, and $V_j(x)/E$ approaches (7.1) with $x = rg$. Thus, for a unit-step input voltage (i.e., $E(s) = 1/s$), the node voltage for a node in the first ladder beyond the input node approaches a finite positive quantity less than 1, but for a node in the second ladder the node voltage approaches 0. This means that the artificial wave settles toward a positive real value at any node in the first ladder, but it eventually becomes infinitesimal at any node in the second ladder. For r and g large enough and for l and c small enough, the response at any node will be monotonic, and we can conclude that the response then remains infinitesimal in the second ladder for all $\mathbf{t}$.

7.6 Conventionally Infinite, Uniform Transmission Lines and Cables and Nonstandard Enlargements

We now consider better models of transmission lines and cables whose electrical parameters are distributed in a uniform fashion along their lengths. During the initial truncation process, we will be reducing the transfinite lines and cables to conventionally infinite ones and will use the known real transient responses of the conventional lines and cables when constructing the hyperreal transient responses of the transfinite lines and cables. So, in this section we summarize some facts about those real responses. We also note how such transient responses can be immediately lifted through the transfer principle into a nonstandard setting. This however is different from the transfinite extensions of transmission lines and cables obtained through cascades of conventional lines and cables. It is the latter that we shall examine in the remaining sections of this chapter.

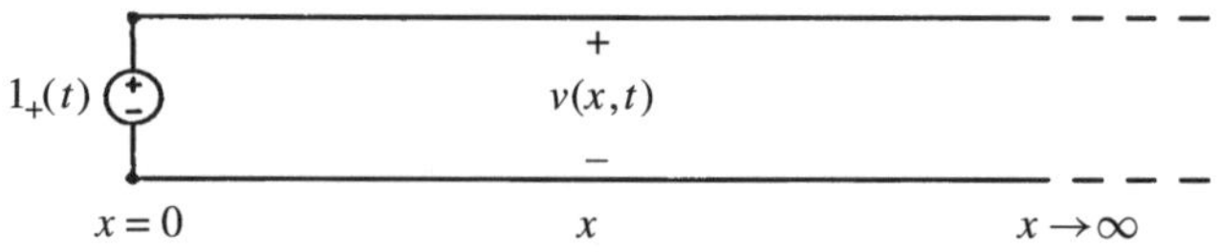

Fig. 7.3. An ω-line or ω-cable.

A conventionally infinite (transmission) line is illustrated in Fig. 7.3. A cable is a special case of a line. For a reason that will become evident later on, we call this an *ω-line*. We assume throughout that the line is uniform with the distributed series resistance and inductance being r ohms/meter and l henries/meter and the distributed shunt conductance and capacitance being g siemens/meter and c farads/meter. The distance along the line from the input in the conventional case is x meters, and the voltage at the distance x and time t seconds is $v(x,t)$. We will have $v(x,t) \in I\!R$ and $x, t \in I\!R_+$. We also assume throughout that at the input to the line $v(0,t) = 1_+(t)$, where 1_+ denotes the unit step function: $1_+(t) = 1$ for $t \geq 0$ and $1_+(t) = 0$ for $t < 0$.

When r, l, g, c are all positive real numbers, we have the general case of an ω-line, and the Laplace transform $V(x,s)$ of $v(x,t)$ is [42, page 379]

$$V(x,s) = \frac{1}{s} \exp\left(-x\sqrt{lc}\sqrt{(s+\delta)^2 - \sigma^2}\right), \quad \mathrm{Re}\, s > 0, \tag{7.9}$$

where

$$\delta = \frac{1}{2}\left(\frac{r}{l} + \frac{g}{c}\right), \quad \sigma = \frac{1}{2}\left(\frac{r}{l} - \frac{g}{c}\right). \tag{7.10}$$

Taking the inverse Laplace transform of (7.9), we get [42, page 383]

$$v(x,t) = f_0(x,t) + f_\sigma(x,t), \tag{7.11}$$

where

$$f_0(x,t) \;=\; e^{-\delta x\sqrt{lc}}\, 1_+(t - x\sqrt{lc}), \tag{7.12}$$

and

$$f_\sigma(x,t) \;=\; \sigma x\sqrt{lc}\int_{x\sqrt{lc}}^{t} \frac{e^{-\delta\tau}}{\sqrt{\tau^2 - x^2 lc}}\, I_1(\sigma\sqrt{\tau^2 - x^2 lc})\; d\tau\; 1_+(t - x\sqrt{lc}). \tag{7.13}$$

Here, I_1 is the modified Bessel function of first kind and order 1 [2, page 374]; it is an entire function. Furthermore, $f_0(x,t)$ is the voltage for the distortionless line occurring when $\sigma = 0$ (i.e., $rc = lg$), and $f_\sigma(x,t)$ is the added distortion occurring when $\sigma \neq 0$. Both $f_0(x,t)$ and $f_\sigma(x,t)$ take on nonnegative values only.

As a special case of the distortionless line, we have the lossless line occurring when $r = g = 0$, $l > 0$, and $c > 0$. Its transient response is simply

$$v(x,t) \;=\; 1_+(t - x\sqrt{lc}). \tag{7.14}$$

A different phenomenon occurs when $g = l = 0$, $r > 0$, and $c > 0$. This corresponds to a cable, for which case the wave response (7.12) is replaced by a diffusion response. Specifically, the response to the unit-step input $v(0,t) = 1_+(t)$ is [42, page 330]

$$v(x,t) \;=\; \operatorname{erfc}\left(\frac{x}{2}\sqrt{\frac{rc}{t}}\right). \tag{7.15}$$

Here, $\operatorname{erfc}(\cdot)$ is the complementary error function [2, page 297]

$$\operatorname{erfc} z \;=\; \frac{2}{\sqrt{\pi}}\int_z^{\infty} e^{-\zeta^2}\, d\zeta. \tag{7.16}$$

For fixed x and as t increases, $v(x,t)$ increases continuously and strictly monotonically from 0 toward its final value 1.

By using the transfer principle of nonstandard analysis, we can lift these expressions for $v(x,t)$ to obtain hyperreal-valued voltages on nonstandard enlargements of the transmission lines and cables. In particular, the transmission line of Fig. 7.3 is replaced by a nonstandard transmission line for which the spatial variable is $\mathbf{x} \in {}^*\mathbb{R}_+$, and the time variable is $\mathbf{t} \in {}^*\mathbb{R}_+$ as well. For $\mathbf{x}$ restricted to limited values, we have the first galaxy of nonnegative $\mathbf{x}$-values corresponding to the transmission line of Fig. 7.3. Moreover, all the subsequent hyperreal values of $\mathbf{x}$ are partitioned into galaxies of positive unlimited hyperreal values. Then, the equations (7.11) through (7.14) can be lifted immediately to obtain an internal function $\mathbf{v}$ mapping $(\mathbf{x},\mathbf{t}) \in {}^*\mathbb{R}_+^2 = {}^*\mathbb{R}_+ \times {}^*\mathbb{R}_+$ into $\mathbf{v}(\mathbf{x},\mathbf{t}) \in {}^*\mathbb{R}_+$, and we obtain thereby a hyperreal voltage $\mathbf{v}(\mathbf{x},\mathbf{t})$ defined upon a nonstandard enlargement of the transmission line.

However, this enlargement is not the transfinite extension of the transmission line of Fig.7.3 indicated in Fig. 7.4. It is instead a rather nebulous concept when

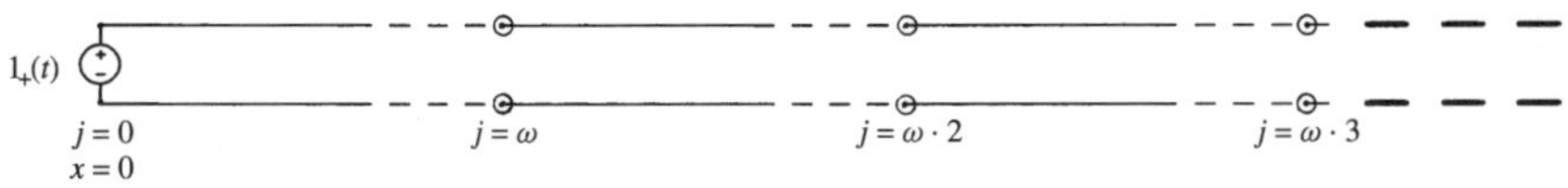

Fig. 7.4. An ω^2-line or ω^2-cable.

compared to our transfinite generalization of conventional networks. Indeed, there is no "next" galaxy after the first one for $\mathbf{x}$ in the hyperreal line $^*\mathbb{R}_+$. Between any two galaxies there is another, and we cannot view the enlargement as a cascade of one-way infinite conventional transmission lines as shown in Fig. 7.4. Our objective in the rest of this chapter is to analyze cascades such as that of Fig. 7.4, and this will require a much different analysis as compared to the simple lifting of the equations (7.11) to (7.14) by means of the transfer principle.

In our last section, Sec. 7.10, we analyze in a similar way transfinite cascades of cables.

7.7 The ω^2-Line

The first transfinite distributed-parameter structure we wish to explore is a transmission line that extends transfinitely with infinitely many (more precisely, ω-many) ω-lines connected in cascade by having the infinite extremity of each ω-line connected to the input of the next ω-line. We call this an *ω^2-line*. It is illustrated in Fig. 7.4. We shall analyze this structure by choosing uniformly spaced sample points Δx meters apart within the entire ω^2-line, taking the input of each ω-line as one of the sample points. We will then determine hyperreal voltage transients $\mathbf{v}(\mathbf{x}_j, \mathbf{t})$ at each sample point. The number of sample points within each ω-line does not depend upon the choice of Δx; that number is always ω (more precisely, the cardinality of the set of such points is always $\omega = \aleph_0$). Thus, we obtain the sample-point numbering at the input of each ω-line as shown in Fig. 7.4, and we therefore number the infinite extremity of the entire ω^2-line as ω^2 (not shown in Fig. 7.4). Moreover, the indices j of the sample points within the entire ω^2-line first traverse the natural numbers within the initial ω-line, then the ordinals from ω through all those below $\omega \cdot 2$ in the next ω-line, and so forth. In general, $\omega \cdot k \leq j < \omega \cdot (k+1)$ for the kth ω-line ($k \in I\!N$) in Fig. 7.4.

Two questions may arise when trying to make sense out of the configuration in Fig. 7.4. First, how can the connection between the infinite extremity of the kth ω-line and the input of the $(k+1)$st ω-line be defined? One way is to think of the line as being "artificial," having discrete series and shunt branches and then to use the definition of 1-nodes. In this way, the small circles in Fig. 7.4 represent 1-nodes that connect the 0-tips of the horizontal paths in any artificial ω-line to the input 0-nodes of the next artificial ω-line. Then, the present distributed structure might be viewed as a limiting case arising when the artificial line smooths out into a distributed one. But, perhaps, such elaboration is not needed if one is willing to accept the idea that the infinite extremity of each distributed ω-line "is connected to" the input terminals of the next ω-line.

The second question concerns the continuous spatial variable x. Within the initial ω-line its values are real numbers, but what are its values in the subsequent ω-lines? Even at the sample points we have a problem in interpreting $x =$ "$j\Delta x''$ when j is a transfinite ordinal and Δx is a real number. This difficulty will be circumvented in the following way: When setting up a restoration sequence of the ω^2-line of Fig. 7.4, we will truncate each ω-line within the ω^2-line into a finite line and will thereby reduce the ω^2-line to an ω-line. This will allow us to use the voltage transients cited in Sec. 7.6 in order to derive hyperreal transients at the sample points of the ω^2-line.

7.8 A Hyperreal Wave on an ω^2-Line

The general idea for analyzing the ω^2-line of Fig. 7.4 is to reduce it to an infinite cascade of finite lines which together comprise an ω-line; then, the finite lines are expanded in steps to "fill out" the ω^2-line. Since the response of an ω-line to a unit step of voltage is known, we will obtain at each sample point a sequence of voltages depending upon time t, which can then be identified as a hyperreal voltage depending upon hyperreal time $\mathbf{t}$.

So, consider the nth sample point ($n \in I\!N$) within each ω-line of the ω^2-line. Remove that part of the ω-line beyond that nth sample point and then connect that nth sample point to the input of the next ω-line. What is left is a cascade of finite lines, each having n sample points (not counting the input node) and together comprising an ω-line because the number of finite lines is ω. We shall refer to this structure as the nth truncation of the ω^2-line. Now, consider the sequence of such nth truncations as $n \to \infty$. Any fixed sample point of the ω^2-line will eventually appear in those nth truncations for all n sufficiently large and will have a voltage in accordance with (7.12). That fixed sample point will be absent in no more than finitely many nth truncations, and in these cases we can set the voltage equal to 0 without disturbing the hyperreal voltage that will arise as $n \to \infty$.

To be more specific, consider the jth sample point in the ω^2-line of Fig. 7.4. We can set $j = \omega \cdot k_1 + k_0$ ($k_0, k_1 \in I\!N$), where k_1 is the number of ω-lines to the left of the ω-line in which the jth sample point appears (but not counting that ω-line) and k_0 is the number of sample points to the left of the jth sample point in the ω-line in which the jth sample point appears. Then, for all n sufficiently large (i.e., for $n \geq k_0$), the jth sample point appears in the nth truncation, and the distance from the input to the jth sample point in the nth truncation is

$$x_{j,n} = (nk_1 + k_0)\Delta x \in I\!R_+,$$

where Δx is the distance between sample points, as before. In this case, the voltage at the jth sample point is

$$v(x_{j,n}, t) = f_0(x_{j,n}, t) + f_\sigma(x_{j,n}, t). \tag{7.17}$$

As $n \to \infty$, we obtain a sequence $\langle x_{j,n} : n \in I\!N \rangle$ of distances to the fixed jth sample point, which we take to be a representative sequence for a hyperreal distance $\mathbf{x}_j = [x_{j,n}]$ to the jth sample point. Furthermore, the voltage wave on the original ω^2-line will require an infinite amount of time in order to reach the ω-lines beyond the initial one. So, in order to examine this, we will need to use hyperreal time $\mathbf{t} = [t_n]$, where $\langle t_n \rangle$ is any representative sequence of nonnegative real time values t_n for $\mathbf{t}$. Altogether then, by replacing x by $\mathbf{x}_j$ and t by $\mathbf{t}$ and using (7.11), (7.12), and (7.13), we obtain the following hyperreal voltage as the response of the ω^2-line at its jth sample point:

$$\mathbf{v}(\mathbf{x}_j, \mathbf{t}) = [v(x_{j,n}, t_n)] = [f_0(x_{j,n}, t_n) + f_\sigma(x_{j,n}, t_n)] . \tag{7.18}$$

Because of the presence of the unit-step function $1_+(t - x\sqrt{lc})$ in (7.12) and (7.13), we see that $\mathbf{v}(\mathbf{x}_j, \mathbf{t})$ equals 0 for $\mathbf{t} = [t_n] < [\sqrt{lc}\, \Delta x\, (nk_1 + k_0)]$ and is positive for $\mathbf{t} = [t_n] \geq [\sqrt{lc}\, \Delta x\, (nk_1 + k_0)]$. Moreover, because of the factor $\exp(-\delta x \sqrt{lc})$ in (7.12), the distortionless term $[f_0(x_{j,n}, t_n)]$ is an infinitesimal for all $\mathbf{t} = [t_n]$ if $k_1 > 0$ (i.e., at all sample points at and beyond $j = \omega$; that $k_1 > 0$ means that the jth sample point is beyond the initial ω-line).

More generally, the total response $\mathbf{v}(\mathbf{x}_j, \mathbf{t})$ is also an infinitesimal when k_1, r, g, l, c are all positive. To show this, first apply the final-value theorem to (7.9):

$$\begin{aligned} \lim_{t\to\infty} v(x, t) &= \lim_{s\to 0+} \exp\left(-x\sqrt{lc}\sqrt{(s+\delta)^2 - \sigma^2}\right) \\ &= \exp\left(-x\sqrt{lc}\sqrt{\delta^2 - \sigma^2}\right) \\ &= \exp\left(-x\sqrt{rg}\right) . \end{aligned} \tag{7.19}$$

Now, at the jth sample point, we have $x_{j,n} = \Delta x(k_1 n + k_0)$. Because of the unit-step function, we need only consider the case where $t_n \geq x_{j,n}\sqrt{lc}$. So, the character of $\mathbf{v}(\mathbf{x}_j, \mathbf{t}) = [v(x_{j,n}, t_n)]$ can be determined by letting $n \to \infty$ but keeping j fixed. From (7.19) we have that, for $k_1 > 0$ again,

$$\lim_{n\to\infty} v(x_{j,n}, t_n) = \lim_{n\to\infty} \exp\left(-\Delta x\, (k_1 n + k_0)\, \sqrt{rg}\right) = 0.$$

Whence our assertion; in fact, $\mathbf{v}(\mathbf{x}_j, \mathbf{t})$ is a positive infinitesimal for $\mathbf{t} \geq [x_{j,n}\sqrt{lc}]$. Note also that, since the distortionless term and the total response are both infinitesimals when $k_1 > 0$, so also is the additional distortion term $[f_\sigma(x_{j,n}, t_n)]$ occurring when $\sigma \neq 0$.

Thus, we can describe the total response $\mathbf{v}(\mathbf{x}_j, \mathbf{t})$ at any sample point $\mathbf{x}_j$ beyond the initial ω-line, as follows: $\mathbf{v}(\mathbf{x}_j, \mathbf{t})$ remains equal to $\mathbf{0}$ for $\mathbf{t} < [x_{j,n}\sqrt{lc}] = \mathbf{x}_j\sqrt{lc}$. Then, there is a positive infinitesimal jump $\exp(-\delta \mathbf{x}_j \sqrt{lc})$ occurring at $\mathbf{t} = \mathbf{x}_j\sqrt{lc}$ due to the distortionless term (7.12). This is followed, for $\mathbf{t} > \mathbf{x}_j\sqrt{lc}$, by a continuous, strictly monotonic, increase toward the larger infinitesimal final value $\exp(-\mathbf{x}_j\sqrt{rg})$. (Indeed, a manipulation shows that $\delta\sqrt{lc} > \sqrt{rg}$.)

However, when $g = 0$ and r, l, c are positive, the same analysis shows that the limit in (7.19) is 1. This means that $\mathbf{v}(\mathbf{x}_j, \mathbf{t})$ is infinitesimally close to 1 when $\mathbf{t}$

is a sufficiently large unlimited hyperreal. Thus, the subsequent increase from the infinitesimal $\exp(-\delta \mathbf{x}_j\sqrt{lc})$ toward 1 is appreciable.

In the still more special case of a lossless line ($r = g = 0, l > 0, c > 0$), $\mathbf{v}(\mathbf{x}_j\mathbf{t}) = 0$ for $\mathbf{t} < \mathbf{x}_j\sqrt{lc}$ and is exactly equal to 1 for $\mathbf{t} \geq \mathbf{x}_j\sqrt{lc}$.

Finally, let us point out that we have obtained a hyperreal transient response for the ω^2-line by specifying a particular way of truncating it into an ω-line and then expanding the ω-line in steps to fill out the ω^2-line. However, there are many ways of doing this by using, say, nonuniformly spaced sample points and different truncations and subsequent expansions among the ω-line. In this way, we can generate many restoration sequences for the ω^2-line. We have presented one method for which the hyperreal transients have known formulas.

7.9 Transfinite Lines of Higher Ranks

By connecting ω-many ω^2-lines in cascade, we will obtain an ω^3-*line*. Recursively, we can construct in this way an ω^μ-*line*, where μ is any natural number; the ω^μ-line consists of ω-many $\omega^{\mu-1}$-lines in cascade. To analyze this by means of a restoration sequence, we again choose sample points with Δx spacing. Then, a typical sample point has the index

$$j = \omega^{\mu-1}\cdot k_{\mu-1} + \omega^{\mu-2}\cdot k_{\mu-2} + \cdots + \omega\cdot k_1 + k_0, \tag{7.20}$$

where $k_{\mu-1}$ is the number of $\omega^{\mu-1}$-lines to the left of the $\omega^{\mu-1}$-line in which the jth sample point appears, $k_{\mu-2}$ is the number of $\omega^{\mu-2}$-lines within the $\omega^{\mu-1}$-line in which the jth sample point appears and to the left of the $\omega^{\mu-2}$-line in which the jth sample point appears, and so on. In general, for $1 \leq \alpha \leq \mu - 1$, k_α is the number of ω^α-lines within the $\omega^{\alpha+1}$-line in which the jth sample point appears and to the left of the ω^α-line in which the jth sample point appears. Finally, k_0 is the number of sample points within the ω-line in which the jth sample point appears and to the left of that sample point.

Next, we create the nth truncation ($n \in I\!N, n \geq 1$) of this ω^μ-line as follows. When $\mu > 1$, each of the ω-lines in the ω^μ-line is truncated, as before, by removing all of the ω-line beyond the nth sample point and connecting that nth sample point to the input of the next ω-line. Let us call this truncation of an ω-line an "n-line." When $\mu > 2$, each of the ω^2-lines in the ω^μ-line thus becomes a cascade of ω-many n-lines. Retain the first n of those n-lines, remove all the others, but maintain the connection to the next ω^2-line. Let us call this truncation of an ω^2-line an "n^2-line." When $\mu > 3$, each of the ω^3-lines in the ω^μ-line becomes a cascade of ω-many n^2-lines. Again, we retain the first n of those n^2, remove the subsequent ones, but maintain the connection to the rest of the ω^μ-line. Continue this way, truncating each ω^ρ-line for each $\rho = 1, 2, \ldots, \mu - 1$ to get an "n^ρ-line." For $\rho \geq 2$, each n^ρ-line consists of n-many $n^{\rho-1}$-lines in cascade.

The final result will be a cascade of ω-many finite lines (namely, $n^{\mu-1}$-lines), each having $n^{\mu-1}$ sample points (not counting the input node of the original

ω^μ-line).[1] We call this the nth truncation of the ω^μ-line. It is simply an ω-line. Its voltage is given in Sec. 7.6. As $n \to \infty$, this ω-line expands to ultimately fill out the original ω^μ-line.

Now, any fixed sample point of index j on the ω^μ-line will eventually appear as a sample point of the nth truncation of the ω^μ-line for all n sufficiently large. When this happens, the distance $x_{j,n}$ from the input of that truncated ω^μ-line to the jth sample point will be

$$x_{j,n} \;=\; \Delta x\,(n^{\mu-1}k_{\mu-1} + n^{\mu-2}k_{\mu-2} + \cdots + nk_1 + k_0) \tag{7.21}$$

where the natural numbers k_p $(p = 0, \ldots, \mu - 1)$ have the same meanings as before in (7.20). Since the nth truncation of the ω^μ-line is an ω-line, we can invoke (7.11), (7.12), and (7.13) to obtain the voltage at $x_{j,n}$. This gives us a sequence $\langle v(x_{j,n}, t) : n \in I\!N\rangle$ of voltage values at the jth sample point for any given time t. In order to obtain a hyperreal time $\mathbf{t} = [t_n]$ when analyzing the original ω^μ-line, we choose a sequence $\langle t_n : n \in I\!N\rangle$ of time values in place of t. All this yields a hyperreal voltage $\mathbf{v}(\mathbf{x}_j, \mathbf{t}) = [v(x_{j,n}, t_n)]$ for this restoration-sequence version of the ω^μ-line. More specifically, we have

$$v(x_{j,n}, t_n) \;=\; f_0(x_{j,n}, t_n) \;+\; f_\sigma(x_{j,n}, t_n) \tag{7.22}$$

where $x_{j,n}$ is given by (7.21). Here, too, $\mathbf{v}(\mathbf{x}_j, \mathbf{t}) = 0$ for $0 \le \mathbf{t} = [t_n] < \mathbf{x}_j\sqrt{lc} = [x_{j,n}]\sqrt{lc}$ and $\mathbf{v}(\mathbf{x}_j, \mathbf{t})$ is a positive hyperreal for $\mathbf{t} \ge \mathbf{x}_j\sqrt{lc}$.

As in the case of an ω^2-line, the presence of the factor $e^{-\delta x\sqrt{lc}}$ in (7.12) in the distortionless case (i.e., $\sigma = 0$) and the expression (7.21) for $x_{j,n}$ insures that the distortionless term $[f_0(x_{j,n}, t_n)]$ is infinitesimal when the jth sample point is beyond the initial ω-line (that is, when at least one of the k_p, $p = 1, \ldots, \mu - 1$, is positive. Also, the argument employing the final-value theorem as in the preceding section shows that the total response $\mathbf{v}(\mathbf{x}_j, \mathbf{t}) = [v(x_{j,n}, t_n)]$ for $\mathbf{t} > \mathbf{x}_j\sqrt{lc}$ and at any sample point $\mathbf{x}_j$ beyond the initial ω-line is also infinitesimal but larger than the initial jump at $\mathbf{t} = \mathbf{x}_j\sqrt{lc}$ whenever r, g, l, c are all positive. Therefore, the added distortion term $[f_\sigma(x_{j,n}, t_n)]$ is also infinitesimal. On the other hand, when $g = 0$ and r, l, c are positive, $\mathbf{v}(\mathbf{x}_j, \mathbf{t})$ is eventually infinitesimally close to 1 at every sample point after the initial ω-line. Altogether then, the voltage response of the ω^μ-line increases monotonically in much the same way as that of the ω^2-line.

The next rank for transfinite transmission lines beyond those of the natural-number ranks is ω^ω, and for this we have the *ω^ω-line*. This can be viewed as the following sequence of nested ω^μ-lines with μ increasing indefinitely. Specifically, we have an ω-line appearing as the first ω-line of an ω^2-line, which in turn is the first ω^2-line of an ω^3-line, ..., which is the first ω^μ-line of an $\omega^{\mu+1}$-line, and so on indefinitely. Now, we can assign uniformly spaced sample points at increments of Δx along the entire structure. Also, we can truncate the ω^ω-line much as before,

[1] For an ω^2-line, we get ω-many finite lines of n sample points each. For an ω^3-line, we get ω-many finite lines of n^2 sample points each. For an ω^4-line, we get ω-many finite lines of n^3 sample points each. Continue this way.

except that now the process continues for $\rho = 1, 2, 3, \ldots$ indefinitely. That is, every ω-line in the ω^ω-line becomes an n-line, every ω^2-line becomes an n^2-line, ...,every ω^ρ-line becomes an n^ρ-line, and so on. We will call this an "n^ω-line," but it is in fact an ω-line with ω-many sample points.

Now, the standard formulas of Sec. 7.6 can be applied once more to get a hyperreal transient for the ω^ω-line. In particular, consider the jth sample point of the ω^ω-line, and let μ be the least natural number for which that sample point appears in an initial ω^μ-line (see (7.20). Within the ω-line that is the truncation of the ω^ω-line, the number of sample points to the left of the jth sample point is

$$N_j(n) \;=\; n^{\mu-1}k_{\mu-1} + n^{\mu-2}k_{\mu-2} + \cdots + nk_1 + k_0 \tag{7.23}$$

where the $k_{\mu-1}, k_{\mu-2}, \ldots, k_0$ are defined exactly as in (7.20). Here, μ depends upon j as stated. Thus, as the chosen jth sample point recedes further from the input, μ will eventually increase through all the natural numbers. So, with $\mathbf{x}_j = [x_{j,n}] = [\Delta x N_j(n)]$, $\mathbf{t} = [t_n]$, $\mathbf{v}(\mathbf{x}_j, \mathbf{t}) = [v(x_{j,n}, t_n)]$, and $v(x_{j,n}, t_n)$ defined again by (7.22), we have the monotonically increasing hyperreal voltage response $\mathbf{v}(\mathbf{x}_j, \mathbf{t})$ of the ω^ω-line at its jth sample point $\mathbf{x}_j$. Again, we can identify $\mathbf{v}(\mathbf{x}_j, \mathbf{t})$ as being either exactly equal to 0, or equal to a positive infinitesimal, or, when $g = 0$, appreciable and eventually infinitesimally close to 1 just as before at any sample point beyond the initial ω-line.

Having treated all the ω^μ-lines ($\mu = 1, 2, 3, \ldots$) and then the ω^ω-line, we can now treat the $\omega^{\omega+\mu}$-lines ($\mu = 1, 2, 3, \ldots$) and then the $\omega^{\omega\cdot 2}$-line in much the same way. So also can the $\omega^{\omega\cdot 2+\mu}$-lines and the $\omega^{\omega\cdot 3}$-lines be treated similarly—and so on.

For example, consider the *$\omega^{\omega+1}$-line*. It is a cascade of ω-many ω^ω-lines. We truncate each ω^ω-line as before to get an n^ω-line. Then, each of the n^ω-lines is truncated by retaining the nth nested initial line (that is, the n^n-line) but removing all of the remainder of the n^ω-line while connecting the output of that n^n-line to the input of the next initial n^n-line. Thus, the $\omega^{\omega+1}$-line becomes a cascade of ω-many n^n-lines. We call this result the nth truncation of the ω^ω-line. The number of sample points $M_j(n)$ up to the jth sample point in that nth truncation is now determined as follows. Let k_ω be the number of n^n-lines to the left of the n^n-line in which the jth sample point appears (for all n large enough, this sample point will so appear). Then,

$$M_j(n) \;=\; n^n k_\omega \;+\; N_j(n) \tag{7.24}$$

where $N_j(n)$ is given by (7.23) for the ω^ω-line in which the jth sample point appears. So, with $\mathbf{x}_j = [x_{j,n}] = [\Delta x M_j(n)]$ and $\mathbf{t} = [t_n]$, we have as before

$$\mathbf{v}(\mathbf{x}_j, \mathbf{t}) \;=\; [f_0(x_{j,n}, t_n)] \;+\; [f_\sigma(x_{j,n}, t_n)]. \tag{7.25}$$

We can identify what kind of hyperreal $\mathbf{v}(\mathbf{x}_j, \mathbf{t})$ is again as before.

Finally, we point out again that many other restoration-sequence models can be constructed by varying the ways we take truncations and subsequent expansions.

7.10 A Hyperreal Diffusion on a Transfinite Cable

For a cable we have $g = l = 0$, $r > 0$, $c > 0$, and the standard voltage response $v(x,t)$ on an ω-cable due to a unit step of voltage at the input to the cable is given by (7.15). This represents a diffusion phenomenon.

For a transfinite cable of rank greater than ω, we truncate it down to an ω-cable, with the truncations depending upon the natural number n, and then let $n \to \infty$ to obtain a representative sequence for the hyperreal voltage response. Thus, for an ω^μ-cable with $\mu \in I\!N$, $\mu > 1$, we choose the nth truncation exactly as in Sec. 7.9. In particular, the index of the jth sample point on the ω^μ-cable is given by (7.20), and for all n sufficiently large the distance $x_{j,n}$ of the jth sample point from the input of the nth truncation of the ω^μ-cable is given by (7.21). Then, with $\mathbf{x}_j = [x_{j,n}]$ being hyperreal distance and $\mathbf{t} = [t_n]$ being hyperreal time as before, we have the hyperreal voltage $\mathbf{v}(\mathbf{x}_j, \mathbf{t})$ on the ω^μ-cable due to a unit step of voltage at the input as

$$\mathbf{v}(\mathbf{x}_j, \mathbf{t}) = \left[\operatorname{erfc} \frac{x_{j,n}}{2} \sqrt{\frac{rc}{t_n}} \right]. \tag{7.26}$$

We can invoke the properties of the complementary error function erfc$(\cdot)$ to assert the following. In the initial ω-cable of the ω^μ-cable ($\mu > 1$), $\mathbf{v}(\mathbf{x}_j, \mathbf{t})$ is infinitesimal (resp. appreciable; resp. unlimited) when $\mathbf{t}$ is infinitesimal (resp. appreciable; resp. unlimited). However, for subsequent ω-cables within the ω^μ-cable, the following properties hold. As $\mathbf{t}$ increases through all limited and then through some initial unlimited values, $\mathbf{v}(\mathbf{x}_j, \mathbf{t})$ increases through some but not all infinitesimal values. Then, as $\mathbf{t}$ increases through larger unlimited values, $\mathbf{v}(\mathbf{x}_j, \mathbf{t})$ continues to increase first through larger infinitesimal values, then through appreciable values, and eventually gets infinitesimally close to 1 but never reaches 1. This variation of $\mathbf{v}(\mathbf{x}_j, \mathbf{t})$ is strictly monotonic.

At the next rank of transfiniteness, we have the ω^ω-cable. This time we construct the nth-truncation of the ω^ω-cable exactly as we constructed the nth-truncation of the ω^ω-line. Then, at the jth sample point of the ω^ω-cable, we have the number $N_j(n)$ of sample points within the nth-truncation up to that jth sample point. $N_j(n)$ is given by (7.23) as before. Finally, we have $\mathbf{v}(\mathbf{x}_j, \mathbf{t})$ given by (7.26) with $x_{j,n} = \Delta x N_j(n)$ now.

Continuing on to the next higher rank of transfiniteness, we have the $\omega^{\omega+1}$-cable. This time we truncate as we did the $\omega^{\omega+1}$-line to get the following expressions for the hyperreal voltage at the jth sample point in the $\omega^{\omega+1}$-cable. Let $M_j(n)$ be given by (7.24). Once again, $\mathbf{v}(\mathbf{x}_j, \mathbf{t})$ is given by (7.26) but with $x_{j,n} = \Delta x M_j(n)$.

This analysis can be continued to still higher ranks of transfinite cables. The voltage $\mathbf{v}(\mathbf{x}_j, \mathbf{t})$ behaves monotonically as before with respect to increasing hyperreal $\mathbf{t}$, but with no jumps. Now, we have a diffusion instead of a wave.

8

Nonstandard Graphs and Networks

Our nonstandard analyses for the determination of hyperreal operating points in transfinite resistive networks in Chapter 6 and of hyperreal transients in transfinite RLC networks in Chapter 7 do not enlarge those networks in a nonstandard way. The restoration processes employed therein merely recover the original standard networks and only provide a means of constructing hyperreal currents and voltages that satisfy Kirchhoff's laws and Ohm's law. Another approach might start with an arbitrary sequence of conventional (finite or infinite) 0-graphs and construct from that a nonstandard 0-graph in much the same way as an internal set in the hyperreal line $^*\mathbb{R}$ is constructed from a given sequence of subsets of the real line $\mathbb{R}$, that is, by means of an ultrapower construction. In this case, the resulting nonstandard graph has nonstandard branches and nonstandard nodes, and it thereby is much different from what we have presented so far.

Our objective in this chapter is to develop this latter approach to nonstandard 0-graphs in the way outlined in Sec. 6.8. The individual elements are chosen to be the nodes of the graphs along with all the natural numbers. These individuals are not sets by assumption and therefore have no members. All the other standard and nonstandard entities are derived from these individuals. For example, a branch is defined to be a pair of nodes, which is one of the conventional ways of defining a graph. Then, certain equivalence classes of sequences of standard nodes are defined to be nonstandard nodes, and these then yield nonstandard branches as certain equivalence classes of sequences of standard branches. In this approach, there are no parallel branches and no self-loops, but isolated nodes can now occur.[1]

After setting up our nonstandard graphs using an ultrapower approach, we invoke the transfer principle to lift several standard graph-theoretic results into a nonstandard setting. For example, the relationship between the number of nodes, the number

[1] Alternatively, we might start with 0-tips along with the natural numbers as the individuals and construct nonstandard branches and nodes from them [58]. In this approach, parallel nonstandard branches and self-loops can occur, but there are no isolated nodes. If we restrict ourselves to graphs having no parallel branches, no self-loops, and no isolated nodes, the two approaches yield exactly the same results.

of branches, and the cyclomatic number of a standard finite connected graph continues to hold for nonstandard "hyperfinite," "connected" graphs except that these numbers are replaced by hypernatural numbers.[2] By virtue of the transfer principle, this only requires that the standard theorems be stated as sentences in symbolic logic, which are then transferred to appropriate sentences for nonstandard graphs. All this is a straightforward application of nonstandard analysis—there are no surprises. Undoubtedly, other standard results for finite graphs can be lifted in this way into a nonstandard setting.

Similarly, standard transfinite ν-graphs can be enlarged, yielding thereby nonstandard ν-graphs. We indicate how this can be accomplished in Sec. 8.8 by doing so for 1-graphs. Finally, we construct in Sec. 8.9 a nonstandard 1-network whose 1-graph is nonstandard and whose electrical branch parameters are hyperreals, and we establish a fundamental theorem for the existence of a hyperreal current-voltage regime. The extension of these results to higher ranks of transfiniteness also appears to be straightforward but more complicated. We do not pursue this matter.

There are other kinds of nonstandard graphs constructed differently. See Sec. 6.8 for a brief survey of them.

Up through Sec. 8.7, we drop the superscript 0 on the symbols for a 0-graph and a 0-node since these are the only graphs and nodes we will be considering. We define nonstandard nodes and branches in the next section and use boldface symbols for them, just as we do for hyperreals and hypernaturals.

Our sentences of symbolic language employ the following conventional symbols: $\exists$ meaning "there exists," $\forall$ "for all," $\wedge$ "and," $\vee$ "or (possibly both)," $\neg$ "it is not true that," $\rightarrow$ "implies," $\leftrightarrow$ "if and only if." Let us mention again that the ideas regarding nonstandard analysis that we use are listed in Appendix A.

8.1 Nonstandard Graphs Defined

Instead of using the definition given in Sec. 2.2, let us now define a (finite or infinite) *graph* G in a more conventional way, namely, $G = \{X, B\}$, where X is the set of its *nodes* and B is the set of its *branches*.[3] Each *branch* $b \in B$ is a two-element set $b = \{x, y\}$ with $x, y \in X$ and $x \neq y$; b and x are said to be *incident* and so, too, are b and y. Also, x and y are said to be *adjacent* through b. In contrast to the 0-graphs of Sec. 2.2, G may now have isolated nodes, that is, nodes having no incident branches. Now, however, G will have neither parallel branches nor self-loops.

Next, let $\langle G_n : n \in \mathbb{N}\rangle$ be a given sequence of graphs. The nonstandard graph we shall construct will depend upon this choice of the sequence $\langle G_n : n \in \mathbb{N}\rangle$. For each n, we have $G_n = \{X_n, B_n\}$, where X_n is the set of nodes and B_n is the set of branches.

---

[2]Similarly, standard theorems concerning Eulerian graphs, Hamiltonian graphs, and a coloring theorem are extended to this nonstandard setting in [58] and [59]. A copy of [59] is available in the website: http://www.arxiv.org under "Zemanian."

[3]For this different definition of a 0-graph we use the italic notation G instead of the calligraphic notation $\mathcal{G}^0$ used in Section 2.2.

We allow $X_n \cap X_m \neq \emptyset$ so that G_n and G_m may be subgraphs of a larger graph. In fact, we can view each G_n as being a subgraph of the union $G = \{\cup X_n, \cup B_n\}$ of all the G_n. As a special case, we may have $X_n = X_m$ and $B_n = B_m$ for all $n, m \in \mathbb{N}$ so that G_n may be the same graph for all $n \in \mathbb{N}$.

Furthermore, let $\mathcal{F}$ be a chosen nonprincipal ultrafilter on $\mathbb{N}$ (Appendix A4).

In the following, $\langle x_n \rangle = \langle x_n : n \in \mathbb{N} \rangle$ will denote a sequence of nodes with $x_n \in X_n$ for all $n \in \mathbb{N}$. A *nonstandard node* $\mathbf{x}$ is an equivalence class of such sequences of nodes, where two such sequences $\langle x_n \rangle$ and $\langle y_n \rangle$ are taken to be equivalent if $\{n : x_n = y_n\} \in \mathcal{F}$, in which case we write "$\langle x_n \rangle = \langle y_n \rangle$ a.e." or say that $x_n = y_n$ "for almost all n." We also write $\mathbf{x} = [x_n]$, where it is understood that the x_n are the members of any one sequence in the equivalence class.

That this truly defines an equivalence class can be shown as follows. Reflexivity and symmetry being obvious, consider transitivity: Given that $\langle x_n \rangle = \langle y_n \rangle$ a.e. and that $\langle y_n \rangle = \langle z_n \rangle$ a.e., we have $N_{xy} = \{n : x_n = y_n\} \in \mathcal{F}$ and $N_{yz} = \{n : y_n = z_n\} \in \mathcal{F}$. By the properties of the ultrafilter, $N_{xy} \cap N_{yz} \in \mathcal{F}$. Moreover, $N_{xz} = \{n : x_n = z_n\} \supseteq (N_{xy} \cap N_{yz})$. Therefore, $N_{xz} \in \mathcal{F}$. Hence, $\langle x_n \rangle = \langle z_n \rangle$ a.e.; transitivity holds. We let ${}^*X$ denote the set of nonstandard nodes.

Next, we define the nonstandard branches: Let $\mathbf{x} = [x_n]$ and $\mathbf{y} = [y_n]$ be two nonstandard nodes. This time, let $N_{xy} = \{n : \{x_n, y_n\} \in B_n\}$ and $N_{xy}^c = \{n : \{x_n, y_n\} \notin B_n\}$. Since $\mathcal{F}$ is an ultrafilter, exactly one of N_{xy} and N_{xy}^c is a member of $\mathcal{F}$. If it is N_{xy}, then $\mathbf{b} = [\{x_n, y_n\}]$ is defined to be a *nonstandard branch*; that is, $\mathbf{b}$ is an equivalence class of sequences $\langle b_n \rangle$ where $b_n = \{x_n, y_n\}$, $n = 0, 1, 2, \ldots$. In this case, we also write $\mathbf{x}, \mathbf{y} \in \mathbf{b}$ and $\mathbf{b} = \{\mathbf{x}, \mathbf{y}\}$. We let ${}^*B$ denote the set of nonstandard branches. On the other hand, if $N_{xy}^c \in \mathcal{F}$, then $[\{x_n, y_n\}]$ is not a nonstandard branch.

We shall now show that this definition is independent of the representatives chosen for the nodes. Let $[\{x_n, y_n\}]$ and $[\{v_n, w_n\}]$ represent the same nonstandard branch. We want to show that, if $\langle x_n \rangle = \langle v_n \rangle$ a.e., then $\langle y_n \rangle = \langle w_n \rangle$ a.e. Suppose $\langle y_n \rangle \neq \langle w_n \rangle$ a.e. Then, $\{n : x_n = v_n\} \cap \{n : y_n \neq w_n\} \in \mathcal{F}$. Thus, there is at least one n for which the three nodes $x_n = v_n$, y_n, and w_n are all incident to the same standard branch—in violation of the definition of a branch. Similarly, if all of $\langle x_n \rangle$, $\langle y_n \rangle$, $\langle v_n \rangle$, $\langle w_n \rangle$ are different a.e., then there would be a standard branch having four incident nodes—again a violation.

Next, we show that we truly have an equivalence relationship for the set of all sequences of standard branches. Reflexivity and symmetry being obvious again, consider transitivity: Let $\mathbf{b} = [\{x_n, y_n\}]$, $\tilde{\mathbf{b}} = [\{\tilde{x}_n, \tilde{y}_n\}]$, $\acute{\mathbf{b}} = [\{\acute{x}_n, \acute{y}_n\}]$, and assume that $\mathbf{b} = \tilde{\mathbf{b}}$ and $\tilde{\mathbf{b}} = \acute{\mathbf{b}}$. We want to show that $\mathbf{b} = \acute{\mathbf{b}}$. We have $N_{b\tilde{b}} = \{n : \{x_n, y_n\} = \{\tilde{x}_n, \tilde{y}_n\}\} \in \mathcal{F}$ and $N_{\tilde{b}\acute{b}} = \{n : \{\tilde{x}_n, \tilde{y}_n\} = \{\acute{x}_n, \acute{y}_n\}\} \in \mathcal{F}$. Moreover, $N_{b\acute{b}} = \{n : \{x_n, y_n\} = \{\acute{x}_n, \acute{y}_n\}\} \supseteq (N_{b\tilde{b}} \cap N_{\tilde{b}\acute{b}}) \in \mathcal{F}$. Therefore, $N_{b\acute{b}} \in \mathcal{F}$. Thus, $\mathbf{b} = \acute{\mathbf{b}}$, as desired.

Finally, we define a *nonstandard graph* ${}^*G$ to be the pair ${}^*G = \{{}^*X, {}^*B\}$; we also write ${}^*G = [G_n]$.

Let us now take note of a special case that arises when all the G_n are the same standard graph $G = \{X, B\}$. In this case, we call ${}^*G = \{{}^*X, {}^*B\}$ an *enlargement* of G, in conformity with an "enlargement" ${}^*A$ of a subset A of $\mathbb{R}$ (see Appendix A15).

If in addition G is a finite graph, each node $\mathbf{x} \in {}^*X$ can be identified with a node x in X because the enlargement of a finite set equals the set itself. Similarly, every branch $\mathbf{b} \in {}^*B$ can be identified with a branch b in B. In short, ${}^*G = G$. On the other hand, if G is a conventionally infinite graph, X is an infinite set and its enlargement ${}^*X$ has more elements, namely, nonstandard nodes that are not equal to standard nodes, (i.e., ${}^*X \setminus X$ is not empty). Similarly, ${}^*B \setminus B$ is not empty too. In short, ${}^*G$ is a proper enlargement of G.

Example 8.1-1. Let G be a one-way infinite path P:

$$P = \langle x_0, b_0, x_1, b_1, x_2, b_2, \ldots \rangle .$$

Also, let $G_n = G$ for all $n \in \mathbb{N}$, and let ${}^*G = [G_n] = \{{}^*X, {}^*B\}$. Next, let $\langle k_n : n \in \mathbb{N}\rangle$ be any sequence of natural numbers, and set $\mathbf{x} = [x_{k_n}]$ and $\mathbf{y} = [x_{k_n+1}]$. Then, $\mathbf{x}$ and $\mathbf{y}$ are two nodes in the enlargement ${}^*G$ of G, and $\mathbf{b} = \{\mathbf{x}, \mathbf{y}\} = [\{x_{k_n}, x_{k_n+1}\}]$ is a branch in ${}^*G$. On the other hand, if $\langle m_n : n \in \mathbb{N}\rangle$ is another sequence of natural numbers with $m_n > 1$, then $\mathbf{z} = [x_{k_n+m_n}]$ is another nonstandard node in ${}^*G$ different from $\mathbf{x}$ and $\mathbf{y}$ and appearing after $\mathbf{y}$ in the enlarged path. Moreover, $[\{x_{k_n}, x_{k_n+m_n}\}]$ is not a nonstandard branch. In this way, no node or branch repeats in the enlarged path. ♣

Another special case arises when almost all the G_n are (possibly different) finite graphs. Again in conformity with the terminology used for hyperfinite internal subsets of ${}^*R$, we will refer to the resulting nonstandard graph ${}^*G$ as a *hyperfinite graph*.[4] As a result, we can lift many theorems concerning finite graphs to hyperfinite graphs. It is just a matter of writing the standard theorem in an appropriate form using symbolic logic and then applying the transfer principle. We let ${}^*G_f$ denote the set of hyperfinite graphs.

8.2 Incidences and Adjacencies between Nodes and Branches

Let us now define these ideas for nonstandard graphs both in terms of an ultrapower construction and by transfer of appropriate symbolic sentences. In the subsequent sections, we will usually confine ourselves to the transfer principle.

Incidence between a node and a branch: Given a sequence $\langle G_n : n \in \mathbb{N}\rangle$ of standard graphs $G_n = \{X_n, B_n\}$, a nonstandard node $\mathbf{x} = [x_n] \in {}^*X$ and a nonstandard branch $\mathbf{b} = [b_n] \in {}^*B$ are said to be *incident* if $\{n : x_n \in b_n\} \in \mathcal{F}$, where as always the nonprincipal ultrafilter $\mathcal{F}$ is understood to be chosen and fixed.

On the other hand, we can define incidence between a standard node $x \in X$ and a standard branch $b \in B$ for the graph $G = \{X, B\}$ through the symbolic sentence

$$(\exists\, x \in X)\ (\exists\, b \in B)\ (x \in b)\,.$$

[4] This should not be confused with a hypergraph—an entirely different object [6]. See also the first paragraph of Sec. 3.3.

By transfer, we have that $\mathbf{x} \in {}^*X$ and $\mathbf{b} \in {}^*B$ are *incident* when the following sentence is true.

$$(\exists\, \mathbf{x} \in {}^*X)\ (\exists\, \mathbf{b} \in {}^*B)\ (\mathbf{x} \in \mathbf{b}) .$$

These are equivalent definitions.

Adjacency between nodes: For a standard graph $G = \{X, B\}$, two nodes $x, y \in X$ are called *adjacent* and we write $x \diamond y$ if the following sentence on the right-hand side of $\leftrightarrow$ is true.

$$x \diamond y \ \leftrightarrow\ (\exists\, x, y \in X)\ (\exists\, b \in B)\ (b = \{x, y\}) .$$

By transfer, this becomes for a nonstandard graph ${}^*G = \{{}^*X,\ {}^*B\}$,

$$\mathbf{x} \diamond \mathbf{y} \ \leftrightarrow\ (\exists\, \mathbf{x}, \mathbf{y} \in {}^*X)\ (\exists\, \mathbf{b} \in {}^*B)\ (\mathbf{b} = \{\mathbf{x}, \mathbf{y}\}) .$$

Alternatively, under an ultrapower construction, we have for ${}^*G = [G_n] = [\{X_n, B_n\}]$ that $\mathbf{x} = [x_n] \in {}^*X$ and $\mathbf{y} = [y_n] \in {}^*X$ are *adjacent* (i.e., $\mathbf{x} \diamond \mathbf{y}$) if there exists a $\mathbf{b} = [b_n] \in {}^*B$ such that $\{n\colon b_n = \{x_n, y_n\}\} \in \mathcal{F}$.

Adjacency between branches: For a standard graph, two branches $b, c \in B$ are called *adjacent* and we write $b \bowtie c$ when the following sentence on the right-hand side of $\leftrightarrow$ is true:

$$b \bowtie c \ \leftrightarrow\ (\exists\, b, c \in B)\ (\exists\, x \in X)\ (x \in b \wedge x \in c) .$$

By transfer, we have for nonstandard branches $\mathbf{b}$ and $\mathbf{c}$,

$$\mathbf{b} \bowtie \mathbf{c} \ \leftrightarrow\ (\exists\, \mathbf{b}, \mathbf{c} \in {}^*B)\ (\exists\, \mathbf{x} \in {}^*X)\ (\mathbf{x} \in \mathbf{b} \wedge \mathbf{x} \in \mathbf{c}) .$$

Under an ultrapower approach, we would have $\mathbf{b} = [b_n]$ and $\mathbf{c} = [c_n]$ are adjacent nonstandard branches when there exists a nonstandard node $\mathbf{x} = [x_n]$ such that

$$\{n\colon x_n \in b_n \text{ and } x_n \in c_n\} \in \mathcal{F}.$$

8.3 Nonstandard Hyperfinite Paths and Loops

Again, we start with a standard graph $G = \{X, B\}$. Remember that since B is a set of two-element subsets of X, there are no parallel branches and no self-loops. A *finite path* P in G is defined by the sentence

$$(\exists\, k \in \mathbb{N} \setminus \{0\})\ (\exists\, x_0, x_1, \ldots, x_k \in X)\ (\exists\, b_0, b_1, \ldots, b_{k-1} \in B) \tag{8.1}$$
$$(x_0 \in b_0 \ \wedge\ b_0 \ni x_1 \ \wedge\ x_1 \in b_1 \ \wedge\ b_1 \ni x_2 \ \wedge\ \cdots \ \wedge\ x_{k-1} \in b_{k-1} \ \wedge\ b_{k-1} \ni x_k) .$$

That all the nodes and branches in P are distinct is implied by the fact that those k nodes in X and those $k-1$ branches in B are perforce all distinct because X and B are sets. The *length* $|P|$ of P is the number of branches in P; thus, $|P| = k$.

We may apply transfer to (8.1) to get the following definition of a *nonstandard path* $^*P$:

$$(\exists\, \mathbf{k} \in {}^*\mathbb{N} \setminus \{0\})\ (\exists\, \mathbf{x}_0, \mathbf{x}_1, \ldots, \mathbf{x}_\mathbf{k} \in {}^*X)\ (\exists\, \mathbf{b}_0, \mathbf{b}_1, \ldots \mathbf{b}_{\mathbf{k}-1} \in {}^*B) \tag{8.2}$$
$$(\mathbf{x}_0 \in \mathbf{b}_0 \wedge \mathbf{b}_0 \ni \mathbf{x}_1 \wedge \mathbf{x}_1 \in \mathbf{b}_1 \wedge \mathbf{b}_1 \ni \mathbf{x}_2 \wedge \cdots \wedge \mathbf{x}_{\mathbf{k}-1} \in \mathbf{b}_{\mathbf{k}-1} \wedge \mathbf{b}_{\mathbf{k}-1} \ni \mathbf{x}_\mathbf{k})\,.$$

In this case, the *length* $|^*P|$ equals $\mathbf{k} \in {}^*\mathbb{N} \setminus \{0\}$; $\mathbf{k}$ is now a positive hypernatural number. We therefore call $^*P$ a *hyperfinite path.* To view this fact in terms of an ultrapower construction of $^*G = [G_n]$, note that the G_n may be finite graphs growing in size or indeed may be conventionally infinite graphs. Thus, $^*P$ may have an unlimited length, that is, its length may be a member of ${}^*\mathbb{N} \setminus \mathbb{N}$.

A standard loop is defined as is a standard path except that the first and last nodes are required to be the same. Also, there will be at least three nodes and three branches because there are no parallel branches. Upon applying transfer, we get the following definition of a *nonstandard loop*:[5]

$$(\exists\, \mathbf{k} \in {}^*\mathbb{N} \setminus \{0, 1\})\ (\exists\, \mathbf{x}_0, \mathbf{x}_1, \ldots, \mathbf{x}_\mathbf{k} \in {}^*X)\ (\exists\, \mathbf{b}_0, \mathbf{b}_1, \ldots \mathbf{b}_\mathbf{k} \in {}^*B) \tag{8.3}$$
$$(\mathbf{x}_0 \in \mathbf{b}_0 \wedge \mathbf{b}_0 \ni \mathbf{x}_1 \wedge \mathbf{x}_1 \in \mathbf{b}_1 \wedge \mathbf{b}_1 \ni \mathbf{x}_2 \wedge \cdots \wedge \mathbf{x}_\mathbf{k} \in \mathbf{b}_\mathbf{k} \wedge \mathbf{b}_\mathbf{k} \ni \mathbf{x}_0)\,.$$

8.4 Connected Nonstandard Graphs

A standard graph $G = \{X, B\}$ is called *connected* if, for every two nodes x and y in G, there is a finite path (8.1) terminating at those nodes, that is, $x_0 = x$ and $x_k = y$. Let $\mathcal{C}$ denote the set of connected standard graphs, and let $\mathcal{P}(G)$ be the set of all finite paths in a given standard graph G. Also, for any $P \in \mathcal{P}(G)$, let $x_0(P)$ and $x_k(P)$ denote the first and last nodes of P in accordance with (8.1); k depends upon P. Then, the connectedness of G is defined symbolically by the truth of the following sentence to the right of $\leftrightarrow$:

$$G \in \mathcal{C}\ \leftrightarrow\ (\forall\, x, y \in X)\ (\exists\, P \in \mathcal{P}(G))\ ((x_0(P) = x)\ \wedge\ (x_k(P) = y))\,. \tag{8.4}$$

By transfer, we obtain the definition of the set ${}^*\mathcal{C}$ of all *connected* nonstandard graphs: For $^*G = \{^*X, {}^*B\}$, for ${}^*\mathcal{P}(^*G)$ being the set of nonstandard paths $^*P \in {}^*\mathcal{P}(^*G)$, and for $\mathbf{x}_0(^*P)$ and $\mathbf{x}_\mathbf{k}(^*P)$ being the first and last nodes of $^*P$, we have

$$^*G \in {}^*\mathcal{C} \leftrightarrow (\forall\, \mathbf{x}, \mathbf{y} \in {}^* X)\ (\exists\, ^*P \in {}^*\mathcal{P}(^*G))\ ((\mathbf{x}_0(^*P) = \mathbf{x}) \wedge (\mathbf{x}_\mathbf{k}(^*P) = \mathbf{y}))\,. \tag{8.5}$$

Here, $\mathbf{k} \in {}^*\mathbb{N} \setminus \{0\}$ as in (8.2).

Let us explicate this still further in terms of an ultrapower construction of $^*G = [G_n]$ from an equivalence class of sequences of (possibly infinite) graphs, $\langle G_n \rangle$ being one of those sequences. With $^*P = [P_n]$ denoting a nonstandard path

[5] An equivalent but more complicated definition of a nonstandard loop is given in Sec. 8.9. It has the advantage that it can be extended to nonstandard transfinite loops.

obtained similarly from a representative sequence $\langle P_n \rangle$ of finite paths, P_n being in G_n, we let $x_0(P_n)$ and $x_k(P_n)$ denote the first and last nodes of P_n. (Of course, k also depends on n through its dependence on P_n.) Then, $\mathbf{x}_0(^*P) = [x_0(P_n)]$ and $\mathbf{x_k}(^*P) = [x_k(P_n)]$ are the first and last nonstandard nodes of $^*P$. Now, $^*G$ is called *connected* if and only if, given any nonstandard nodes $\mathbf{x} = [x_n]$ and $\mathbf{y} = [y_n]$ in $^*G$, we have that, for almost all n, there exists a finite path P_n terminating at x_n and y_n. This can be restated by saying that there exists a hyperfinite path $^*P$ in $^*G$ terminating at $\mathbf{x}$ and $\mathbf{y}$.

Later on, we will need a special case of $^*\mathcal{C}$: Let $\mathcal{C}_f$ denote the subset of $\mathcal{C}$ consisting of all finite connected standard graphs $G = \{X, B\}$, where $|X| \in \mathbb{N} \setminus \{0, 1\}$, $|B| \in \mathbb{N} \setminus \{0\}$. Then, $^*\mathcal{C}_f$ is the subset of $^*\mathcal{C}$ obtained by lifting $\mathcal{C}_f$ through transfer to a subset of $^*\mathcal{C}$. In this case,

$$^*G \in^* \mathcal{C}_f \;\leftrightarrow (^*G = \{^*X, ^*B\} \in^* \mathcal{C}) \wedge (|^*X| \in^* \mathbb{N} \setminus \{0, 1\} \wedge |^*B| \in^* \mathbb{N} \setminus \{0\}) \,. \quad (8.6)$$

We call such a $^*G$ a *nonstandard hyperfinite connected* graph.

8.5 Nonstandard Subgraphs

If A and C are sets of nodes with $A \subseteq C$, we get upon transfer the following definition for sets of nonstandard nodes:

$$^*A \subseteq {}^*C \;\leftrightarrow\; (\forall\, \mathbf{x} \in {}^*A)\; (\mathbf{x} \in {}^*C)\,.$$

Our purpose in this section is to define a nonstandard subgraph $^*G_s$ of a given nonstandard graph $^*G$. We let $\mathcal{G}$ denote the set of all standard graphs. By definition, $G_s = \{X_s, B_s\}$ is a (node-induced) subgraph of $G = \{X, B\} \in \mathcal{G}$ if $X_s \subseteq X$ and B_s is the set of those branches in B that are each incident to two nodes in X_s. Let us denote the set of all such subgraphs of G by $\mathcal{G}_s(G)$. Then, symbolically G_s is defined by

$$G_s \in \mathcal{G}_s(G) \;\leftrightarrow$$

$$(\exists\, G_s = \{X_s, B_s\} \in \mathcal{G})\; (\exists\, G = \{X, B\} \in \mathcal{G})$$

$$((X_s \subseteq X) \;\wedge\; (B_s = \{b = \{x, y\} \in B : x, y \in X_s\}))\,.$$

By transfer, we get the definition of a *nonstandard subgraph* $^*G_s$ of a given nonstandard graph $^*G$. In this case, $^*\mathcal{G}$ denotes the set of all nonstandard graphs, and $^*\mathcal{G}_s(^*G)$ denotes the set of all nonstandard subgraphs of a given $^*G \in^* \mathcal{G}$.

$$^*G_s \in {}^*\mathcal{G}_s(^*G) \;\leftrightarrow$$

$$(\exists\, ^*G_s = \{^*X_s, {}^*B_s\} \in {}^*\mathcal{G})\; (\exists\, ^*G = \{^*X, ^*B\} \in {}^*\mathcal{G})$$

$$((^*X_s \subseteq {}^*X) \;\wedge\; (^*B_s = \{\mathbf{b} = \{\mathbf{x}, \mathbf{y}\} \in {}^*B : \mathbf{x}, \mathbf{y} \in {}^*X_s\}))\,.$$

8.6 Nonstandard Trees

The symbols $\mathcal{G}$, $\mathcal{G}_s(G)$, $\mathcal{C}$, $\mathcal{C}_f$, and their nonstandard counterparts have been defined above. Now, let $\mathcal{L}(G)$ be the set of all loops in the standard graph G, and let $\mathcal{T}$ be the set of all standard trees. Then, a *tree* $T = \{X_T, B_T\}$ can be defined symbolically by

$$T \in \mathcal{T} \;\leftrightarrow\; (\exists\, T \in \mathcal{C})\,(\neg(\exists\, L \in \mathcal{L}(T)))\,. \tag{8.7}$$

To transfer this, we let ${}^*\mathcal{L}({}^*G)$ be the set of all nonstandard loops in a given nonstandard graph ${}^*G$, as defined by (8.3), and we let ${}^*\mathcal{T}$ denote the set of *nonstandard trees* ${}^*T$, defined as follows:

$${}^*T \in {}^*\mathcal{T} \;\leftrightarrow\; (\exists\, {}^*T \in {}^*\mathcal{C})\,(\neg(\exists\, {}^*L \in {}^*\mathcal{L}({}^*T)))\,. \tag{8.8}$$

Next, let $\mathcal{T}_{sp}(G)$ be the set of all spanning trees in a given finite connected standard graph $G = \{X, B\}$. That T is such a *spanning tree* can be expressed symbolically as follows:

$$T \in \mathcal{T}_{sp}(G) \;\leftrightarrow \tag{8.9}$$

$$(\exists\, G = \{X, B\} \in \mathcal{C}_f)\,(\exists\, T = \{X_T, B_T\} \in \mathcal{T})\,((T \in \mathcal{G}_s(G)) \;\wedge\; (|X| = |X_T|))\,.$$

By transfer, we have the set ${}^*\mathcal{T}_{sp}({}^*G)$ of all *spanning trees* of a given hyperfinite connected nonstandard graph ${}^*G$, defined as follows:

$$\begin{aligned} &{}^*T \in {}^*\mathcal{T}_{sp}({}^*G) \;\leftrightarrow \\ &(\exists\, {}^*G = \{{}^*X, {}^*B\} \in {}^*\mathcal{C}_f)\,(\exists\, {}^*T = \{{}^*X_T, {}^*B_T\} \in {}^*\mathcal{T}) \\ &(({}^*T \in {}^*\mathcal{G}_s({}^*G)) \;\wedge\; (|{}^*X| = |{}^*X_T|))\,. \end{aligned} \tag{8.10}$$

8.7 Some Numerical Formulas

With these symbolic definitions in hand, we can now lift some standard formulas regarding numbers of nodes and branches into a nonstandard setting. For example, if p, q, and r are the number of nodes, the number of branches, and the cyclomatic number respectively of a given connected finite graph, then $r = q - p + 1$. Symbolically, this can be stated as follows. Again, we use the notation $G = \{X, B\}$ for a graph and $T = \{X_T, B_T\}$ for a tree:

$$(\forall\, p, q, r \in \mathbb{N})\,(\forall\, G \in \mathcal{C}_f)\,(\forall\, T \in \mathcal{T}_{sp}(G))$$

$$((p = |X| \;\wedge\; q = |B| \;\wedge\; r = |B| - |B_T|) \;\rightarrow\; (r = q - p + 1))\,.$$

By transfer, we obtain the following formula in hypernatural numbers, where we identify natural numbers with their images in ${}^*\mathbb{N}$:

$$(\forall\, \mathbf{p}, \mathbf{q}, \mathbf{r} \in {}^*\mathbb{N})\ (\forall\, {}^*G \in {}^*\mathcal{C}_f)\ (\forall\, {}^*T \in {}^*\mathcal{T}_{sp}({}^*G))$$

$$((\mathbf{p} = |{}^*X| \,\wedge\, \mathbf{q} = |{}^*B| \,\wedge\, \mathbf{r} = |{}^*B| - |{}^*B_T|) \;\rightarrow\; (\mathbf{r} = \mathbf{q} - \mathbf{p} + 1))\,.$$

Another standard formula [5, pages 3 and 53] for a connected finite graph having no parallel branches is that $p - 1 \le q \le p(p-1)/2$. Symbolically, we have

$$(\forall\, p, q \in \mathbb{N})\ (\forall\, G \in \mathcal{C}_f)\ ((p = |X| \,\wedge\, q = |B|) \;\rightarrow$$

$$(p - 1 \le q \le p(p-1)/2))\,.$$

So, by transfer, we have

$$(\forall\, \mathbf{p}, \mathbf{q} \in {}^*\mathbb{N})\ (\forall\, {}^*G \in {}^*\mathcal{C}_f)\ ((\mathbf{p} = |{}^*X| \,\wedge\, \mathbf{q} = |{}^*B|) \;\rightarrow$$

$$(\mathbf{p} - 1 \le \mathbf{q} \le \mathbf{p}(\mathbf{p} - 1)/2))\,.$$

Still another example of such a lifting concerns the "radius" R and "diameter" D of a finite connected graph G. It is a fact that $R \le D \le 2R$ [9, page 37], [14, pages 20–21]. Again, we need to express ideas symbolically.

Let $A \subset \mathbb{N}$ be such that $|A| \in \mathbb{N}$ (i.e., A is a finite subset of $\mathbb{N}$). We use the symbols $\overline{a} = \max A$ and $\underline{a} = \min A$ as abbreviations for the following sentences:

$$\overline{a} = \max A \;\leftrightarrow\; (\forall\, c \in A)\ (\exists\, \overline{a} \in A)\ (\overline{a} \ge c),$$

$$\underline{a} = \min A \;\leftrightarrow\; (\forall\, c \in A)\ (\exists\, \underline{a} \in A)\ (\underline{a} \le c)\,.$$

This transfers to

$$\overline{\mathbf{a}} = \max {}^*A \;\leftrightarrow\; (\forall\, \mathbf{c} \in {}^*A)\ (\exists\, \overline{\mathbf{a}} \in {}^*A)\ (\overline{\mathbf{a}} \ge \mathbf{c}),$$

$$\underline{\mathbf{a}} = \min {}^*A \;\leftrightarrow\; (\forall\, \mathbf{c} \in^* A)\ (\exists\, \underline{\mathbf{a}} \in {}^*A)\ (\underline{\mathbf{a}} \le \mathbf{c})$$

where now ${}^*A$ is a hyperfinite subset of ${}^*\mathbb{N}$ and $\overline{\mathbf{a}}$ and $\underline{\mathbf{a}}$ are hypernatural numbers. That is, ${}^*A = [A_n]$, where A_n is a finite subset of $\mathbb{N}$ for almost all n. ${}^*A$ does have a maximum element and a minimum element so that these definitions are valid (see Appendix A15).

Now, for a given finite connected graph $G \in \mathcal{C}_f$ and with x and y being any two nodes in G, let $\mathcal{P}_{x,y}$ denote the set of all paths in G terminating at x and y. The length $|P_{x,y}|$ of any $P_{x,y} \in \mathcal{P}_{x,y}$ is the number of branches in $P_{x,y}$. Then, the distance between any two nodes x and y of $G = \{X, B\}$ is the natural number $\min\{|P_{x,y}| \colon P_{x,y} \in \mathcal{P}_{x,y}\}$. So, with $\mathcal{D}(G)$ denoting the set of distances in G, we may write

$$d(x, y) \in \mathcal{D}(G) \;\leftrightarrow$$

$$(\exists x, y \in X)(\forall P_{x,y} \in \mathcal{P}_{x,y})(d(x, y) = \min\{|P_{x,y}| \colon P_{x,y} \in \mathcal{P}_{x,y}\})\,.$$

If $x = y$, we set $d(x, x) = 0$.

Now, let $^*G = \{^*X, ^*B\}$ be a hyperfinite connected graph (i.e., $^*G \in {}^*\mathcal{C}_f$, as defined in Sec. 8.4). By transfer, we have the set $^*\mathcal{P}_{\mathbf{x},\mathbf{y}}$ consisting of all the nonstandard paths in $^*G$ terminating at the nonstandard nodes **x** and **y**. We also have the internal set $^*\mathcal{D}(^*G)$ of all hypernatural distances in $^*G$:

$$\mathbf{d}(\mathbf{x}, \mathbf{y}) \in {}^*\mathcal{D}(^*G) \leftrightarrow$$

$$(\exists \mathbf{x}, \mathbf{y} \in {}^*X)(\forall {}^*P_{\mathbf{x},\mathbf{y}} \in {}^*\mathcal{P}_{\mathbf{x},\mathbf{y}})(\mathbf{d}(\mathbf{x}, \mathbf{y}) = \min\{|^*P_{\mathbf{x},\mathbf{y}}| : {}^*P_{\mathbf{x},\mathbf{y}} \in {}^*\mathcal{P}_{\mathbf{x},\mathbf{y}}\}).$$

Here, too, $\mathbf{d}(\mathbf{x}, \mathbf{x}) = \mathbf{0}$.

Furthermore, two nonstandard nodes $\mathbf{x}, \mathbf{y} \in {}^*X$ are said to be *limitedly distant* if $\mathbf{d}(\mathbf{x}, \mathbf{y})$ is a limited hypernatural (i.e., a natural number). This defines an equivalence relation on $^*X$; indeed, reflexivity and symmetry are obvious, and the triangle inequality follows by transfer from the triangle inequality for distances in G. So, by analogy with galaxies for hyperreals, we might refer to the equivalence classes induced by limited distances as "nodal galaxies"; these partition $^*X$.

Next, we have the eccentricity e_x of any node of $G = \{X, B\} \in \mathcal{C}_f$ as the maximum of the distances from x. With $\mathcal{E}(G)$ being the set of eccentricities in G and with $\mathcal{D}_x$ being the set of distances from x, we have symbolically

$$e_x \in \mathcal{E}(G) \leftrightarrow (\exists x \in X)(e_x = \max\{d(x) : d(x) \in \mathcal{D}_x\}).$$

So, for a hyperfinite connected graph $^*\mathcal{G} = \{^*X, ^*B\} \in {}^*\mathcal{C}_f$, we have by transfer

$$\mathbf{e}_\mathbf{x} \in {}^*\mathcal{E}(^*G) \leftrightarrow (\exists \mathbf{x} \in {}^*X)(\mathbf{e}_\mathbf{x} = \max\{\mathbf{d}(\mathbf{x}) : \mathbf{d}(\mathbf{x}) \in {}^*\mathcal{D}_\mathbf{x}\}),$$

where now $^*\mathcal{E}(^*G)$ is the set of hypernatural eccentricities for nonstandard nodes in $^*G$ and $^*\mathcal{D}_\mathbf{x}$ is the set of hypernatural distances starting at the nonstandard node **x**.

Then, for any $G = \{X, B\} \in \mathcal{C}_f$, the *radius* $R(G)$ is defined by

$$(\forall\, e_x \in \mathcal{E}(G))\ (\exists\ R(G) \in \mathbb{N})\ (R(G) = \min\{e_x : x \in X\}),$$

which by transfer gives the following definition of the hypernatural radius $^*R(^*G) \in {}^*\mathbb{N}$ of any $^*G = \{^*X, ^*B\} \in {}^*\mathcal{C}_f$:

$$(\forall\, \mathbf{e}_\mathbf{x} \in {}^*\mathcal{E}(G))\ (\exists\ {}^*R(^*G) \in {}^*\mathbb{N})\ ({}^*R(^*G) = \min\{\mathbf{e}_\mathbf{x} : \mathbf{x} \in {}^*X\}).$$

Similarly, the *diameter* $D(G)$ of G is defined by

$$(\forall\, e_x \in \mathcal{E}(G))\ (\exists\ D(G) \in \mathbb{N})\ (D(G) = \max\{e_x : x \in X\}),$$

which by transfer gives the hypernatural diameter $^*D(^*G) \in {}^*\mathbb{N}$ of $^*G$:

$$(\forall\, \mathbf{e}_\mathbf{x} \in {}^*\mathcal{E}(^*G))\ (\exists\ {}^*D(^*G) \in {}^*\mathbb{N})\ ({}^*D(^*G) = \max\{\mathbf{e}_\mathbf{x} : \mathbf{x} \in {}^*X\}).$$

Both $^*R(^*G)$ and $^*D(^*G)$ exist because $\mathcal{E}(^*G)$ is a hyperfinite set of hypernaturals (Appendix A15).

We have the following sentence for a standard result:

$$(\forall\ G \in \mathcal{C}_f)\ (R(G) \le D(G) \le 2R(G)).$$

By transfer this yields the nonstandard result:

$$(\forall\ {}^*G \in {}^*\mathcal{C}_f)\ ({}^*R(^*G) \le {}^*D(^*G) \le 2^*R(^*G)).$$

8.8 Nonstandard 1-Graphs

Our objective in this section is to introduce transfiniteness into our construction of nonstandard graphs, but we shall do so only for the first rank of transfiniteness, that is, we construct a nonstandard 1-graph. We shall employ an ultrapower construction based upon a given sequence $\langle G_n^1 : n \in \mathbb{N}\rangle$ of 1-graphs $G_n^1 = \{X_n^0, B_n, X_n^1\}$, where now $G_n^0 = \{X_n^0, B_n\}$ is a 0-graph defined as in the first paragraph of Sec. 8.1 and X_n^1 is a set of 1-nodes defined from a partition of the 0-tips augmented possibly by 0-nodes exactly as in Sec. 2.3. We could define the nonstandard 1-nodes from sequences of 1-nodes taken from the G_n^1 in the same way as are the nonstandard 0-nodes obtained in Sec. 8.1 from the 0-nodes of the G_n^0. However, we wish to relate those nonstandard 1-nodes to the nonstandard 0-tips and nonstandard 0-nodes obtained from the sequence $\langle G_n^0 : n \in \mathbb{N}\rangle$ of 0-graphs G_n^0. So, we shall proceed in a more detailed fashion. As always, we choose and fix upon a nonprincipal ultrafilter $\mathcal{F}$ on $\mathbb{N}$. Given the sequence $\langle \mathcal{G}_n^1 : n \in \mathbb{N}\rangle$ of 1-graphs, the "individuals" are now the 0-nodes and 0-tips of all the $\mathcal{G}_n^1$ along with the natural numbers.

For each n, the set $\mathcal{T}_n^0$ of all the 0-tips of the 0-graph $G_n^0 = \{X_n^0, B_n\}$ is partitioned in some arbitrarily chosen fashion into subsets $\{\mathcal{T}_{n,k}^0 : k \in K\}$, where K is an index set for the partition $\mathcal{T}_n^0 = \bigcup_{k\in K} \mathcal{T}_{n,k}^0$. Each $\mathcal{T}_{n,k}^0$ may be, but need not be, augmented by exactly one 0-node of X_n^0, but no 0-node augments more than one set $\mathcal{T}_{n,k}^0$. Each such possibly augmented set is a 1-node $x_{n,k}^1$ of G_n^1; that is, $x_{n,k}^1 = \mathcal{T}_{n,k}^0 \cup \mathcal{Z}_{n,k}^0$, where $\mathcal{Z}_{n,k}^0$ is either the empty set or is a singleton whose element is a 0-node, with $\mathcal{Z}_{n,k}^0 \cap \mathcal{Z}_{n,l}^0 = \emptyset$ if $k \neq l$. We refer to the 0-tips and the augmenting 0-nodes (those occurring in 1-nodes) as the *extremities*[6] of the 0-graph $\{X_n^0, B_n\}$. If e_n and f_n are extremities in the same 1-node $x_{n,k}^1$ of G_n^1, we will say that e_n and f_n are *shorted together* by $x_{n,k}^1$ and will write $e_n \asymp f_n$.

Our next step is to make an ultrapower construction to get the nonstandard 1-nodes. We consider sequences of extremities of the G_n^0, $\langle e_n\rangle$ being one such sequence and e_n being an extremity of G_n^0. Two such sequences $\langle e_n\rangle$ and $\langle f_n\rangle$ are taken to be equivalent if $e_n = f_n$ for almost all n. This partitions the set of all such sequences into equivalence classes. Indeed, reflexivity and symmetry are obvious, and transitivity follows as usual (i.e., if $e_n = f_n$ a.e. and if $f_n = g_n$ a.e., then $e_n = g_n$ a.e.). Each equivalence class is taken to be a *nonstandard extremity* $\mathbf{e} = [e_n]$ where $\langle e_n\rangle$ is any sequence in that equivalence class.

Given any sequence $\langle e_n\rangle$, let $N_{t^0} = \{n : e_n \text{ is a 0-tip}\}$ and $N_{x^0} = \{n : e_n \text{ is a 0-node}\}$. Thus, $N_{t^0} \cap N_{x^0} = \emptyset$ and $N_{t^0} \cup N_{x^0} = \mathbb{N}$. So, exactly one of N_{t^0} and N_{x^0} is a member of $\mathcal{F}$. If it is N_{t^0} (resp. N_{x^0}), $\langle e_n\rangle$ is a representative of a *nonstandard 0-tip* (resp. a *nonstandard 0-node*).

Now, let $\mathbf{e} = [e_n]$ and $\mathbf{f} = [f_n]$ be two nonstandard extremities, and let $\mathbb{N}_{ef} = \{n : e_n \asymp f_n\}$ and $\mathbb{N}_{ef}^c = \{n : e_n \not\asymp f_n\}$. Exactly one of $\mathbb{N}_{ef}$ and $\mathbb{N}_{ef}^c$ is a member of $\mathcal{F}$. If it is $\mathbb{N}_{ef}$ (resp. $\mathbb{N}_{ef}^c$), we say that $\mathbf{e}$ is *shorted* to $\mathbf{f}$ (resp. $\mathbf{e}$ is *not shorted* to $\mathbf{f}$),

[6] Any 0-node in a 1-node is nonmaximal in G_n^1 and might be viewed as an "extremity" because it coincides with the 1-node so far as its position in the 1-graph is concerned.

and we write $\mathbf{e} \asymp \mathbf{f}$ (resp. $\mathbf{e} \not\asymp \mathbf{f}$). Furthermore, we take it that every $\mathbf{e}$ is shorted to itself: $\mathbf{e} \asymp \mathbf{e}$. This shorting is an equivalence relation for the set of all nonstandard extremities, as can be shown much as before; indeed, for transitivity, assume $\mathbf{e} \asymp \mathbf{f}$ and $\mathbf{f} \asymp \mathbf{g}$. Since $\{n\colon e_n \asymp f_n\} \cap \{n\colon f_n \asymp g_n\} \subseteq \{n\colon e_n \asymp g_n\}$, we have $\mathbf{e} \asymp \mathbf{g}$. The resulting equivalence classes are the *nonstandard* 1*-nodes*.

This definition can be shown to be independent of the representative sequences chosen for the nonstandard extremities. To be specific, let $\mathbf{e} = [e_n] = [\tilde{e}_n]$ and $\mathbf{f} = [f_n] = [\tilde{f}_n]$. Set $\mathbb{N}_e = \{n : e_n = \tilde{e}_n\} \in \mathcal{F}$ and $\mathbb{N}_f = \{n : f_n = \tilde{f}_n\} \in \mathcal{F}$. Assume $[e_n] \asymp [f_n]$. Thus, $\mathbb{N}_{ef} = \{n : e_n \asymp f_n\} \in \mathcal{F}$. We want to show that $\mathbb{N}_{\tilde{e}\tilde{f}} = \{n : \tilde{e} = \tilde{f}\} \in \mathcal{F}$ and thus $[\tilde{e}_n] \asymp [\tilde{f}_n]$. We have $(\mathbb{N}_e \cap \mathbb{N}_f \cap \mathbb{N}_{ef}) \subseteq \mathbb{N}_{\tilde{e}\tilde{f}}$, whence our conclusion.

Altogether we have defined a nonstandard 1-node $\mathbf{x}^1$ to be any set in the partition of the set of nonstandard extremities induced by the shorting $\asymp$, with every nonstandard 1-node having at least one nonstandard 0-tip. $^*X^1$ will denote the set of nonstandard 1-nodes. Finally, we define the *nonstandard* 1*-graph* $^*G^1$ to be the triplet $^*G^1 = \{^*X^0, {}^*B, {}^*X^1\}$.

Let us observe that each nonstandard 1-node $\mathbf{x}^1$ contains no more than one nonstandard 0-node, and, if it does contain such a 0-node, it does not share that nonstandard 0-node with any other nonstandard 1-node. Indeed, if $\mathbf{x}^1$ had two nonstandard 0-nodes, then, for at least one n, two 0-nodes in G_n^1 would have to be shorted together within a 1-node of G_n^1, a violation of the definition of standard 1-nodes. Our second observation follows in the same way because for no n will a 1-node in G_n^1 share a 0-node with another 1-node in G_n^1.

8.9 A Fundamental Theorem for Nonstandard 1-Networks

Several fundamental theorems[7] concerning the existence of a current-voltage regime in a 1-network can be lifted into a nonstandard setting. Such liftings involve essentially the same manipulations except for minor differences in details. Let us perform one of them to illustrate how this can be done.

We now start with a sequence $\langle \mathbf{N}_n^1 \rangle$ of 1-networks, each 1-network $\mathbf{N}_n^1$ having $G_n^1 = \{X_n^0, B_n, X_n^1\}$ as its 1-graph and with each branch b_n of G_n^1 being assigned an orientation and electrical parameters r_{b_n} and e_{b_n} connected in the Thevenin form as shown in Fig. 6.6(b). The individuals are now the 0-nodes and 0-tips in all the $\mathbf{N}_n^1$ along with the real numbers.

We will first present a fundamental theorem asserting a unique current-voltage regime for each $\mathbf{N}_n^1$. To do this, we need to construct for each $\mathbf{N}_n^1$ a solution space $\mathcal{L}_n$ that will be searched for a unique branch-current vector satisfying a form of Tellegen's equation (given by (8.12) below). In the following, we let $\sum_{b_n \in B_n}$ denote the summation over all the branches b_n in B_n. $\mathcal{I}_n$ will denote the linear space over the

[7] Such as [50, Theorems 3.3-1 and 3.5-2], [51, Theorems 5.2-8 and 5.2-14], [54, Theorem 5.1-4], and Theorem 5.13-3 in this book.

field $\mathbb{R}$ of all finite-powered branch-current vectors in $\mathbf{N}_n^1$; that is, $i_n = \{i_{b_n} : b_n \in B_n\}$ is a member of $\mathcal{I}_n$ whenever

$$\|i_n\|^2 = \sum_{b_n \in B_n} i_{b_n}^2 r_{b_n} < \infty.$$

We assign the norm $\|i_n\|$ to the members of $\mathcal{I}_n$ and make $\mathcal{I}_n$ a real Hilbert space with the inner product $(i_n, s_n) = \sum_{b_n \in B_n} i_{b_n} s_{b_n} r_{b_n}$.

The (unit) *loop current* for a given oriented loop L in $\mathbf{N}_n^1$ is a vector $i_n = \{i_{b_n} : b_n \in B_n\}$ of branch currents i_{b_n} such that $i_{b_n} = 1$ (resp. $i_{b_n} = -1$, resp. $i_{b_n} = 0$) if the branch b_n is in L with the same orientation as L (resp. b_n is in L with the opposite orientation, resp. b_n is not in L). If L is a 0-loop, the loop current for L is a member of $\mathcal{L}_n$, but, if L is a 1-loop, the loop current for L will be in $\mathcal{I}_n$ if and only if the sum of branch resistances r_{b_n} for the branches in L is finite.

Let $\mathcal{L}_n^o$ be the span of all the loop currents in $\mathcal{I}_n$. Finally, let $\mathcal{L}_n$ be the closure of $\mathcal{L}_n^o$ in $\mathcal{I}_n$. $\mathcal{L}_n$ is a subspace of $\mathcal{I}_n$ and is a Hilbert space by itself with the same norm and inner product as those of $\mathcal{I}_n$.

Next, let $e_n = \{e_{b_n} : b_n \in B_n\}$ be the vector of branch voltage sources in $\mathbf{N}_n^1$. We say that e_n is of *finite total isolated power* if

$$\sum_{b_n \in B_n} e_{b_n}^2 g_{b_n} < \infty, \tag{8.11}$$

where $g_{b_n} = 1/r_{b_n}$. We let $\mathcal{E}_{f,n}$ denote the set of all e_n satisfying (8.11).

We have the following fundamental theorem for $\mathbf{N}_n^1$.

Theorem 8.9-1. *If $e_n \in \mathcal{E}_{f,n}$, then there exists a unique branch-current vector $i_n \in \mathcal{L}_n$ such that*

$$\sum_{b_n \in B_n} r_{b_n} i_{b_n} s_{b_n} = \sum_{b_n \in B_n} e_{b_n} s_{b_n} \tag{8.12}$$

for every $s_n = \{s_{b_n} : b_n \in B_n\} \in \mathcal{L}_n$.

The proof of this theorem is much the same as that of Theorem 5.13-3 and will not be repeated. Note that the branch voltages $v_{b_n} = r_{b_n} i_{b_n} - e_{b_n}$ are also determined by this theorem.

We wish to obtain a nonstandard version of this theorem that is applicable to a nonstandard 1-network ${}^*\mathbf{N}^1$ obtained from a given sequence $\langle \mathbf{N}_n^1 \rangle$ of 1-networks through an ultrapower construction. Upon constructing the nonstandard 1-graph ${}^*G^1 = \{{}^*X^0, {}^*B, {}^*X^1\}$ from $\langle G_n^1 \rangle$ as in the preceding section, the branch parameters undergo an ultrapower construction as well to become hyperreal parameters. Thus, each nonstandard branch $\mathbf{b}$ of ${}^*G^1$ has a hyperreal positive resistor $\mathbf{r_b}$ and possibly a nonzero hyperreal branch voltage source $\mathbf{e_b}$. In particular, for $\mathbf{b} = [\{x_n^0, y_n^0\}] \in {}^*B$, we have $\mathbf{r_b} = [r_{b_n}]$, where $r_{b_n} > 0$ is the resistance of the branch $b_n = \{x_n^0, y_n^0\}$ for almost all n, and similarly $\mathbf{e_b} = [e_{b_n}]$, where $e_{b_n} \in \mathbb{R}$ is the branch voltage source for b_n for almost all n. All this yields the *nonstandard 1-network* ${}^*\mathbf{N}^1 = [\mathbf{N}_n^1]$

whose 1-graph is ${}^*G^1$ and whose branch parameters are the hyperreals $\mathbf{r_b}$ and $\mathbf{e_b}$. Furthermore, i_{b_n} and v_{b_n} denote the branch current and branch voltage for the branch $b_n \in B_n$, and these yield the hyperreal branch current $\mathbf{i_b} = [i_{b_n}]$ and hyperreal branch voltage $\mathbf{v_b} = [v_{b_n}]$ for each $\mathbf{b} = [b_n] \in {}^*B$.

Consider next any space $\mathcal{L}_n$ ($n \in \mathbb{N}$). Every member $i_n = \{i_{b_n} : b_n \in \mathcal{B}_n\}$ of $\mathcal{L}_n$ determines a function mapping the set $\mathcal{B}_n$ of branches in $\mathbf{N}_n^1$ into $\mathbb{R}$, and thus by means of an ultrapower construction of $[i_n] = \mathbf{i} = \{\mathbf{i_b} : b \in {}^*B\}$ determines an internal function mapping ${}^*B$ into ${}^*\mathbb{R}$ with regard to the nonstandard network ${}^*\mathbf{N}_n^1$. In particular, $\mathbf{i_b} = [i_{b_n}]$, where $\{i_{b_n} : b_n \in B_n\}$ is a member of $\mathcal{L}_n$ for almost all n. All this yields a solution space ${}^*\mathcal{L} = [\mathcal{L}_n]$ consisting of the nonstandard current vectors $\mathbf{i} = \{\mathbf{i_b} : \mathbf{b} \in {}^*B\}$.

In order to invoke Theorem 8.9-1, we also assume that, for almost all n, the branch voltage sources e_{b_n} together have finite total isolated power (i.e., (8.11) is satisfied for almost all n). We let ${}^*\mathcal{E}_f$ denote the set of such nonstandard branch-voltage-source vectors; that is, each member of ${}^*\mathcal{E}_f$ is a vector $\mathbf{e} = \{\mathbf{e_b} : \mathbf{b} \in {}^*B\}$, where $\mathbf{e_b} = [e_{b_n}]$ and the e_{b_n} satisfy (8.11) for almost all n.

Then, Theorem 8.9-1 holds again for almost all n. This can be restated as follows for the nonstandard 1-network ${}^*\mathbf{N}^1$.

Theorem 8.9-2. *If* $\mathbf{e} \in {}^*\mathcal{E}_f$, *then there exists a unique branch-current vector* $\mathbf{i} = \{\mathbf{i_b} : \mathbf{b} \in {}^*B\} \in {}^*\mathcal{L}$ *such that*

$$\sum_{\mathbf{b} \in {}^*B} \mathbf{r_b i_b s_b} = \sum_{\mathbf{b} \in {}^*B} \mathbf{e_b s_b} \tag{8.13}$$

for every $\mathbf{s} = \{\mathbf{s_b} : \mathbf{b} \in {}^*B\} \in {}^*\mathcal{L}$.

Each side of (8.13) is well-defined as the hyperreal having as a representative sequence the sequence of real numbers given by the corresponding side of (8.12).

Note that the uniquely determined branch-current vector $\mathbf{i}$ determines a unique branch-voltage vector $\mathbf{v} = \{\mathbf{v_b} : \mathbf{b} \in {}^*B\}$ by means of Ohm's law:

$$\mathbf{v} = \mathbf{r_b i_b} - \mathbf{e_b}\,.$$

Theorem 8.9-2 could also have been obtained from Theorem 8.9-1 by appending asterisks in accordance with the transfer principle, but the latter procedure would require that we express all the conditions inherent in Theorem 8.9-1 symbolically.

Kirchhoff's laws can also be lifted in a nonstandard way for the current-voltage regime dictated by Theorem 8.9-2. First, consider Kirchhoff's current law. The nonstandard 0-node $\mathbf{x}^0 = [x_n^0]$ is called *maximal* if x_n^0 is maximal in $\mathbf{N}_n^1$ for almost all n. In the following, $\sum_{b_n \ni x_n^0}$ will denote a summation over all branches b_n that are incident at x_n^0. As in Sec. 5.14, x_n^0 is called *restraining* if the sum of the conductances g_b for the branches incident at x_n^0 is finite (i.e., in symbols, if $\sum_{b_n \ni x_n^0} g_{b_n} < \infty$. Then, we say that $\mathbf{x}^0$ is *restraining* if x_n^0 is restraining for almost all n.

Now, under the assumptions on $\mathbf{N}_n^1$ required for Theorem 8.9-1, Kirchhoff's current law is satisfied at every restraining maximal 0-node x_n^0 as follows:

$$\sum_{b_n \ni x_n^0} \pm i_{b_n} = 0 \tag{8.14}$$

where the plus (resp. minus) sign is used if b_n is incident away from (resp. toward) x_n^0. Furthermore, the summation in (8.14) converges absolutely. The proof of all this is virtually the same as that of Theorem 5.14-1.

Turning to the nonstandard case, we first observe that every branch b_n in $\mathbf{N}_n^1$ has an orientation. So, for $\mathbf{x}^0 = [x_n^0]$ and $\mathbf{b} = [b_n]$, every branch b_n incident at x_n^0 is either oriented away from x_n^0 a.e. or is oriented toward x_n^0 a.e. Thus, $\mathbf{b}$ acquires an orientation either away from $\mathbf{x}^0$ or toward $\mathbf{x}^0$. We also have that (8.14) holds as stated for almost all n. We set

$$\sum_{\mathbf{b} \ni \mathbf{x}^0} \pm \mathbf{i_b} = \left[\sum_{b_n \ni x_n^0} \pm i_{b_n} \right].$$

In this way, we have Kirchhoff's current law for ${}^*\mathbf{N}^1$ as follows.

Theorem 8.9-3. *If $\mathbf{x}^0$ is a restraining maximal 0-node in ${}^*\mathbf{N}^1$, then under the regime dictated by Theorem 8.9-2,*

$$\sum_{\mathbf{b} \ni \mathbf{x}^0} \pm \mathbf{i_b} = \mathbf{0} \tag{8.15}$$

where the summation converges absolutely (i.e., $\sum_{\mathbf{b} \ni \mathbf{x}^0} |\mathbf{i_b}| < \infty$).

Next, we discuss a nonstandard version of Kirchhoff's voltage law for ${}^*\mathbf{N}^1 = [\mathbf{N}_n^1]$. For this purpose we need to define nonstandard loops. Nonstandard 0-loops were defined in Sec. 8.3, but now we also need nonstandard 1-loops. We could base their definition on the recursive construction of standard 1-loops given in Sec. 2.3, but it is awkward to extend the latter to a nonstandard setting. There is a simpler way based upon the degrees of nodes, which works for both 0-loops and 1-loops. This idea was developed in Sec. 2.7 for both paths and loops and for all ranks up to ω. Here, we only need it for loops of ranks 0 and 1, so let us summarize it for this simpler case using a somewhat different notation.

Let ${}^*\mathbf{N}^1 = [\mathbf{N}_n^1]$ be a nonstandard 1-network. A (branch-induced) subgraph $G_{s,n}$ and its corresponding reduced graph $G_{r,n}$ of the standard 1-graph of $\mathbf{N}_n^1$ can be defined as was done in Sec. 2.3. Then, the *relative degree*[8] $d_x(G_{s,n})$ of a node x (0-node or 1-node) in $G_{s,n}$ is the cardinality of the set of tips (elementary tips or 0-tips) embraced by x and traversed by $G_{s,n}$. Moreover, a node x of $G_{s,n}$ is called *relatively maximal with respect to* $G_{s,n}$ if its corresponding reduced node is maximal in the reduced graph $G_{r,n}$ corresponding to $G_{s,n}$. Finally, a loop L (0-loop or 1-loop) in $\mathbf{N}_n^1$ is a connected subgraph $G_{s,n}$ having at least three branches and whose every node x that is relatively maximal with respect to $G_{s,n}$ has a relative degree equal to 2 (i.e., $d_x(G_{s,n}) = 2$ for every relatively maximal node x in $G_{s,n}$). This definition is entirely

[8]This was denoted by $d_{x|R_n}$ in Sec. 2.7, where $R = G_{r,n}$.

equivalent to those used in Secs. 2.2 and 2.3 when parallel branches, self-loops, and isolated nodes are disallowed.

Now, consider any sequence $\langle G_{s,n}\rangle$ of subgraphs $G_{s,n}$ in the $\mathbf{N}_n^1$. This determines a nonstandard subgraph ${}^*G_s$ of the nonstandard graph ${}^*G^1$ of ${}^*\mathbf{N}^1$ in the same way as $\langle G_n^1\rangle$ determines the nonstandard 1-graph ${}^*G^1$ of ${}^*\mathbf{N}^1$. (A nonstandard branch $\mathbf{b} = [b_n]$ is in ${}^*G_s$ if and only if b_n is in $G_{s,n}$ for almost all n, and similarly for 0-nodes, 0-tips, and 1-nodes.) Then, ${}^*G_s$ is a nonstandard loop (0-loop or 1-loop) if, for almost all n, $G_{s,n}$ is connected, has at least three branches, and the relative degrees of all its relatively maximal nodes equal 2. In this case, we write $\mathbf{L} = {}^*G_s = [L_n]$, where $L_n = G_{s,n}$ is a loop in $\mathbf{N}_n^1$ for almost all n.

In the following, $\sum_{b_n \dashv L_n}$ denotes a sum over all the branches in the standard loop L_n. L_n is called *permissive* if $\sum_{b_n \dashv L_n} r_{b_n} < \infty$. Furthermore, we assign an orientation to each loop L_n. Under the regime dictated by Theorem 8.9-1, Kirchhoff's voltage law is satisfied around every permissive loop L_n in $\mathbf{N}_n^1$. In symbols,

$$\sum_{b_n \dashv L_n} \pm v_{b_n} = 0 \tag{8.16}$$

where the plus (resp. minus) sign is used if the orientations of b and L_n agree (resp. disagree). The proof of this is much the same as that of Theorem 5.14-2.

With regard to the nonstandard case, the nonstandard loop $\mathbf{L} = [L_n]$ is called *permissive* if L_n is permissive for almost all n. Also, $\mathbf{L}$ acquires an orientation with regard to its nonstandard branches $\mathbf{b} = [b_n]$ in the following way. For almost all n, b_n is in L_n, and the orientation of b_n either agrees a.e or disagrees a.e with the orientation of L_n. So, if $\mathbf{v_b}$ is the hyperreal voltage of the nonstandard oriented branch $\mathbf{b}$ in $\mathbf{L}$, we have unambiguously the voltage $+\mathbf{v_b}$ or $-\mathbf{v_b}$ measured with respect to this implicitly defined orientation of $\mathbf{L}$. Upon setting

$$\sum_{\mathbf{b} \dashv \mathbf{L}} \pm \mathbf{v_b} = \left[\sum_{b_n \dashv L_n} \pm v_{b_n} \right],$$

we obtain the following nonstandard version of Kirchhoff's voltage law.

Theorem 8.9-4. *If* $\mathbf{L}$ *is an oriented permissive loop in* ${}^*\mathbf{N}^1$, *then under the regime dictated by Theorem 8.9-2,*

$$\sum_{\mathbf{b} \dashv \mathbf{L}} \pm \mathbf{v_b} = \mathbf{0} \tag{8.17}$$

where the summation converges absolutely (i.e., $\sum_{\mathbf{b} \dashv \mathbf{L}} |\mathbf{v_b}| < \infty$*)).*

Finally, let us note an immediate corollary. If x_n^0 is a 0-node of finite degree for almost all n, then $\mathbf{x}^0 = [x_n^0]$ is restraining. Also, if L_n is a finite 0-loop for almost all n, then $\mathbf{L} = [L_n]$ is permissive. It follows that Kirchhoff's laws will always hold for nonstandard 0-networks having hyperfinite graphs, as defined in Sec. 8.4.

Corollary 8.9-5. *If the nonstandard 0-network* ${}^*\mathbf{N}^0$ *has a hyperfinite graph, then Kirchhoff's laws are satisfied at all its nodes and around all its loops.*

This result can be obtained much more easily by transferring the sentences for Kirchhoff's laws for finite networks.

In general, however, the possible nonsatisfaction of Kirchhoff's current law at nonrestraining 0-nodes of infinite degree and the possible nonsatisfaction of Kirchhoff's voltage law around nonpermissive transfinite loops in standard transfinite 1-networks is reflected in the nonstandard 1-networks of this chapter. So, here is another way the use of nonstandard analysis in this chapter differs radically from that of Chapter 6. There, nonstandard analysis reestablished Kirchhoff's laws at nonrestraining nodes of infinite degree and around nonpermissive transfinite loops, without the graphs of those networks being enlarged in a nonstandard way. Here, such an enlargement takes place, but the possible nonsatisfaction of Kirchhoff's laws persists.

Finally, let us simply take note of the following nonlinear result obtained by transferring Duffin's theorem (see [20] or [54, Sec. 6.4]). Let ${}^*\mathbf{N}^0 = [\mathbf{N}_n^0]$ be any nonstandard nonlinear 0-network such that its nonstandard graph ${}^*G^0 = [G_n^0]$ is hyperfinite and, for almost all n, the resistance characteristic $R_{b_n} : i_{b_n} \mapsto v_{b_n}$ for each branch $b_n \in B_n$ in G_n^0 is a continuous, strictly monotonically increasing bijection of $\mathbb{R}$ onto $\mathbb{R}$.[9] Then, the hyperreal current-voltage regime for ${}^*\mathbf{N}^0$ (i.e., the hyperreal operating point) is determined by Kirchhoff's current law at each nonstandard 0-node $\mathbf{x}^0 = [x_n^0]$, by Kirchhoff's voltage law around each nonstandard 0-loop, and by the replacement of Ohm's law by the expression $\mathbf{v_b} = \mathbf{R_b}(\mathbf{i_b})$ for each nonstandard branch $\mathbf{b} = [b_n]$, where now (in contrast to the last paragraph of Sec. 6.4) $\mathbf{R_b} = [R_{b_n}]$ is the internal nonlinear resistance characteristic, that is, $v_{b_n} = R_{b_n}(i_{b_n})$ for almost all n.

[9] See the the end of Sec. 6.4 regarding these conditions; the resistance characteristic R_j for the jth branch b_j given there is now denoted by R_{b_n} for each branch $b_n \in B_n$.

A

Some Elements of Nonstandard Analysis

A.1 The Purpose of This Appendix

In this appendix, we present an outline of some definitions and results needed for an understanding of those parts of nonstandard analysis that we use in this book. This is by no means a tutorial exposition but is instead a listing of ideas interspersed with some explanations. Moreover, no proofs are given. Substantial expositions of nonstandard analysis can now be found in quite a number of books and monographs. The textbook by R. Goldblatt [24] published in 1998 lists 41 of them. Particular books, parts of whose expositions we borrow, are [4], [26], and [29]. Moreover, there is some divergence in the notation and terminology throughout these sources. Another purpose of this appendix is to specify our symbols and terms, which follow almost entirely those of Goldblatt.

There are two approaches to nonstandard analysis, the axiomatic and the constructive. Actually, each subsumes the other, the only difference being the order in which ideas are introduced. The axiomatic approach is based upon symbolic logic and is the method used by Robinson in his seminal works [34], [35]. The constructive approach is probably more accessible to those accustomed to mathematical analysis rather than mathematical logic, and this is the approach we follow up to Sec. A20. After that, we turn to the sentences of symbolic logic to arrive at the powerful "transfer principle," through which results concerning hyperreals and much more can be obtained more directly.

A.2 Infinitesimals and Hyperreals

So, how can one construct an infinitesimal? Well, how are the real numbers constructed? Answering the latter question may give us a clue about the former one. The construction of the rational numbers as ratios of integers was accomplished by the ancients, but it was disconcerting to the Pythagoreans when they discovered that the diagonal of the unit square is not such a ratio [30, pages 104–105]. This situation remained unresolved from ancient times to the late 19th century, when at last Cantor and Dedekind [30, page 179] proposed different but equivalent definitions for the real numbers. The definition most pertinent for us is the one whereby a real number is defined as an equivalent class of Cauchy sequences of rational numbers, two Cauchy sequences being considered equivalent if their terms approach each other. In this way, the real numbers expand and fill out the set of rational numbers.

This idea can be reworked to define infinitesimals as certain equivalence classes of sequences of real numbers that approximate 0 in a certain way but—roughly speaking—with different asymptotic rates of convergence, yielding thereby different infinitesimals. More generally, all real sequences can be partitioned into equivalence classes, called *hyperreal numbers*, and they include not only infinitesimals but also numbers infinitesimally close to any real number, as well as infinitely large numbers. In this way, the real line is enlarged into the hyperreal line, and this in turn empowers simplifications of real analysis, new concepts and methods, and substantial generalizations of conventional ideas.

A.3 Other Nonstandard Entities

There is even more generality available. So far, the sequences we have mentioned take the real numbers as their elements. We can expand our purview by allowing those elements to be members of any set A that contains the set $\mathbb{R}$ of real numbers as a subset. The elements of A are called *individuals* (or "atoms" or "urelements") and are assumed to be not sets, that is, they have no elements—even though in another context they may be sets. For example, A could be the union of $\mathbb{R}$ and the set $\{\mathbf{N}_f\}$ of all finite electrical networks, where two finite electrical networks are identified as being the same network if there exists a graphical isomorphism between them that preserves the electrical parameters of the branches. As another example, A could be $\mathbb{R} \cup \mathcal{N}$, where $\mathcal{N}$ is the set of nodes in a given transfinite graph; this case is explored in Chapter 8.[1]

Turning to A itself, two sequences $\langle a_n \rangle$ and $\langle b_n \rangle$ with $a_n, b_n \in A$ for all n are defined as being "equivalent" if $\{n : a_n = b_n\}$ is a "large enough" subset of $\mathbb{N}$, an idea we explicate shortly. This partitions such sequences into equivalence classes that are taken to be nonstandard entities, and are in particular the hyperreal numbers when $a_n, b_n \in \mathbb{R}$. This is called an *ultrapower construction* of those nonstandard entities.

Still more generality can be achieved if the set A of individuals is expanded into a "superstructure" by applying recursively the power-set operation, that is, by taking the set of all subsets repeatedly in a certain way. This more general approach is described in Secs. A17 through A20.

[1] Actually, we only need $\mathbb{N}$ in place of $\mathbb{R}$ in Secs. 8.1 through 8.8. It is only in Sec. 8.9 that we need the real numbers as some of the individuals.

A.4 Nonprincipal U ltrafilters

With $\mathbb{N} = \{0, 1, 2, \dots\}$ being the set of natural numbers, as always, a "large enough" set is taken to be any member of a chosen and fixed *nonprincipal ultrafilter*[2] $\mathcal{F}$ *on* $\mathbb{N}$ defined as follows: $\mathcal{F}$ is a collection of nonempty subsets of $\mathbb{N}$ satisfying the following axioms:

1. If $A, B \in \mathcal{F}$, then $A \cap B \in \mathcal{F}$.
2. If $A \in \mathcal{F}$ and $A \subseteq B \subseteq \mathbb{N}$, then $B \in \mathcal{F}$.
3. For any $A \subseteq \mathbb{N}$, either $A \in \mathcal{F}$ or $A^c \in \mathcal{F}$ but not both. Here, $A^c = \mathbb{N} \backslash A$ denotes the complement of A in $\mathbb{N}$.
4. No finite subset of $\mathcal{N}$ is a member of $\mathcal{F}$; in particular, $\emptyset \notin \mathcal{F}$.

As a result of these assumptions, we have the following properties:

a. $\mathbb{N} \in \mathcal{F}$.
b. If $\{A_1, A_2, \dots, A_k\}$ is a finite collection of mutually disjoint subsets of $\mathcal{F}$, then no more than one of them is a member of $\mathcal{F}$. If in addition $\cup_{j=1}^{k} A_j = \mathbb{N}$, then exactly one of the A_j is a member of $\mathcal{F}$.
c. Every cofinite set (i.e., the complement of a finite set) in $\mathbb{N}$ is a member of $\mathcal{F}$.
d. $\mathcal{F}$ is a maximal filter in the following sense: A *proper filter* on $\mathbb{N}$ is by definition a collection $\mathcal{H}$ of subsets of $\mathbb{N}$ with $\emptyset \notin \mathcal{H}$ and satisfying Conditions 1 and 2 above. There is no proper filter $\mathcal{H}$ that is larger than $\mathcal{F}$ in the sense that $\mathcal{F}$ is a proper subset of $\mathcal{H}$.

(Actually, an ultrafilter can be defined on any infinite set $\mathbb{I}$ by replacing $\mathbb{N}$ by i in the axioms 1 to 4 above; indeed, every infinite set possesses a nonprincipal ultrafilter on it.)

A.5 Hyperreal Numbers, More Explicitly

There are many nonprincipal ultrafilters on $\mathbb{N}$. Let us choose and fix our attention on one of them, say, $\mathcal{F}$. Also, let $\langle x_n \rangle$ and $\langle y_n \rangle$ be two sequences of real numbers.[3] We call these sequences *equivalent modulo* $\mathcal{F}$—or simply *equivalent* when a chosen and fixed $\mathcal{F}$ is understood—if $\{n \in \mathbb{N} \colon x_n = y_n\} \in \mathcal{F}$. This is truly an equivalence relation. As a result, the set of all sequences of real numbers is partitioned into equivalence classes, each of which is defined to be a *hyperreal number*. It is conventional to say "hyperreal" instead of "hyperreal number," and "real" instead of "real number." The set of all hyperreals is denoted by $^*\mathbb{R}$. Each member of an equivalence class of sequences defining a hyperreal is called a *representative* of that hyperreal, and that hyperreal is denoted by $\mathbf{x} = [\langle x_n \rangle]$ or $\mathbf{x} = [\langle x_0, x_1, x_2, \dots \rangle]$. It is convenient to use the abbreviated notation $\mathbf{x} = [x_n]$ or $\mathbf{x} = [x_0, x_1, x_2, \dots]$, where it is understood that $\langle x_n \rangle$ is one (i.e., any one) of the representative sequences in the equivalence class for

[2] Also called a *free ultrafilter*

[3] Remember that $\langle x_n \rangle$ denotes $\langle x_0, x_1, x_2, \dots \rangle$, which is also represented by $\langle x_n \colon n \in \mathbb{N} \rangle$.

the hyperreal $\mathbf{x}$.[4] Every real $x \in \mathbb{R}$ has a hyperreal version $[x] = [x, x, x, \dots]$, which we also refer to as the *image* $\mathbf{x}$ in $^*\mathbb{R}$ of x. In this way, we view $\mathbb{R}$ as being a subset of the set $^*\mathbb{R}$ of all hyperreals, in which case it is convenient to use the same symbol for the real and the hyperreal. For example, 2 is a real, and 2 also denotes the corresponding hyperreal [2, 2, 2, ...], which we also denote [2] or by the boldface symbol **2**.

A.6 The Hyperreals Form an Ordered Field

If a condition depending upon n holds for all n in some set $F \in \mathcal{F}$, we will simply say that it holds "for almost all n" or simply "a.e.".[5] For example, the hyperreals $\mathbf{x} = \langle x_n \rangle$ and $\mathbf{y} = \langle y_n \rangle$ are defined to be *equal* (i.e., $\mathbf{x} = \mathbf{y}$) if $\langle n \in \mathbb{N}\colon x_n = y_n \} = F \in \mathcal{F}$, and we say in this case that $x_n = y_n$ a.e. For example, $[1, 1/2, 1/3, 1/4, 1/5, 1/6, \dots]$ and $[1, 0, 1/3, 0, 1/5, 0, \dots]$ denote the same hyperreal if $\mathcal{F}$ is so chosen that the set of even natural numbers is a member of $\mathcal{F}$. Similarly, addition, multiplication, inequality, and absolute value are defined componentwise on the representatives of hyperreals. That is, if $\mathbf{x} = [x_n]$ and $\mathbf{y} = [y_n]$, then $\mathbf{x}+\mathbf{y} = [x_n+y_n]$, $\mathbf{x}-\mathbf{y} = [x_n-y_n]$, $\mathbf{xy} = [x_n y_n]$, and $\mathbf{x}/\mathbf{y} = [x_n/y_n]$ if $\mathbf{y} \neq [0]$. Also, $\mathbf{x} < \mathbf{y}$ means $x_n < y_n$ a.e., and $\mathbf{x} \leq \mathbf{y}$ is defined similarly. It can be shown that these definitions are independent of the representatives chosen for the hyperreals. Altogether, the usual arithmetic laws hold in $^*\mathbb{R}$; in fact, $^*\mathbb{R}$ is an ordered field with the zero element [0] and the unit element [1]. Therefore, we can manipulate the hyperreals as we do the real numbers. The absolute value $|\mathbf{x}|$ of a hyperreal $\mathbf{x}$ is $\mathbf{x}$ if $\mathbf{x} > [0]$, is $-\mathbf{x}$ if $\mathbf{x} < [0]$, and is [0] if $\mathbf{x} = [0]$. Finally, $^*\mathbb{R}_+$ will denote the set of all nonnegative hyperreals: $\mathbf{x} = [x_n] \in {}^*\mathbb{R}_+$ if and only if $x_n \geq 0$ a.e.

A.7 Types of Hyperreals

The hyperreal $[x_n]$ is called *infinitesimal* if, for every positive real ϵ, we have $\{n \in \mathbb{N}\colon |x_n| < \epsilon\} \in \mathcal{F}$, that is, if $|x_n| < \epsilon$ a.e. Also, $[x_n]$ is called *unlimited* if $|x_n| > \epsilon$ a.e. for every positive real ϵ. Thus, the reciprocal $[x_n^{-1}]$ of an infinitesimal $[x_n]$ is unlimited, and conversely. A *limited* hyperreal is one that is not unlimited. Thus, $\mathbf{x} = [x_n]$ is limited if and only if there is a $\gamma \in \mathbb{R}_+$ such that $|x_n| < \gamma$ a.e. A hyperreal that is neither infinitesimal nor unlimited is called *appreciable*. Thus, $[x_n]$ is appreciable if, for some ϵ and γ with $0 < \epsilon < \gamma < \infty$, we have that $\epsilon < x_n < \gamma$ a.e.

---

[4]Later on in Secs. A18 and A19, we will also use the notation $\mathbf{x} = \langle x_n \rangle_{\mathcal{F}}$ in order to display the chosen nonprincipal ultrafilter $\mathcal{F}$.

[5]The abbreviation "a.e." stands for "almost everywhere". Although brief and convenient, "a.e." is rather a misnomer, for the set of those n for which the condition holds can be a very sparse subset of $\mathbb{N}$.

Two hyperreals $\mathbf{x}$ and $\mathbf{y}$ are said to be *infinitesimally close*[6] (resp. *limitedly close*) if $|\mathbf{x} - \mathbf{y}|$ is an infinitesimal (resp. $|\mathbf{x} - \mathbf{y}|$ is a limited hyperreal).

For each hyperreal $\mathbf{x}$ (limited or unlimited), there is a set of infinitesimally close hyperreals called the *halo* or (*monad*) for $\mathbf{x}$, and there is a larger set of limitedly close hyperreals $\mathbf{y}$ called the *galaxy* for $\mathbf{x}$. In fact, the hyperreal line is partitioned by the halos and more coarsely by the galaxies. Furthermore, for each limited hyperreal $\mathbf{x}$ there is a unique real number called the *shadow* (also, *standard part*) of $\mathbf{x}$ and denoted by sh$\mathbf{x}$ (also, st$\mathbf{x}$), such that $\mathbf{x}$ is infinitesimally close to sh$\mathbf{x}$.

Since every cofinite set is a member of every nonprincipal ultrafilter, any of the adjectives: infinitesimal, appreciable, limited, and unlimited holds for $\mathbf{x} = [x_n]$ whenever the corresponding inequality on x_n holds for all n in a cofinite subset of $\mathbb{N}$. Moreover, we are free to change the values of x_n in $\mathbf{x} = [x_n]$ for all n in any subset of $\mathbb{N}$ not in $\mathcal{F}$ and in particular in any finite subset of $\mathbb{N}$; this will not alter $\mathbf{x}$.

A.8 How Does the Choice of the Ultrafilter $\mathcal{F}$ Affect the Hyperreal Line $^*\mathbb{R}$?

Let us now discuss the apparent arbitrariness arising from the many possible choices for the ultrafilter $\mathcal{F}$. Consider a sequence $\langle a_n \rangle$, where $a_n = 0$ for n even and $a_n = 1$ for n odd. If we so choose $\mathcal{F}$ to have the set of even natural numbers as a member of $\mathcal{F}$, then $[a_n] = \mathbf{0}$ (modulo $\mathcal{F}$). But, if we choose another ultrafilter, say, $\mathcal{F}'$ to have the set of odd natural numbers as a member of $\mathcal{F}'$, then $[a_n] = \mathbf{1}$ (modulo $\mathcal{F}'$), and $\mathbf{0}$ will now be $[b_n]$, where $b_n = 0$ for n odd and $b_n = 1$ for n even. It appears that, given any choice of $\mathcal{F}$, all the hyperreals can be obtained by appropriate choices of representative sequences for them. There is some justification for this idea. Nonstandard analysis, like mathematics in general, is based upon the theory of sets, which in turn is founded upon the Zermelo–Fraenkel axioms along with the axiom of choice [1], [17]. If the truth of the continuum hypothesis is assumed as an additional axiom, then the following is obtained: There is an isomorphism between the two ordered fields of hyperreals corresponding to any two choices of $\mathcal{F}$ [22].

Let us take note of a special case where the choice of $\mathcal{F}$ does not affect the shadow of a limited hyperreal. Assume $a_n \in \mathbb{R}$ for all n and $a \in \mathbb{R}$ too. Then, $[a_n]$ is in the shadow of a for every choice of $\mathcal{F}$ if and only if $\lim_{n\to\infty} a_n = a$.

Another such result occurs when $a_n > 0$ for all n. Then, $[a_n]$ is positive unlimited for every choice of $\mathcal{F}$ if and only if $\lim_{n\to\infty} a_n = \infty$.

A.9 A Hyperreal Can Be Determined by an Arbitrarily Sparse Subset of $\mathbb{N}$

Another peculiarity arises from the fact that, for any finite partition of $\mathbb{N}$, exactly one of the sets in that partition is a member of $\mathcal{F}$ (see item A4b above). To view

[6] Synonymously, it is also said that $\mathbf{x}$ and $\mathbf{y}$ are *infinitely close*.

this in another way, consider the nested subsets of $\mathbb{N}$ of the form $\mathbb{N}_p = \{2^p k : k = 0, 1, 2, \ldots\}$, where p is a positive natural number. Thus, if $p_j < p_l$, then, $\mathbb{N}_{p_l} \subset \mathbb{N}_{p_j}$. We can choose $\mathcal{F}$ such that $\mathbb{N}_p \in \mathcal{F}$ for all p. Also, for any fixed p, let $a_{p,n} = 1$ for $n = 2^p k$, and let $a_{p,n} = 0$ for $n \neq 2^p k$. Then, the hyperreal $\mathbf{1} = [1, 1, 1, \ldots]$ can be written as $[a_{p,n}]$ for each $p = 1, 2, 3, \ldots$. For example,

$$\begin{aligned} \mathbf{1} &= [1, 0, 1, 0, 1, 0, 1, 0, 1, 0, \ldots] \quad (p = 1) \\ &= [1, 0, 0, 0, 1, 0, 0, 0, 1, 0, \ldots] \quad (p = 2) \end{aligned}$$

and so on. In short, each member of a sequence of nested, progressively sparser samplings of a representative sequence for a hyperreal still determines that hyperreal so long as each sampling comprises a set in $\mathcal{F}$. It is because of this that the "almost all" criterion for a subset of $\mathbb{N}$ might be called a misnomer.

A.10 A First Glimpse of the Transfer Principle

It was noted above that, with the arithmetic operations transferred onto $^*\mathbb{R}$ by means of componentwise definitions, $^*\mathbb{R}$ is a field in much the same way as $\mathbb{R}$ is a field. Consequently, rational functions of hyperreals and equations between such rational functions can be set up in the hyperreal realm. More specifically, consider two rational functions that involve some real coefficients a_k and some real variables x_j, which we may display together by means of a vector $\mathbf{s} = (a_1, \ldots, a_K, x_1, \ldots, x_J)$. Then, an equation between the rational functions will be true in the real realm when and only when $\mathbf{s}$ lies in some nonempty subset $\mathcal{S}$ of thc Cartesian product $\mathbb{R}^{K+J}$ of $\mathbb{R}$ taken $K + J$ times. It is an important property of nonstandard analysis that that equation will again be true in the subset $^*\mathcal{S}$ of $^*\mathbb{R}^{K+J}$ consisting of all $[\mathbf{s}_n]$ such that $\{n : \mathbf{s}_n \in \mathcal{S}\} \in \mathcal{F}$. $^*\mathcal{S}$ is called an "internal set" in $^*\mathbb{R}^{K+J}$.[7] All this is a result of the "transfer principle" of nonstandard analysis (see A22 below).

As a simple example, consider $a_1 x = a_2$ in the real realm. This equation is true whenever $a_1 \in \mathbb{R}\backslash\{0\}$, $a_2 \in \mathbb{R}$, and $x = a_2/a_1$. Thus, the set $\mathcal{S}$ mentioned above is given by

$$\mathcal{S} = \{\mathbf{s} = (a_1, a_2, x) : a_1 \in \mathbb{R}\backslash\{0\}, a_2 \in \mathbb{R}, x = a_2/a_1\}.$$

Consequently, any $[\mathbf{s}_n]$ of the form $[\mathbf{s}_n] = ([a_{1n}], [a_{2n}], [x_n])$, where $a_{1n} \in \mathbb{R}\backslash\{0\}$, $a_{2n} \in \mathbb{R}$, and $x_n = a_{2n}/a_{1n}$ for all n in a set of $\mathcal{F}$, will be a member of $^*\mathcal{S}$. We can succinctly restate this by setting $\mathbf{a}_1 = [a_{1n}]$, $\mathbf{a}_2 = [a_{2n}]$, and $\mathbf{x} = [x_n]$ and asserting that $\mathbf{a}_1\mathbf{x} = \mathbf{a}_2$ has the solution $\mathbf{x} = \mathbf{a}_2/\mathbf{a}_1 \in {}^*\mathbb{R}$ when $\mathbf{a}_1, \mathbf{a}_2 \in {}^*\mathbb{R}$ and $\mathbf{a}_1 \neq [0]$. Here, $\mathbf{a}_1$ and $\mathbf{a}_2$ need not be the images of real numbers.

This is a critically important result, for it allows us to find solutions in the hyperreal realm that do not have preimages in the real realm. For instance, Kirchhoff's current law need not hold at a node of infinite degree in a conventionally infinite network, but it may hold at that node in a certain restricted way when real currents are

[7] There are more general kinds of "internal sets" in $^*\mathbb{R}^{K+J}$ (see A12 and A20 below).

replaced by hyperreal currents. A similar restoration of Kirchhoff's voltage law may occur at a transfinite loop. In general, a hyperreal voltage-current regime satisfying Kirchhoff's laws and Ohm's law may be found for a transfinite network when no such real voltage-current regime exists for it. Chapters 6 and 7 exploit this facility.

A.11 Infinite Series of Hyperreals

An infinite series of hyperreals can be summed under certain circumstances. For instance, consider $\sum_{k=0}^{\infty} \mathbf{x}_k$, where each $\mathbf{x}_k \in {}^*\mathbb{R}$. Let us choose a particular representative $\langle x_{k,n} : n \in \mathbb{N} \rangle$ for each $\mathbf{x}_k$. Thus, we may write

$$\sum_{k=0}^{\infty} \mathbf{x}_k = \left[\sum_{k=0}^{\infty} x_{k,n} \right]$$

$$\begin{aligned} = \quad & [\, x_{0,0},\ x_{0,1},\ x_{0,2},\ \ldots\,] \\ & + [\, 0,\quad x_{1,1},\ x_{1,2},\ \ldots\,] \\ & + [\, 0,\quad 0,\quad x_{2,2},\ \ldots\,] \\ & + \quad \cdots . \end{aligned}$$

In conformity with how finitely many hyperreals are added together, we would like to sum the $\mathbf{x}_k$ componentwise; that is, we wish to sum along columns in the array. This will be possible if every column has only finitely many nonzero entries. One way of insuring this condition is to set $x_{k,n} = 0$ for all $n < k$, as in the array. Any hyperreal $\mathbf{x}_k = [x_{k,n}]$ can be placed in this form without changing that hyperreal. The resulting sum is then the hyperreal

$$\sum_{k=0}^{\infty} \mathbf{x}_k = [x_{0,0},\ x_{0,1} + x_{1,1},\ x_{0,2} + x_{1,2} + x_{2,2},\ \ldots\,].$$

However, another problem now arises if no further restriction is placed on the choice of the representative $\langle x_{k,n} \rangle$ for each $\mathbf{x}_k$. The sum is not uniquely determined, and in fact can be made equal to any hyperreal $[y_n]$. This can be done, for instance, by choosing the main diagonal entries as

$$x_{n,n} = y_n - \sum_{k=0}^{n-1} x_{k,n} .$$

(This, too, does not alter $\mathbf{x}_k$ for every k.)

On the other hand, $\sum_{k=0}^{\infty} \mathbf{x}_k$ does become a unique hyperreal if we specify the representative $\langle x_{k,n} \rangle$ for each $\mathbf{x}_k$ along with the requirement that every column in the array have only finitely many entries. (This is exactly what happens in Sec. 6.4 when the representatives of the branch voltages around any transfinite loop or the representatives of the branch currents on any infinite cut are determined by the finite networks that fill out a transfinite network.)[8]

As an example, consider the infinite series $\sum_{k\in\mathbb{N}} \mathbf{x}_k$, where $\mathbf{x}_k$ is the hyperreal image of the real x_k. We wish that sum to be the hyperreal having as a representative the partial sums of the standard series $\sum_{k\in\mathbb{N}} x_k$, assuming the series converges. To do so, we can use the above array wherein we set $x_{k,n} = 0$ for $n < k$ and $x_{k,n} = x_k$ for $n \geq k$. Upon summing columnwise, we obtain the desired result:

$$\sum_{k\in\mathbb{N}} \mathbf{x}_k = [x_0,\ x_0 + x_1,\ x_0 + x_1 + x_2,\ \dots].$$

This hyperreal will reside in the halo of the limit of the convergent series. Here, too, we can obtain any other hyperreal sum by changing the diagonal entries of the array as stated above without changing the $\mathbf{x}_k$. Thus, in order to obtain the correct hyperreal determined by the partial sums of a convergent series of reals by using the above array, we must specify appropriately which representative to use for each $\mathbf{x}_k$. If, on the other hand, the standard sum $\sum_{k\in\mathbb{N}} x_k$ does not converge, the nonstandard sum $\sum_{k\in\mathbb{N}} \mathbf{x}_k$ still has a meaning in this way as a hyperreal.

A.12 Internal and External Sets

Again let A be any set containing $\mathbb{R}$, and let $\mathcal{F}$ be a chosen nonprincipal ultrafilter on A. As was indicated in A3, we can define nonstandard entities as equivalence classes of sequences taking their elements in A, where two such sequences $\langle a_n \rangle$ and $\langle b_n \rangle$ are defined as being "equivalent" if $\{n : a_n = b_n\} \in \mathcal{F}$. The set of such equivalence classes is denoted by $^*A$. As before, $\mathbf{a} = [a_n]$ denotes such an equivalence class.

Next, let $\langle A_n \rangle$ be a sequence of subsets of A. A subset $[A_n]$ of $^*A$ can be defined by identifying its elements $\mathbf{a} = [a_n]$ as follows: $\mathbf{a} = [a_n] \in [A_n]$ if and only if $\{n : a_n \in A_n\} \in \mathcal{F}$. A set $[A_n]$ formed in this way is called an *internal set*. In the special case where $A = \mathbb{R}$, $[A_n]$ is a subset of $^*\mathbb{R}$.

There are subsets of $^*A$ that are not internal sets. These sets are called *external sets*. An example of an external subset of $^*\mathbb{R}$ is $\mathbb{R}$, as we shall note again in A16. Internal and external sets are discussed in greater generality in A20.

A.13 Hyperfinite Sets

Comprising a special case of internal sets $[A_n]$ are the hyperfinite ones. They occur when the A_n are finite sets for almost all $n \in \mathbb{N}$. With $|A_n|$ denoting the cardinality of A_n, we define the *internal cardinality* of the hyperfinite set $[A_n]$ as the "hypernatural number" $[|A_n|]$.

---

[8]This specialized way of summing is roughly analogous to the use of specialized summation techniques to convert a divergent series into a convergent one, such as Cesàro's summation. When the series at hand is convergent, Cesàro's summation yields the same sum, but, if the series is divergent, Cesàro's method may yield a unique finite sum. [44, page 155].

A.14 Internal Functions

Furthermore, let $\langle f_n \rangle$ be a sequence of functions $f_n : A_n \rightsquigarrow \mathbb{R}$ with $A_n \subseteq A$. Then, $[f_n]$ is defined as an ${}^*\mathbb{R}$-valued function on the internal set $[A_n]$ by $[f_n]([a_n]) = [f_n(a_n)]$, where $[a_n] \in [A_n]$. More specifically, another sequence $\langle g_n \rangle$ of functions $g_n : A_n \rightsquigarrow \mathbb{R}$ is taken to be *equivalent* to $\langle f_n \rangle$ if $\{n : f_n = g_n\} \in \mathcal{F}$. Then, the set of all sequences of such functions that are equivalent to f_n is denoted by $\mathbf{f} = [f_n]$, and we have $\mathbf{f}(\mathbf{a}) = [f_n(a_n)] \in {}^*\mathbb{R}$, where $\mathbf{a} = [a_n] \in [A_n]$. Thus, the domain of $\mathbf{f}$ is $[A_n]$. Moreover, it is a fact that the image $\mathbf{f}([A_n])$ of an internal set $[A_n]$ under the internal function $\mathbf{f}$ is also an internal set $[f_n(A_n)] \subseteq {}^*\mathbb{R}$.

A.15 Enlargements of Sets

With A and A_n still being as in A12, let B be any subset of A. When $A_n = B$ for all n, we have the *enlargement* ${}^*B$ of B as the internal set $[A_n]$. If B is a finite set, it turns out that ${}^*B = B$, where each $\mathbf{b} = [b] \in {}^*B$ is identified with a $b \in B$. If, on the other hand, B is an infinite set, then ${}^*B$ contains B as a proper subset; in this case, the elements of B are called the *standard elements* of ${}^*B$, and those of ${}^*B \setminus B$ are called the *nonstandard elements* of ${}^*B$.

The enlargement ${}^*\mathbb{R}$ of $\mathbb{R}$ (discussed in A5) consists of the hyperreals; $\mathbb{R}$ is the set of standard real numbers. Similarly, the members of the enlargement ${}^*\mathbb{N}$ of $\mathbb{N}$ are called the *hypernaturals*. If A_n is a subset of $\mathbb{N}$ for almost all n with k_n being the minimum member of A_n, then the set $[A_n]$ of hypernatural numbers has the minimum member $[k_n]$. Similarly, $[A_n]$ has a maximum member $[m_n]$ if A_n is a finite set of natural numbers with maximum member m_n for almost all n.

Note that ${}^*\mathbb{R}$ is an internal set containing nonstandard elements. In fact, every infinite subset of $\mathbb{R}$ has an enlargement containing nonstandard hyperreals. It follows that $\mathbb{R}$ itself is an external set because it neither is finite nor contains nonstandard elements. Similarly, $\mathbb{N}$ is an external set, too.

A.16 Overflow and Underflow

Here we list three versions of the "permanence principle," the idea that internal sets that contain the hyperreals in certain sets must contain hyperreals in larger sets as well. In the following, ${}^*A$ is an internal set of hyperreals (thus, ${}^*A \subseteq {}^*\mathbb{R}$).

(i) Overflow: If there exists a real $a \in \mathbb{R}$ such that ${}^*A$ contains all reals b with $b \geq a$, then there exists a positive unlimited hyperreal $\mathbf{c}$ such that ${}^*A$ contains all hyperreals $\mathbf{d}$ with $a \leq \mathbf{d} \leq \mathbf{c}$.

Because of this, we can again conclude that $\mathbb{R}$ is an external set because it does not "overflow."

(ii) Underflow: If there exists a positive unlimited hyperreal $\mathbf{c} \in {}^*\mathbb{R}$ such that ${}^*A$ contains all positive unlimited hyperreals $\mathbf{d} \leq \mathbf{c}$, then there exists a real $a \in \mathbb{R}$ such that A contains all hyperreals $\mathbf{b}$ with $a \leq \mathbf{b} \leq \mathbf{c}$.

(iii) Overflow again: Two other versions of overflow are the following: If $^*A$ contains all hyperreals that are infinitesimally close to some $\mathbf{a} \in {}^*\mathbb{R}$, then there exists a positive real c such that $^*A$ contains all hyperreals $\mathbf{b}$ with $|\mathbf{b} - \mathbf{a}| \le c$. Similarly, if $^*A$ contains all positive infinitesimals, then there exists a positive real $c \in \mathbb{R}$ such that $^*A$ contains all positive hyperreals $\mathbf{b}$ such that $\mathbf{b} \le c$.

A.17 Superstructures

So far, we have considered nonstandard analysis based upon equivalence classes of sequences taking their elements from a set A that contains $\mathbb{R}$, and from these we have defined internal sets and internal functions. However, there are many occasions in analysis when the entities of interest are not only elements of A and the subsets and functions involving them but more generally subsets of the set of subsets of A, subsets of the set of those subsets, and so on. For example, an ordered pair $\langle a, b\rangle$, where $a, b \in A$, can be defined as the set of sets $\{\{a\}, \{a, b\}\}$, and a finite sequence $\langle a_1, \ldots, a_k\rangle$ can be defined as the set $\{\langle 1, a_1\rangle, \ldots, \langle k, a_k\rangle\}$ of ordered pairs, that is, a set of sets of sets.

To accommodate such generality, the set A is expanded into a "superstructure" $V(A)$ defined as follows. The *power set* $\mathcal{P}(A)$ of A is the set of all subsets of A. (A itself and the empty set $\emptyset$ are members of $\mathcal{P}(A)$.) For $k \in \mathbb{N}$, the kth *cumulative power set* $V_k(A)$ is defined recursively by

$$V_0(A) = A, \quad V_{k+1}(A) = V_k(A) \cup \mathcal{P}(V_k(A)), \quad k = 0, 1, 2, \ldots.$$

Then, the *superstructure over* A is the union

$$V(A) = \cup_{k=0}^{\infty} V_k(A).$$

Note that, for each k, $V_k(A) \subseteq V_{k+1}(A)$ and $V_k(A) \in V_{k+1}(A)$ simultaneously. The elements of $V(A)$ are called *entities*, and the entities in A are the *individuals* mentioned above. The *superstructure rank* of an entity $a \in V(A)$ is the least k for which $a \in V_k(A)$.

With $^*A$ denoting the set of all nonstandard entities obtained from ultrapower constructions using only the elements of A, a superstructure $V(^*A)$ over $^*A$ is constructed in exactly the same way. Any real number r is identified with its image $\mathbf{r} = [r, r, r, \ldots]$ in $^*\mathbb{R}$, and so in this way $\mathbb{R} \subset {}^*\mathbb{R} \subseteq {}^*A$. In the same way, A is identified as a subset of $^*A$, and we set $^*a = [a, a, a, \ldots] = \mathbf{a}$ for every $a \in A$. However, A is a proper subset of $^*A$, as is $\mathbb{R}$ a proper subset of $^*\mathbb{R}$.

A.18 Bounded Ultrapowers

Next, equivalence classes of sequences of elements from a superstructure, say, $V(A)$ can be defined with respect to a chosen and fixed nonprincipal ultrafilter $\mathcal{F}$. But, instead of selecting the elements of those sequences arbitrarily from $V(A)$, an additional restriction is imposed. A sequence $a = \langle a_0, a_1, a_2 \ldots \rangle$ of elements a_n of $V(A) = \cup_{k=0}^{\infty} V_k(A)$ is called *bounded* if there is a fixed k such that $a_n \in V_k(A)$ for all n. Two bounded sequences a and $b = \langle b_0, b_1, b_2, \ldots \rangle$ are taken to be *equivalent with respect to* $\mathcal{F}$ if $\{n \in \mathbb{N} : a_n = b_n\} \in \mathcal{F}$, in which case we write $a =_{\mathcal{F}} b$. For the sake of specifying $\mathcal{F}$, the class of sequences equivalent to a will now be denoted by $a_{\mathcal{F}}$ (instead of **a** or $[a_n]$). Then, the set of all such equivalence classes of bounded sequences is called a *bounded ultrapower* and is denoted by $V(A)^{\mathbb{N}}/\mathcal{F}$. Given any bounded sequence $a = \langle a_0, a_1, a_2 \ldots \rangle$ in $a_{\mathcal{F}}$, there will be only finitely many ranks for the members a_n of a. Consequently, exactly one rank k will correspond to a according to item b of A4. Moreover, it is easily shown that that rank k will belong to every bounded sequence equivalent to a. Hence, that rank is assigned to $a_{\mathcal{F}}$ as well, and we can write $a_{\mathcal{F}} \in V_k(A)^{\mathbb{N}}/\mathcal{F}$.

A membership relation $\in_{\mathcal{F}}$ between equivalence classes of differing ranks in this bounded ultrapower is defined as follows: $a_{\mathcal{F}} \in_{\mathcal{F}} b_{\mathcal{F}}$ if $\{n \in \mathbb{N} : a_n \in b_n\} \in \mathcal{F}$. In this case the rank of $a_{\mathcal{F}}$ will be less than the rank of $b_{\mathcal{F}}$. It can be shown that this membership relation does not depend upon the choice of the representatives $a \in a_{\mathcal{F}}$ and $b \in b_{\mathcal{F}}$.

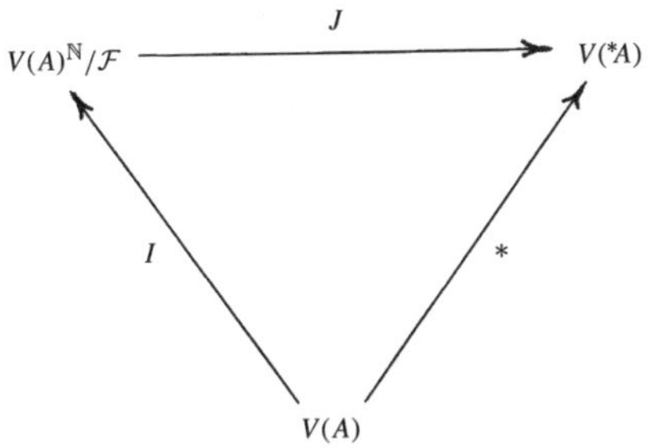

Fig. A.1. The mappings J, I, and $*$ between a superstructure $V(A)$, the bounded ultrapower $V(A)^{\mathbb{N}}/\mathcal{F}$, and the superstructure $V({}^*A)$.

Here too, each member $a_0 \in V(A)$ determines a member of $V(A)^{\mathbb{N}}/\mathcal{F}$ through the equivalence class $\langle a_0, a_0, a_0, \ldots \rangle_{\mathcal{F}}$ of the constant sequence $\langle a_0, a_0, a_0 \ldots \rangle$. This constitutes an injection I of $V(A)$ into $V(A)^{\mathbb{N}}/\mathcal{F}$, as is indicated on the left-hand side of Fig. A.1. We may in fact identify $a_0 \in V(A)$ with its image $\langle a_0, a_0, a_0, \ldots \rangle_{\mathcal{F}}$ and thus view I as an identity injection. Of course, there are many other members of $V(A)^{\mathbb{N}}/\mathcal{F}$ corresponding to bounded sequences that are not equivalent to constant ones.

A.19 The Injection J of $V(A)^{\mathbb{N}}/\mathcal{F}$ into $V({}^*A)$

There is a bijection J from $A^{\mathbb{N}}/\mathcal{F} = V_0(A)^{\mathbb{N}}/\mathcal{F}$ onto ${}^*A = V_0({}^*A)$ defined as in A5; namely, each ${}^*a$ in ${}^*A$ is by definition the equivalence class $a_{\mathcal{F}} \in A^{\mathbb{N}}/\mathcal{F}$. So, in this case, J is the identity mapping. However, for each rank $k > 0$, only some, but not all, of the members of $V_k({}^*A)$ can be obtained from members of $V_k(A)^{\mathbb{N}}/\mathcal{F}$ (as was noted in A12 and A16 and will again be seen in A20).

The mapping J is extended onto all of $V(A)^{\mathbb{N}}/\mathcal{F}$ as follows. Assume $a_{\mathcal{F}} = \langle a_0, a_1, a_2, \ldots \rangle_{\mathcal{F}} \in V_{k+1}(A)^{\mathbb{N}}/\mathcal{F}$, and assume $a_{\mathcal{F}} \notin_{\mathcal{F}} {}^*A$. Consider $c_{\mathcal{F}} = \langle c_0, c_1, c_2, \ldots \rangle_{\mathcal{F}} \in_{\mathcal{F}} a_{\mathcal{F}}$. This means that $\{n \in \mathbb{N} : c_n \in a_n\} \in \mathcal{F}$. Since $a_n \in V_{k+1}(A)$ a.e., we have that $c_n \in V_k(A)$ a.e.; there will be such c_n according to the definition of the $V_{k+1}(A)$. Therefore, $c_{\mathcal{F}} \in V_k(A)^{\mathbb{N}}/\mathcal{F}$.

We now set

$$J(a_{\mathcal{F}}) = \{J(c_{\mathcal{F}}) : c_{\mathcal{F}} \in_{\mathcal{F}} a_{\mathcal{F}}\}.$$

By this definition, J is defined recursively as a mapping on $V(A)^{\mathbb{N}}/\mathcal{F}$ as follows. For $a_{\mathcal{F}} \in V_0(A)^{\mathbb{N}}/\mathcal{F}$, J is the identity mapping giving $J(a_{\mathcal{F}}) \in {}^*A = V_0({}^*A)$.

Next, for any $k \geq 0$, assume $a_{\mathcal{F}} \in V_{k+1}(A)^{\mathbb{N}}/\mathcal{F}$ and $a_{\mathcal{F}} \notin {}^*A$. As was noted just above, there is at least one $c_{\mathcal{F}} \in_{\mathcal{F}} a_{\mathcal{F}}$ with $c_{\mathcal{F}} \in V_k(A)^{\mathbb{N}}/\mathcal{F}$. Moreover, $J(c_{\mathcal{F}})$ is determined at the lower rank k. Hence, by the last displayed equation, J is defined on $V_{k+1}(A)^{\mathbb{N}}/\mathcal{F}$ for each $k = 0, 1, 2, \ldots$ in turn, and thus on $V(A)^{\mathbb{N}}/\mathcal{F} = \cup_{k=0}^{\infty} V_k(A)^{\mathbb{N}}/\mathcal{F}$. This too is indicated in Fig. A.1.

It turns out that J is an injection of $V_k(A)^{\mathbb{N}}/\mathcal{F}$ into $V_k({}^*A)$ and that $*(\cdot) = J(I(\cdot))$ is an injection of $V_k(A)$ into $V_k({}^*A)$ for each $k = 0, 1, 2, \ldots$. Hence, $*(\cdot) = J(I(\cdot))$ is an injection of $V(A)$ into $V({}^*A)$. We set ${}^*a = *(a)$ for each $a \in V(A)$.[9] In general, however, we use boldface notation to indicate that an entity **a** is a member of $V({}^*A)$ but not necessarily of $V(A)$ or of $V(A)^{\mathbb{N}}/\mathcal{F}$.

A.20 Standard, Internal, and External Sets

Let $a \in V(A)$. Then, $a_{\mathcal{F}} = \langle a, a, a, \ldots \rangle_{\mathcal{F}} = I(a)$ is the equivalence class of a constant sequence, and ${}^*a = *(a) = J(I(a))$ is the image of a under the injection $*$. Any member ${}^*a$ of $V({}^*A)$ that is obtained in this way from a constant sequence is called *standard*. On the other hand, if $\mathbf{b} \in V({}^*A)$ is an element of a standard member ${}^*a$, then $\mathbf{b}$ is called *internal*. Otherwise, $\mathbf{c} \in V({}^*A)$ is called *external* when $\mathbf{c}$ is not internal. In symbols, ${}^*a = J(I(a))$ is standard when $a \in V(A)$; $\mathbf{b}$ is internal if $\mathbf{b} \in {}^*a$ for some $a \in V(A)$; $\mathbf{c}$ is external if $\mathbf{c} \notin {}^*a$ for any $a \in V(A)$. Every standard element ${}^*a$ is internal because $a \in V_k(A)$ for some k, and thus ${}^*a \in *(V_k(A))$. Similarly, $\mathbf{b}$ is internal if and only if $\mathbf{b} \in *(V_k(A))$ for some k. Another way of comparing standard and internal sets is to note that a standard set ${}^*a$ is the image under J of the equivalence class $\langle a, a, a, \ldots \rangle_{\mathcal{F}}$ in $V(A)^{\mathbb{N}}/\mathcal{F}$ of a constant sequence. On the other hand, an internal set $\mathbf{a}$ is the image under J of an equivalence class $\langle a_0, a_1, a_2, \ldots \rangle_{\mathcal{F}}$ in $V(A)^{\mathbb{N}}/\mathcal{F}$ of a possibly nonconstant, bounded sequence.

Some other consequences of these definitions are the following:

[9] In much of the literature the same symbol is used for both a and ${}^*a$. Thus, for example, $\mathbb{R}$ is imply viewed as a subset of ${}^*\mathbb{R}$.

(i) Every element of an internal set is internal.
(ii) Every nonempty internal subset of $^*N$ has a least element.
(iii) Every nonempty internal subset of $^*\mathbb{R}$ with an upper bound (resp. lower bound) has a least upper bound (resp. greatest lower bound).

In fact, all the ideas discussed in A12 through A16 including hyperfinite sets, enlargements of sets, overflow, and underflow are subsumed under this more general construction based upon superstructures. For example, a function f defined on a subset D of A with values in $\mathbb{R}$ is a subset F of the Cartesian product $D \times \mathbb{R}$ with the following property: Each member of that subset is an ordered pair $\langle d, r\rangle$ with $d \in D$ and $r \in \mathbb{R}$ such that, if $\langle d, r\rangle$ and $\langle d, r'\rangle$ are in F, then $r = r'$. But, that ordered pair can be defined as a set of sets of elements in A according to $\langle d, r\rangle = \{\{d\}, \{d, r\}\}$ and is thereby an element of $V_2(A)$. Therefore, f can be identified as a member of $V_3(A)$. So, an equivalence class $\langle f_0, f_1, f_2, \ldots\rangle_{\mathcal{F}}$ of a sequence of functions $f_n : A_n \rightsquigarrow \mathbb{R}$ is a member of $V_3(A)^{\mathbb{N}}/\mathcal{F}$, and it is mapped by J to an internal element $\mathbf{f}$ of $V_3(^*A)$. Thus, $\mathbf{f}$ is the internal function mapping $[A_n] \in V_1(A)^{\mathbb{N}}/\mathcal{F}$ into $^*\mathbb{R}$ that was discussed in A14.

Here are some examples of standard, internal, and external entities in $V(^*\mathbb{R})$. $^*\mathbb{R}$ and $^*\mathbb{N}$ are standard. Each $\mathbf{r} \in {}^*\mathbb{R} \setminus \mathbb{R}$ is internal but not standard. The interval $[\mathbf{a}, \mathbf{b}]$ of hyperreals, where $\mathbf{a}, \mathbf{b} \in {}^*\mathbb{R} \setminus \mathbb{R}$ and $\mathbf{a} < \mathbf{b}$ is also internal but not standard, but, if $\mathbf{a}$ and $\mathbf{b}$ are images of reals $a, b \in \mathbb{R}$ with $a < b$, then that interval of hyperreals is standard. As external sets in $V(^*\mathbb{R})$ we have $\mathbb{R}$, $\mathbb{N}$, the halo of all infinitesimals, and the galaxy of all limited hyperreals.

A.21 A Symbolic Language

We have noted in A10 how an equation involving real numbers can be transferred to an equation involving hyperreal numbers, thereby expanding the domain of the first equation. This is an example of the fundamental "transfer principle" of nonstandard analysis, which states in a precise way that a "sentence" involving real numbers is true if and only if it is true when reals are replaced by hyperreals. To extend this facility to the superstructures $V(A)$ and $V(^*A)$ and implement it rigorously, the "sentence" must be stated in a language articulating symbolic logic.

A "language" $\mathcal{L}_A$ for $V(A)$ consists of a set of symbols appropriately combined to make sentences that have truth values (either "true" or "false"). The symbols are the following.

Constant symbols: Each element $a \in V(A)$ is assigned one or more symbols such as $\overline{a}$. Note that the element $a \in V(A)$ and a symbol $\overline{a}$ for it are different ideas, but we will identify them and always use a for both the symbol and the element. (Similarly, a language $\mathcal{L}_{^*A}$ for $V(^*A)$ has its constant symbols representing elements of $V(^*A)$.)

Variable symbols: A denumerable collection of symbols, such as w, x, y, z. These may be assigned meanings by specifying sets of which they are members.

Connectives: Five symbols to which are assigned meanings; namely, $\wedge$ meaning "and," $\vee$ meaning "or (possibly both)," $\neg$ meaning "it is not true that," $\rightarrow$ meaning "implies," $\leftrightarrow$ meaning "if and only if."

Quantifiers: $\exists$ meaning "there exists," and $\forall$ meaning "for all."

Equality symbol: $=$ meaning "equals." (It indicates identity for members of $V_0(A)$ and set equality for members of $V(A) \setminus V_0(A)$.)

Predicate symbol: $\in$ meaning "is a member of" some specified set.

Parentheses and brackets: The symbols $(\ldots)$ and $[\ldots]$ are used to bracket combinations of symbols to clarify the constructions of sentences. Also, the angle brackets $\langle\cdot,\cdot,\ldots,\cdot\rangle$ are used for finite sequences such as $\langle x, y, \ldots, z\rangle$; any symbol within the angle brackets may appear more than once.

The " sentences" in $\mathcal{L}_A$ employ these symbols and are special cases of *formulas* in $\mathcal{L}_A$. The latter are defined recursively through the following steps.

1. If $x_1, \ldots, x_k, x, y, z$ are either constants or variables, the following six combinations of symbols are formulas:

$$x = y,\ x \in y,$$
$$\langle x_1, \ldots, x_k\rangle = y,\ \langle x_1, \ldots, x_k\rangle \in y,$$
$$\langle\langle x_1, \ldots, x_k\rangle, y\rangle = z,\ \langle\langle x_1, \ldots, x_k\rangle, y\rangle \in z\,.$$

2. If ϕ and ψ are formulas, then so too are $\neg\phi$, $\phi \wedge \psi$, $\phi \vee \psi$, $\phi \rightarrow \psi$, and $\phi \leftrightarrow \psi$.
3. If x is a variable symbol, if y is either a variable symbol or a constant symbol, and ϕ is a formula that does not contain a combination of the form $(\exists x \in z)$ or $(\forall x \in z)$ with that variable symbol x, then $(\exists x \in y)\phi$ and $(\forall x \in y)\phi$ are formulas.

"Sentences" are particular kinds of formulas defined as follows: A variable x is said to *occur in the scope of a quantifier* if whenever x occurs in a formula ϕ, then x is contained in a formula ψ that occurs in ϕ along with the combination of symbols $(\exists x \in y)\psi$ or $(\forall x \in y)\psi$, where y is either a constant or a variable. In this case, x is said to be *bound*. Otherwise, x is said to be *free* when it is not bound. A *sentence* is a formula in which every variable x is bound whenever it occurs.

For example, with a and b being constants and x, y, and z being variables, neither $(\forall x \in a)(x \in y)$ nor $(\forall x \in y)(\exists z \in b)(x = z)$ is a sentence because y is not bound. On the other hand, $(\exists y \in a)(\forall x \in y)(\exists z \in b)(x = z)$ is a sentence.

Conventional notation is commonly used for some formulas. For example, the ordered doublet $\langle x, y\rangle$ can be defined as the set $\{\{x\}, \{x, y\}\} \in V_2(A)$. Then, with regard to a constant $f \in V_2(A)$ that represents a function, the notation $f(x) = y$ is used in place of the formula $\langle x, y\rangle \in f$. Similarly, $\langle\langle x_1, x_2\rangle, y\rangle \in +$, $\langle x, y\rangle \in <$, and $\langle x, y\rangle \in \leq$ are replaced by $x_1 + x_2 = y$, $x < y$, and $x \leq y$, respectively. Also, the ordered k-tuple $\langle x_1, \ldots, x_k\rangle$ can be defined to be $\{\langle 1, x_1\rangle, \ldots, \langle k, x_k\rangle\} \in V_3(A)$; then, the constant f representing a function of k variables can be written as $\langle\{\langle 1, x_1\rangle, \ldots, \langle k, x_k\rangle\}, y\rangle \in f$, and this is replaced simply by $f(x_1, \ldots, x_k) = y$. Moreover, the formula $(x_1 \in d) \wedge (x_2 \in d) \wedge \ldots \wedge (x_k \in d)$ may be abbreviated to $x_1, x_2, \ldots x_k \in d$.

A sentence in $\mathcal{L}_A$ concerning the entities of the superstructure V(A) may be assigned "true" or "false" values. We now list some basic cases for which a sentence is *true*. The truth of more complicated sentences is determined by induction on the complexity of the sentence. If it is impossible to establish the truth of a sentence by such induction, then the sentence is *false*.

1. Let a, a_k, b, and c be constants. The sentences $a = b$, $\langle a_1, \ldots, a_m \rangle = b$, and $\langle \langle a_1, \ldots, a_m \rangle, c \rangle = b$ are true when the indicated formulas on the left of "=" are identical to b or equal as sets to b. Similarly, $a \in b$, $\langle a_1, \ldots, a_m \rangle \in b$, and $\langle \langle a_1, \ldots, a_m \rangle, c \rangle \in b$ are true when those left-hand sides of "$\in$" are elements of b.
2. If ϕ and ψ are sentences, then
   (i) $\neg\phi$ is true if ϕ is not true,
   (ii) $\phi \wedge \psi$ is true if both ϕ and ψ are true,
   (iii) $\phi \vee \psi$ is true if at least one of ϕ and ψ is true,
   (iv) $\phi \rightarrow \psi$ is true if either ψ is true or ϕ is not true,
   (v) $\phi \leftrightarrow \psi$ is true if ϕ and ψ are both true or if ϕ and ψ are both not true.
3. Let $\phi = \phi(x)$ be a formula containing x as the only free variable, and let b be a constant. Then,
   (i) $(\forall x \in b)\phi$ is true, if, for all $a \in b$, the substitution of a for each occurrence of x in $\phi(x)$ yields a sentence $\phi(a)$ that is true.
   (ii) $(\exists x \in b)\phi$ is true if there is an $a \in b$ such that the same substitution of a for x yields a true sentence $\phi(a)$

A.22 The Transfer Principle

We are at last ready to state the "transfer principle" of nonstandard analysis. Let ϕ be any sentence in the language $\mathcal{L}_A$ for $V(A)$. The *$*$-transform* ${}^*\phi$ of ϕ is the sentence in the language $\mathcal{L}_{*A}$ for $V({}^*A)$ obtained by replacing each constant symbol $a \in V(A)$ appearing in ϕ by the constant symbol ${}^*a = J(I(a)) \in V({}^*A)$. (See Fig. A.1.)

The Transfer Principle. *ϕ is true if and only if ${}^*\phi$ is true.*

It turns out that this principle can assert true sentences about internal sets in $V({}^*A)$ but may fail to do so for external sets. As an example, consider item (iii) in A20. Every nonempty internal subset ${}^*x$ of ${}^*\mathbb{R}$ with an upper bound $\mathbf{z}$ has a least upper bound $\mathbf{w}$ in ${}^*\mathbb{R}$; this result can be obtained by transferring the corresponding sentence for upper-bounded subsets of $\mathbb{R}$. On the other hand, $\mathbb{R}$ is an upper-bounded subset of ${}^*\mathbb{R}$, but $\mathbb{R}$ has no least upper bound in ${}^*\mathbb{R}$; thus, $\mathbb{R}$ must be an external set.

Let us examine this transfer symbolically. That a nonempty upper-bounded subset x of $\mathbb{R}$ has a least upper bound w can be asserted by a single symbolic sentence, but it is more comprehensible to break that sentence into three sentences. (See Fig. A.2 for an illustration of some of the following entities.) First, we let $\mathcal{X}$ be the set of all nonempty upper-bounded subsets of $\mathbb{R}$. We can define $\mathcal{X}$ symbolically by specifying its members x as follows:

$$x \in \mathcal{X} \;\leftrightarrow\; (\exists x \in \mathcal{P}(\mathbb{R}) \setminus \emptyset)(\exists z \in \mathbb{R})(\forall y \in x)(z \geq y)\,.$$

Then, for any $x \in \mathcal{X}$, there is a unique interval u_x of real numbers each of whose members is larger than every member of x. The members of u_x are specified symbolically as follows:

$$v \in u_x \;\leftrightarrow\; (\exists x \in \mathcal{X})(\exists u_x \in \mathcal{P}(\mathbb{R}) \setminus \emptyset)(\forall y \in x)(\forall v \in u_x)$$
$$((v \geq y) \wedge (v \notin x \rightarrow v \in u_x))\,.$$

Finally, the fact that u_x has a least member w is asserted by the following sentence:

$$(\forall x \in \mathcal{X})(\exists w \in u_x)(\forall v \in u_x)(w \leq v)\,.$$

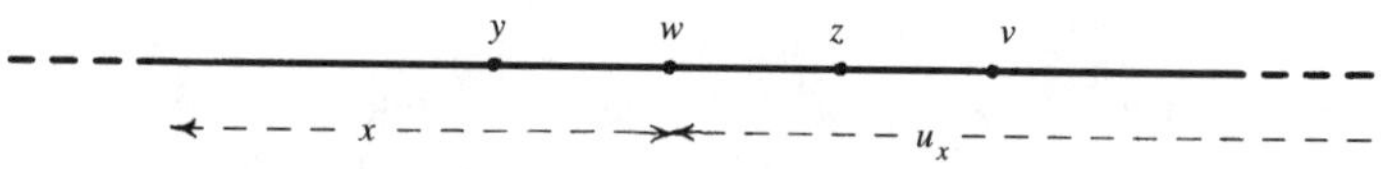

Fig. A.2. Illustration of the example for the transfer principle.

By transfer, the corresponding result for internal sets of hyperreals is obtained. We now use the facts that every element of ${}^*\mathcal{P}(\mathbb{R})$ is an internal set and that ${}^*\emptyset = \emptyset$.

$${}^*x \in {}^*\mathcal{X} \;\leftrightarrow\; (\exists\, {}^*x \in {}^*\mathcal{P}(\mathbb{R}) \setminus \emptyset)(\exists \mathbf{z} \in {}^*\mathbb{R})(\forall \mathbf{y} \in {}^*x)(\mathbf{z} \geq \mathbf{y})\,.$$

Also,

$$\mathbf{v} \in {}^*u_{{}^*x} \;\leftrightarrow\; (\exists\, {}^*x \in {}^*\mathcal{X})(\exists\, {}^*u_{{}^*x} \in {}^*\mathcal{P}(\mathbb{R}) \setminus \emptyset)(\forall \mathbf{y} \in {}^*x)(\forall \mathbf{v} \in {}^*u_{{}^*x})$$
$$((\mathbf{v} \geq \mathbf{y}) \wedge (\mathbf{v} \notin {}^*x \rightarrow \mathbf{v} \in {}^*u_{{}^*x}))\,.$$

Finally,

$$(\forall\, {}^*x \in {}^*\mathcal{X})(\exists \mathbf{w} \in {}^*u_{{}^*x})(\forall \mathbf{v} \in {}^*u_{{}^*x})(\mathbf{w} \leq \mathbf{v})\,.$$

In words again, a nonempty internal set of hyperreals having an upper bound has a least upper bound.

B

The Fibonacci Numbers

The Fibonacci numbers $F(k)$ comprise a sequence defined recursively by setting $F(-1) = 0$, $F(0) = 1$, and

$$F(k) = F(k-1) + F(k-2)$$

for $k = 1, 2, 3, 4, \ldots$. See, for example, [38, page 144]. Thus, the next several values are $F(1) = 1$, $F(2) = 2$, $F(3) = 3$, $F(4) = 5$, $F(5) = 8$, $F(6) = 13, \ldots$. A formula can be derived for any $F(k)$ by solving this linear difference equation with constant coefficients in the standard way [37, pages 167–168]. This gives

$$F(k) = \frac{\lambda_1^{k+1} - \lambda_2^{k+1}}{\sqrt{5}},$$

where $\lambda_1 = (1+\sqrt{5})/2 = 1.618\cdots$ and $\lambda_2 = (1-\sqrt{5})/2 = -0.618\cdots$. (The value λ_1 happens to be the so-called "golden ratio" [10].) As $k \to \infty$, the asymptotic behavior of $F(k)$ is

$$F(k) \sim \lambda_1^{k+1}/\sqrt{5}.$$

C

A Laplace Transform for an Artificial RC Cable

The voltage transfer ratio across one el-section of a one-way-infinite artificial RC cable is $V_j(s)/V_{j-1}(s) = Z_d(s)/(r+Z_d(s))$, where Z_d is the characteristic impedance as seen at a point just after a series resistor r and just before the next shunting capacitor c. By a customary manipulation, we have

$$Z_d(s) \;=\; -\frac{r}{2} + \sqrt{\left(\frac{r}{2}\right)^2 + \frac{r}{cs}}.$$

So, with $a = rc/2$, we get, after some manipulation involving the completion of a square,

$$\frac{V_j(s)}{V_{j-1}(s)} \;=\; as + 1 - \sqrt{(as+1)^2 - 1}.$$

Then, for $V_0(s) = 1/s$, we get

$$V_j(s) \;=\; \frac{1}{s}\left(as + 1 - \sqrt{(as+1)^2 - 1}\right)^j.$$

By Formula 90 of [47, page 354], we get the following voltage transient at node j due to a unit step of voltage at the input:

$$v_j(t) \;=\; j\int_0^t \tau^{-1}\, I_j(b\tau)\, e^{-b\tau}\, d\tau$$

where I_j is the modified Bessel function of first kind and order j and $b = 1/a = 2/rc$. $I_j(bt)$ is positive for all $t > 0$, and $I_j(bt)/t$ is asymptotic to $(b/2)^j t^{j-1}/j!$ as $t \to 0+$ for each $j \geq 1$. (See page 374 et seq. of [2].) Thus, $v_j(t)$ is continuous and strictly monotonically increasing for $t > 0$. We are justified in applying the final-value theorem. (See [47, Theorem 8.7-1] and [2, Formula 9.7.1].) This gives $\lim_{t\to\infty} v_j(t) = \lim_{s\to 0+} sV_j(s) = 1$. This establishes all the properties of $v_{j,\infty}(t)$ asserted in Sec. 7.4.

Let us note in passing that $v_j(t)$ can be lifted into a nonstandard setting by means of the transfer principle. This will yield a hyperreal voltage $\mathbf{v_j(t)}$ for an enlargement of the artificial cable where $\mathbf{j} \in {}^*\mathbb{N}$, $\mathbf{t} \in {}^*\mathbb{R}_+$, and $\mathbf{v_j}(\mathbf{t}) \in {}^*\mathbb{R}_+$. However, this enlargement is not the kind of transfinite extension of an artificial RC cable that is discussed in Sec. 7.10.

References

1. A. Abian, *The Theory of Sets and Transfinite Arithmetic*, W.B. Saunders, Philadelphia (1965).
2. M. Abramowitz and I.A. Stegun, *Handbook of Mathematical Functions*, National Bureau of Standards, Washington, DC (1964).
3. A.D. Aczel, *The Mystery of the Aleph*, Four Walls Eight Windows, New York (2000).
4. S. Albeverio, J.E. Fenstad, R. Hoegh-Krohn, and T. Lindstrom, *Nonstandard Methods in Stochastic Analysis and Mathematical Physics*, Academic Press, New York (1986).
5. M. Behzad and G. Chartrand, *Introduction to the Theory of Graphs*, Allyn and Bacon Inc., Boston (1971).
6. C. Berge, *Graphs and Hypergraphs*, North-Holland Publishing Co., Amsterdam (1973).
7. B. Bollobas, *Modern Graph Theory*, Springer, New York (1998).
8. F. Buckley (Editor), Centrality Concepts in Network Location, *Networks*, **34** (1999), Issue No. 4.
9. F. Buckley and F. Harary, *Distance in Graphs*, Addison-Wesley Publishing Co., New York (1990).
10. D.A. Burton, *The History of Mathematics*, Allyn and Bacon, Boston (1985).
11. B.D. Calvert, Infinite nonlinear resistive networks, after Minty, *Circuits Syst. Signal Process.*, **15** (1996), 727–733.
12. B.D. Calvert, Unicursal resistive networks, *Circuits, Systems, and Signal Processing*, **16** (1997), 307–324.
13. B.D. Calvert and A.H. Zemanian, Operating points in infinite nonlinear networks approximated by finite networks, *Trans. Amer. Math. Soc.*, **352**, (2000),753–780.
14. G. Chartrand and L. Lesniak, *Graphs and Digraphs*, Third Edition, Chapman & Hall, New York (1996).
15. L. DeMichele and P.M. Soardi, A Thomson's principle for infinite, nonlinear, resistive networks, *Proc. Amer. Math. Soc.*, **109** (1990), 461–468.
16. L. DeMichele and P.M. Soardi, Nonlinear infinite networks with nonsymmetric resistances, *Partial Differential Equations and Applications*, Lecture Notes in Pure and Applied Mathematics, **177**, Dekker, New York (1996).
17. K. Devlin, *The Joy of Sets*, Springer-Verlag, New York (1993).
18. V. Dolezal, *Nonlinear Networks*, Elsevier, New York (1977).
19. V. Dolezal, *Monotone Operators and Applications in Control and Network Theory*, Elsevier, New York (1979).
20. R.J. Duffin, Nonlinear networks. IIa, *Bull. Amer. Math. Soc., 53* (1947), 963–971.

21. *Encyclopedia Britannica.*
22. P. Erdos, L. Gillman, and M. Henriksen, An isomorphism theorem for real-closed fields, *Annals of Mathematics*, **61** (1955), 542–554.
23. H. Flanders, Infinite networks: I – Resistive networks, *IEEE Trans. Circuit Theory*, **CT-18** (1971), 326–331.
24. R. Goldblatt, *Lectures on the Hyperreals*, Springer, New York (1998).
25. R. Halin, Über unendliche Wege in Graphen, *Math. Ann.*, **157** (1964), 125–137.
26. A.E. Hurd and P.E. Loeb, *An Introduction to Nonstandard Real Analysis*, Academic Press, New York (1985).
27. H.A. Jung, Connectivity in infinite graphs, in *Studies in Pure Mathematics* (L. Mirsky, Ed.), Academic Press, New York (1971).
28. T. Kayano and M. Yamasaki, Dirichlet finite solutions of Poisson equations on an infinite network, *Hiroshima Math. J.*, **12** (1982), 569–579.
29. H.J. Keisler, *Foundations of Infinitesimal Calculus*, Prindle, Weber and Schmidt, Boston (1976).
30. M. Kline, *Mathematics*, Oxford University Press, New York (1980).
31. C.St.J.A. Nash-Williams, Random walks and electric currents in networks, *Proc. Cambridge Philos. Soc.*, **55** (1959), 181–194.
32. C.St.J.A. Nash-Williams, Infinite graphs – A survey, *Journal of Combinatorial Theory*, **3** (1967), 286–301.
33. N. Polat, Aspects topologiques de la séparation dans les graphes infinis. I, *Math. Z.*, **165** (1979), 73–100.
34. A. Robinson, Nonstandard analysis, *Proc. Royal Academy of Sciences, Series A*, Amsterdam, The Netherlands, **64** (1961), 432–440.
35. A. Robinson, *Nonstandard Analysis*, North-Holland Publ. Co., Amsterdam, The Netherlands (1966).
36. R. Rucker, *Infinity and the Mind*, Birkhäuser, Boston (1982).
37. T.L. Saaty, *Modern Nonlinear Equations*. New York: Dover (1981).
38. E.R. Scheinerman, *Mathematics, A Discrete Introduction*. Pacific Grove, CA, USA: Brooks/Cole Publishing Co. (2000).
39. P.M. Soardi, *Potential Theory on Infinite Networks*, Lecture Notes in Mathematics 1590, Springer-Verlag, New York (1994).
40. F.J. Thayer, Nonstandard Analysis on Graphs, *Houston Journal of Mathematics*, **29** (2004), 403–436.
41. C. Thomassen, Resistances and currents in infinite electrical networks, *J. Combinatorial Theory, Series B*, **49** (1990), 87–102.
42. E. Weber, *Linear Transient Analysis*, Vol. II, Wiley, New York (1956).
43. L. Weinberg, *Network Analysis and Synthesis*, McGraw-Hill Book Co., New York (1962).
44. E.T. Whittaker and G.N. Watson, *A Course of Modern Analysis*, 4th Edition, Cambridge University Press, New York (1963).
45. R.J. Wilson, *Introduction to Graph Theory*, Academic Press, New York (1972).
46. M. Yamasaki, Nonlinear Poisson equations on an infinite network, *Mem. Fac. Sci. Shimane Univ.* **23** (1989), 1–9.
47. A.H. Zemanian, *Distribution Theory and Transform Analysis*, McGraw-Hill Book Co., New York (1965); republished by Dover Publications, New York (1987).
48. A.H. Zemanian, The connections at infinity of a countable resistive network, *Int. J. Circuit Theory Appl.*, **3** (1975), 333–337.

49. A.H. Zemanian, Operator-valued transmission lines in the analysis of two-dimensional anomalies imbedded in a horizontally layered earth under transient polarized electromagnetic excitation, *SIAM J. Appl. Math.*, **45** (1985), 591–620.
50. A.H. Zemanian, *Infinite Electrical Networks*, Cambridge University Press, New York (1991).
51. A.H. Zemanian, *Transfiniteness – for Graphs, Electrical Networks, and Random Walks*, Birkhäuser, Boston (1996).
52. A.H. Zemanian, Nonstandard electrical networks and the resurrection of Kirchhoff's laws, *IEEE Trans. Circuits and Systems – I: Fund. Theory and Appl.*, **44** (1997), 221–233.
53. A.H. Zemanian, Transfinite electrical networks, *IEEE Trans. Circuits Syst. – I: Fund. Theory Appl.*, **46** (1999), 59–70.
54. A.H. Zemanian, *Pristine Transfinite Graphs and Permissive Electrical Networks*, Birkhäuser, Boston (2001).
55. A.H. Zemanian, Nonstandard Versions of Conventionally Infinite Networks, *IEEE Trans. Circuits and Systems – I: Fund. Theory and Appl.*, **48** (2001), 1261–1265.
56. A.H. Zemanian, *Nonstandard Transfinite Graphs and Random Walks on Them*, CEAS Technical Report 785, State University of New York at Stony Brook, January (2002).
57. A.H. Zemanian, *Nonstandard Transfinite Electrical Networks*, CEAS Technical Report 795, State University of New York at Stony Brook, January (2002).
58. A.H. Zemanian, *Nonstandard Graphs Based on Tips as the Individual Elements*, CEAS Technical Report 802, State University of New York at Stony Brook, August (2002).
59. A.H. Zemanian, *Nonstandard Graphs*, CEAS Technical Report 803, State University of New York at Stony Brook, August (2002).
60. A.H. Zemanian and P.E. Fleischer, On the transient responses of ladder networks, *IRE Transactions of Circuit Theory*, **CT-5** (1958), 197–201.

Index of Symbols

Subject Index